OXFORD IB STUDY GUIDES

Garrett Nagle
Briony Cooke

Geography

FOR THE IB DIPLOMA

2nd edition

OXFORD
UNIVERSITY PRESS

Introduction

Aims

This second edition Study Guide (for first assessment in 2019) will help you to:

- develop an understanding of how people, places, spaces and the environment are connected and interdependant

- develop an awareness of global environments, and an understanding of how these are shaped and sustained

- develop a global perspective of diversity and change, and an appreciation of the issues and challenges that face the world today.

Study Guide

This Study Guide provides a comprehensive coverage of all the Options and Units required for the updated SL and HL syllabus. It also provides a series of questions and answers, so that you can "Check Your Understanding", and exam-style "Exam practice" questions and mark schemes to help you prepare for the final examinations.

You will learn geographic skills throughout this course, many of which you will be able to apply across a variety of topics. You will develop core research and writing skills, as well as learning to interpret and analyse maps, infographics and data.

Contents

 Answers can be found online at
www.oxfordsecondary.co.uk/9780198396079

The drainage basin

DEFINITIONS

The **drainage basin** is an area that is drained by a river and its tributaries. Drainage basins have inputs, stores, processes and outputs. The inputs and outputs cross the boundary of the drainage basin, hence the drainage basin is an open system. The main input is precipitation, which is regulated by various means of storage. The outputs include evaporation and transpiration. Flows include infiltration, throughflow, overland flow and base flow, and stores include vegetation, soil, aquifers and the cryosphere (snow and ice).

Drainage basin hydrology

Precipitation is the transfer of moisture to the earth's surface from the atmosphere. It includes dew, hail, rain, sleet and snow. Precipitation total and intensity are important for overland flow.

Interception refers to the capture of raindrops by plant cover that prevents direct contact with the soil. If rain is prolonged, the retaining capacity of leaves will be exceeded and water will drop to the ground (throughfall). Some will trickle along branches and down the stems or trunk (stemflow). Some is retained on the leaves and later evaporated.

Evaporation is the physical process by which a liquid becomes a gas. It is a function of:
- vapour pressure
- air temperature
- wind
- rock surface, for example, bare soils and rocks have high rates of evaporation compared with surfaces which have a protective tilth where rates are low.

Transpiration is the loss of water from vegetation. **Evapotranspiration** is the combined loss of water from vegetation and water surfaces to the atmosphere. **Potential evapotranspiration** is the rate of water loss from an area if there were no shortage of water.

FLOWS

Infiltration is the process by which water sinks into the ground. **Infiltration capacity** refers to the amount of moisture that a soil can hold. By contrast, the **infiltration rate** refers to the speed with which water can enter the soil. **Throughflow** refers to water moving in soil, laterally following natural pipes (percolines) or between horizons. **Overland run-off occurs** when precipitation intensity exceeds the infiltration rate, or when the infiltration capacity is reached and the soil is saturated. **Percolation** refers to water moving deep into the groundwater zone. **Baseflow** refers to the movement of groundwater – for groundwater to flow the water table must rise above the river level to provide the hydraulic gradient needed for water movement.

STORES

There are many stores including vegetation, soils, aquifers and the cryosphere. **Aquifers** are rocks that hold water. They provide the most important store of water, regulate the hydrological cycle and maintain river flow.

Soil moisture varies with porosity (the amount of pore spaces) in a soil, and with permeability (the ability to transmit water).

The **cryosphere** is the largest store of freshwater, and water may be stored for millennia. **Vegetation** is another important store – trees store more water than grasses or crops.

EXAM TIP

You may be asked to draw a diagram of a drainage basin hydrological cycle. A systems diagram – with inputs, stores, flows and outputs – is a much better diagram that a diagram that tries to show trees, clouds, rainfall, glaciers, rivers, lakes and oceans, for example.

CHECK YOUR UNDERSTANDING

1. Identify five forms of storage in a drainage basin.
2. Briefly explain why drainage basins can be considered as open systems.

River discharge

DEFINITIONS

Discharge refers to the volume of water passing a certain point per unit of time. It is usually expressed in cubic metres per second (cumecs). Normally, discharge increases downstream as shown by the Bradshaw model.

BRADSHAW MODEL OF CHANNEL VARIABLES

Bradshaw's model shows changes to channel characteristics over the course of a river. Water velocity and discharge increase downstream while channel bed roughness and load particle size decrease.

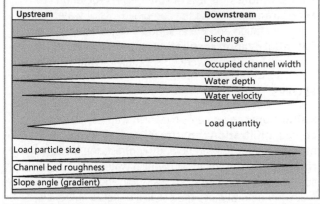

THE MAIN TYPES OF EROSION

Abrasion (or **corrasion**) is the wearing away of the bed and bank by the load carried by a river.

Attrition is the wearing away of the load carried by a river. It creates smaller, rounder particles.

Hydraulic action is the force of air and water on the sides of rivers and in cracks.

Solution (or **corrosion**) is the removal of chemical ions, especially calcium, which causes rocks to dissolve.

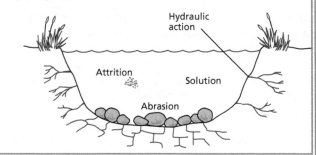

FACTORS AFFECTING EROSION

- **Load** – the heavier and sharper the load the greater the potential for erosion.
- **Velocity and discharge** – the greater the velocity and discharge the greater the potential for erosion.
- **Gradient** – increased gradient increases the rate of erosion.
- **Geology** – soft, unconsolidated rocks, such as sand and gravel, are easily eroded.
- **pH** – rates of solution are increased when the water is more acidic
- **Human impact** – deforestation, dams, and bridges interfere with the natural flow of a river and frequently end up increasing the rate of erosion.

TRANSPORT

The main types of transportation include:
- **Suspension** – small particles are held up by turbulent flow in the river.
- **Saltation** – heavier particles are bounced or bumped along the bed of the river.
- **Solution** – the chemical load is dissolved in the water.
- **Traction** – the heaviest material is dragged or rolled along the bed of the river.
- **Floatation** – leaves and twigs are carried on the surface of the river.

THEORY OF RIVER CHANNEL LOAD

The **capacity** of a stream refers to the largest amount of debris that a stream can carry; its **competence** refers to the diameter of the largest particle that can be carried. The **critical erosion velocity** is the lowest velocity at which grains of a given size can be moved. The relationship between these variables is shown by means of a **Hjulström curve**.

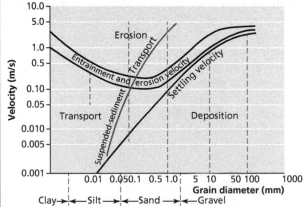

There are three important features on Hjulström curves:
- the smallest and largest particles require high velocities to lift them
- higher velocities are required for entrainment than for transport
- when velocity falls below a certain level (settling or fall velocity) particles are deposited.

CHECK YOUR UNDERSTANDING
3. Define the terms hydraulic action, attrition and abrasion.
4. Outline the ways in which a river transports its load.

Temporal variations in processes: River landforms (1)

RIVER REGIME

A **river regime** is the annual variation in the flow of a river.

The character or **regime** of the resulting stream or river is influenced by several variable factors:
- the amount and nature of precipitation
- the local rocks, especially porosity and permeability
- the amount and type of vegetation cover.

When there is more precipitation, rivers are more likely to erode. Certain rocks, such as chalk and limestone, are likely to be affected by solution.

The regime of the River Shannon, Ireland

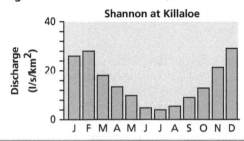

WATERFALLS

Waterfalls frequently occur on horizontally bedded rocks. The soft rock is undercut by hydraulic action and abrasion. The weight of the water and the lack of support cause the waterfall to collapse and retreat. Over thousands of years, the waterfall may retreat enough to form a gorge of recession.

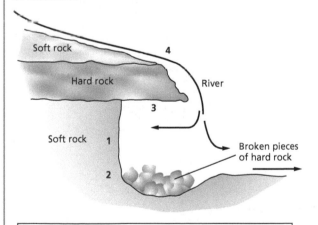

1 Hydraulic impact.
2 Abrasion of soft rock by hard fragments.
3 Lack of support by soft rock.
4 Weight of water causes unsupported hard rock to collapse.

DEPOSITION

Deposition occurs as a river slows down and loses its energy. Typically, this occurs as a river floods across a floodplain, enters the sea or behind a dam. It is also more likely during low flow conditions (such as in a drought) than during high flow (flood) conditions – as long as the river is carrying sediment. The larger, heavier particles are deposited first, the smaller, lighter ones later. Features of deposition include floodplains, levees and deltas.

FLOODPLAINS

Floodplains are flat areas found in the lower parts of a river, comprising of clay, silt or alluvium deposited when the river is in flood.

LEVEES

When a river floods, its speed is reduced, slowed down by friction caused by contact with the floodplain. As its velocity is reduced the river has to deposit some of its load. It drops the coarser, heavier material first to form raised banks, or **levees** at the edge of the river. This means that over centuries the levees are built up of coarse material, such as sand and gravel, while the floodplain consists of fine silt and clay.

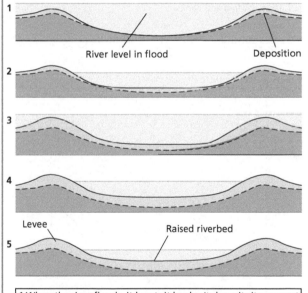

1 When the river floods, it bursts it banks. It deposits its coarsest load (gravel and sand) closer to the bank and the finer load (silt and clay) further away.
2, 3, 4. This continues over a long time, for centuries.
5 The river has built up raised banks called levees, consisting of coarse material, and a floodplain of fine material.

EXAM TIP

Make sure that you use units – it would be easy here to just refer to high discharge and low discharge (or high flow and low flow). A scale is provided – please make sure that you make use of it.

CHECK YOUR UNDERSTANDING

5. Describe the regime of the River Shannon.
6. Suggest reasons for the variation in flow between December and July.

River landforms (2)

MEANDERS

Meandering is the normal behaviour of fluids and gases in motion. Meanders can occur on a variety of materials from ice to solid rock. Meander development occurs in conditions where channel slope, discharge and load combine to create a situation where meandering is the only way that the stream can use up the energy it possesses equally throughout the channel reach.

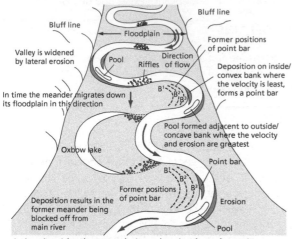

A river is said to be meandering when its **sinuosity** ratio exceeds 1.5. The **wavelength** of meanders is dependent on three major factors: channel width, discharge, and the nature of the bed and banks.

Development of a meander through time

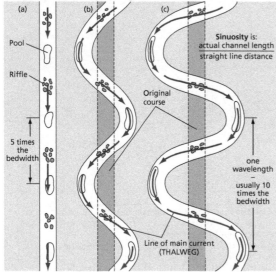

Meanders

DELTAS

Deltas are formed as river sediments are deposited when a river enters a standing body of water such as a lake, lagoon, or ocean. Deposition occurs because velocity is reduced. A number of factors affect the formation of deltas:

* the amount and size of load – rivers must be heavily laden, and coarse sediments will be deposited first
* salinity – salt-water causes clay particles to stick together, they get heavier and are deposited
* gradient of coastline – delta formation is more likely on gentle coastlines
* vegetation – plant waters will slow waters and so increase deposition
* low energy river discharge and/or low energy wave or tidal energy.

Deltas occur in three main forms:

* arcuate – many distributaries which branch out radially, for example, the Nile Delta
* cuspate – a pointed delta formed by a dominant channel
* bird's foot – long, projecting fingers which grow at the end of distributaries, for example, the Mississippi Delta.

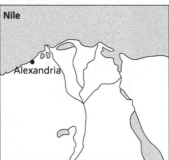

Different forms of deltas

EXAM TIP

For most landforms (of erosion and deposition) you should learn an annotated diagram, and ensure that you can explain how the landform is formed. For many features, for example, waterfalls and levees, you may need to learn a sequence of diagrams.

CHECK YOUR UNDERSTANDING

7. Outline the main factors required for the formation of deltas.
8. Briefly explain how waterfalls are formed by rivers.

Hydrographs

DEFINITION

A **storm** or **flood hydrograph** is a graph that shows how a river changes over a short period, such as a day or a couple of days. It shows how a river channel responds to the key processes of the hydrological cycle. It measures the speed at which rainfall falling on a drainage basin reaches the river channel. It is a graph on which river discharge during a storm or run-off event is plotted against time.

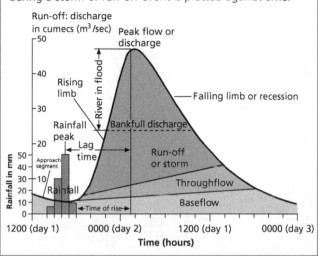

CHARACTERISTICS

- **Discharge** (Q) is the volume of flow passing through a cross-section of the river during a given period of time (usually measured in cumecs, that is, m³/s).
- The **rising limb** indicates the amount of discharge and the speed at which it is increasing. It is very steep in a flash flood or in small drainage basins where the response is rapid. It is generally steep in urbanized catchments.
- **Peak flow or discharge** is higher in larger basins. Steep catchments will have lower infiltration rates; flat catchments will have high infiltration rates, so more throughflow and lower peaks.
- **Lag time** is the time interval between peak rainfall and peak discharge. It is influenced by basin shape, steepness, and stream order.
- The **run-off** curve reveals the relationship between overland flow and throughflow. Where infiltration is low, high antecedent moisture, impermeable surface and rainfall strong overland flow will dominate.
- **Baseflow** is the seepage of groundwater into the channel – this can be very important where rocks have high pore spaces. Baseflow is a slow movement and is the main, long-term supplier of the river's discharge.
- The **recessional limb** is influenced by basin size, geological composition and behaviour of local aquifers.
- Larger catchments, flatter gradients and permeable rocks have gentler recessional limbs.

HYDROGRAPH SIZE (AREA UNDER THE GRAPH)

Generally, the higher the rainfall, the greater the discharge, and the larger the basin size, the greater the discharge.

VARIATION IN HYDROGRAPHS

A number of factors affect flood hydrographs:
- climate (rainfall total, intensity, seasonality)
- soils (impermeable clay soils create more flooding)
- vegetation (vegetation intercepts rainfall and so flooding is less likely)
- infiltration capacity (soils with a low infiltration capacity cause much overland flow)
- rock type (permeable rocks will allow water to infiltrate, thereby reducing the flood peak)
- slope angle (on steeper slopes there is greater run-off)
- drainage density (the more stream channels there are the more water that gets into rivers
- dams disrupt the flow of water; afforestation schemes increase interception
- basin size, shape, and relief (small, steep basins reduce lag time, while basin shape influences where the bulk of the floodwaters arrive).

URBAN HYDROLOGY AND THE STORM HYDROGRAPH

Urban hydrographs are different to rural ones. They have:
- a shorter lag time
- a steeper rising limb
- a higher peak flow (discharge)
- a steeper recessional limb.

This is because there are more impermeable surfaces in urban areas (roofs, pavements, roads, buildings) as well as more drainage channels (gutters, drains, sewers).

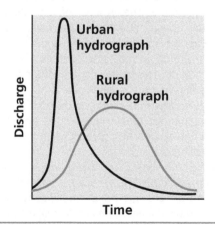

CHECK YOUR UNDERSTANDING

9. Briefly explain how climate and soils influence hydrographs.
10. Explain how an urban hydrograph differs from a rural one.

Land-use change and flood risk

POTENTIAL HYDROLOGICAL EFFECTS OF URBANIZATION

Urbanizing influence	Potential hydrological response
Removal of trees and vegetation	Decreased evapotranspiration and interception; increased stream sedimentation
Initial construction of houses, streets and culverts	Decreased infiltration and lowered groundwater table; increased storm flows and decreased baseflows during dry periods
Complete development of residential, commercial and industrial areas	Decreased porosity, reducing time of run-off concentration, thereby increasing peak discharges and compressing the time distribution of the flow; greatly increased volume of run-off and flood damage potential
Construction of storm drains and channel improvements	Local relief from flooding; concentration of floodwaters may aggravate flood problems downstream

Urbanization has a greater impact on processes in the lower part of a drainage basin than the upper course. This is because more urban areas are found in the lower parts of drainage basins.

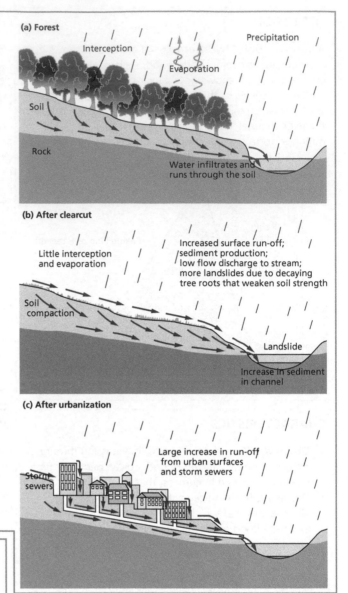

DEFORESTATION

Deforestation can have a similar impact to urbanization on flood hydrographs. The presence of vegetation increases interception, reduces overland flow, and increases evapotranspiration. In contrast, deforestation reduces interception, increases overland flow, and reduces evapotranspiration. This causes flood hydrographs to have shorter time lags and higher peak flows.

In deforested areas flood risk is increased. The risk of higher magnitudes, greater frequencies, and reduced recurrence intervals becomes greater when the vegetation cover is removed.

Deforestation is likely to occur over a much broader area than urbanization. This is because deforestation may occur for land-use changes (for example, conversion to agriculture), industrial development, to make way for tourist developments and to allow urbanization to occur. Hence it likely to have a more widespread impact on hydrological processes.

CHANNEL MODIFICATIONS

Channel modifications include channelization, enlargement and straightening. Channelization may create new channels. These are likely to be quite straight. This speeds up water movement and so time lags are likely to be reduced. Enlarging channels through levees (raised banks) enables rivers to carry more water. Thus the peak flow may be higher. However, the purpose of channelization and straightening is to remove water from an area, and so reduce the threat of a flood.

COMMON MISTAKE

✗ *All floods have the same impact.*

✓ *An annual flood may be only a few centimetres deep whereas a 50-year flood could be a metre deep. The low-frequency high-magnitude flood events are much more powerful and damaging than the high-frequency low-magnitude flood events.*

CHECK YOUR UNDERSTANDING

11. Describe the hydrological impacts of the removal of trees and vegetation.
12. Describe the impact of urbanization on channels in an urban area.

Flood prediction and mitigation

FORECASTING AND WARNING

According to the United Nations Environment Programme's publication *Early warning and assessment* there are a number of things that could be done to improve flood warnings. These include:

- improved rainfall and snow-pack estimates, better and longer forecasts of rainfall
- better gauging of rivers, collection of meteorological information and mapping of channels
- better and more-current information about human populations, and infrastructure, elevation and stream channels need to be incorporated into flood risk assessment models
- better sharing of information is needed between forecasters, national agencies, relief organizations and the general public
- more complete and timely sharing of information of meteorological and hydrological information is needed among countries within international drainage basins
- technology should be shared among all agencies involved in flood forecasting and risk assessment both in the basins and throughout the world.

EMERGENCY MEASURES

Emergency action includes the removal of people and property, and flood-fighting techniques, such as sandbags. Much depends on the efficiency of forecasting and the time available to warn people and clear the area. Flood-proofing includes sealing walls, sewer adjustment by the use of valves, covering buildings and machinery

PREVENTION AND AMELIORATION OF FLOODS

Loss-sharing adjustments include disaster aid and insurance. **Disaster aid** refers to any aid, such as money, equipment, staff and technical assistance that are given to a community following a disaster. In developed countries **insurance** is an important loss-sharing strategy. However not all flood-prone households have insurance and many of those that are insured may be underinsured. Its lack of availability in many poor countries makes it of limited use.

Event modification adjustments include environmental control and hazard-resistant design. Physical control of floods depend on two measures – flood abatement and flood diversion.

Flood abatement involves decreasing the amount of run-off, thereby reducing the flood peak in a drainage basin. There are a number of ways of reducing flood peaks. These include:

- reforestation
- reseeding of sparsely vegetated areas to increase evaporative losses
- treatment of slopes such as contour ploughing or terracing to reduce the run-off coefficient
- comprehensive protection of vegetation from wildfires, overgrazing, and clearcutting of forests
- clearance of sediment and other debris from headwater streams
- construction of small water and sediment holding areas
- preservation of natural water storage zones, such as lakes.

Flood diversion measures, by contrast, include the construction of levees, reservoirs, and the modification of river channels. **Levees** are the most common form of river engineering. **Reservoirs** store excess rainwater in the upper drainage basin. However, this may only be appropriate in small drainage networks.

CASE STUDY

PROTECTING THE MISSISSIPPI

For over a century the Mississippi has been mapped, protected and regulated. The river drains one-third of the USA and it affects some of the USA's most important agricultural regions. A number of methods have been used to control flooding and the effects of flooding including:

- stone and earthen levees to raise the banks of the river
- dams to hold back water in times of flood
- straightening of the channel to remove water speedily.

Altogether over $10 billion has been spent on controlling the Mississippi, and annual maintenance costs are nearly $200 million.

In 1993, following heavy rain, many of the levees collapsed allowing the river to flood its floodplain. The damage was estimated to be over $12 billion yet only 43 people died. Over 25,000 km^2 of land were flooded. In 2005 more than 50 breaches in New Orleans's hurricane surge protection occurred during Hurricane Katrina leading to death and destruction on a massive scale.

According to some geographers, if the Mississippi were left to its own devices a new channel would have been created by the mid-1970s, so much so that the ports at New Orleans and Baton Rouge would be defunct. However, river protection schemes have prevented this. For example, at New Orleans 7-metre levees flank the river. New Orleans is 1.5 metres below the average river level and 5.5 metres below flood level.

CHECK YOUR UNDERSTANDING
13. Distinguish between a levee and a flood relief channel.
14. Explain the term "flood abatement".

Flood control – protective measures along flood channels

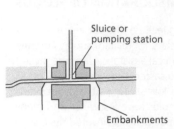

1 Flood embankments with sluice gates. The main problem with this is it may raise flood levels upstream and downstream.

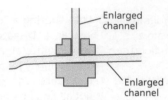

2 Channel enlargement to accommodate larger discharges. One problem with such schemes is that as the enlarged channel is only rarely used it becomes clogged with weed.

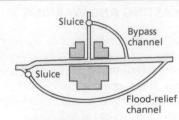

3 Flood-relief channel. This is appropriate where it is impossible to modify original channel as it tends to be rather expensive, e.g. the flood-relief channels around Oxford UK.

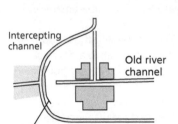

4 Intercepting channels. These divert only part of the flow away, allowing flow for town and agricultural use, e.g. the Great Ouse Protection Scheme in the UK.

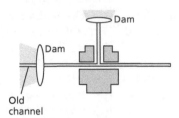

5 Flood storage reservoirs. This solution is widely used, especially as many reservoirs created for water-supply purposes may have a secondary flood control role, such as the intercepting channels along the Loughton Brook, UK.

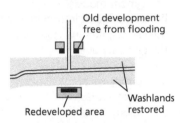

6 The removal of settlements. This is rarely used because of cost, although many communities were forced to leave as a result of the 1993 Mississippi floods, e.g. Valmeyer, Illinois.

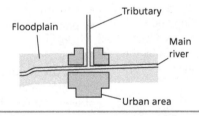

CHANNEL MODIFICATION

Modification of river channels includes raising the banks (to hold more water), straightening the river (to speed up flow and remove the water as quickly as possible) and creating new channels (flood-relief channels) to carry water when the river is in flood. Channels can also be strengthened with steel or concrete to make them less vulnerable to erosion.

Artificial **levees** are the most common form of river engineering. This is when the banks of the river are increased in height so that the river can carry more water and sediment. Levees can also be used to divert and restrict water to low-value land on the flood plain. Over 4,500 km of the Mississippi River has levees. Channel improvements such as dredging the river bed of sediment will increase the carrying capacity of the river.

> **EXAM TIP** ✓
> When writing about floods, make sure you use a named, located example, with dates of the floods and details such as the loss of life, economic loss or number of people made homeless.

> **COMMON MISTAKE** ❗
> ✗ *Flood protection measures will protect a settlement against all floods.*
> ✓ Flood protection measures protect against a flood of a given magnitude/recurrence interval. For example, levees on the Mississippi at New Orleans are designed to withstand a 100-year flood. However, the floodwaters caused by Hurricane Katrina were a 150–200-year flood so the levees were breached.

> **CHECK YOUR UNDERSTANDING** ❓
> **15.** Identify two ways in which flood warnings could be improved.
> **16.** Explain how revegetation of an area could help reduce flooding.

Water scarcity

SCARCITY THAT AFFECTS LICs

Where water supplies are inadequate, two types of **water scarcity** affect LICs in particular:
- **Physical water scarcity** occurs where water consumption exceeds 60% of the usable supply. To help meet water needs some countries such as Saudi Arabia and Kuwait import much of their food and invest in desalinization plants.

- **Economic water scarcity** occurs where a country physically has sufficient water resources to meet its needs, but additional storage and transport facilities are required – this will mean embarking on large and expensive water-development projects as in many in sub-Saharan countries.

In addition, in LICs, access to adequate water supplies is affected by the exhaustion of traditional sources, such as wells and seasonal rivers.

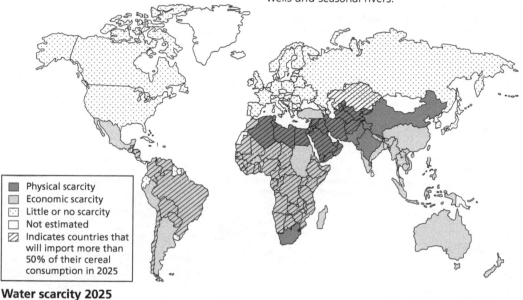

Key:
- Physical scarcity
- Economic scarcity
- Little or no scarcity
- Not estimated
- Indicates countries that will import more than 50% of their cereal consumption in 2025

Water scarcity 2025

DROUGHT

Drought is an extended period of dry weather leading to conditions of extreme dryness. Absolute drought is a period of at least 15 consecutive days with less than 0.2 mm of rainfall. Partial drought is a period of at least 29 consecutive days during which the average daily rainfall does not exceed 0.2 mm.

The severity of a drought depends upon the length of the drought and how severe the water shortage is. The impacts of drought can include reduced crop yields, increased animal mortality, an increase in illnesses in humans (linked to dehydration), an increase in forest fires, hosepipe bans, a ban on watering private gardens or washing cars.

QUALITY

Water also needs to be of an adequate quality for consumption. In developing countries too many people lack access to safe and affordable water supplies and sanitation. The World Health Organization (WHO) estimates that around 4 million deaths each year are from water-related disease, particularly cholera, hepatitis, malaria and other parasitic diseases. Water quality may be affected by organic waste from sewage, fertilizers and pesticides from farming, and by heavy metals and acids from industrial processes and transport. Factors affecting access to safe drinking water include:
- water availability
- water infrastructure
- cost of water.

WATER QUANTITY AND WATER QUALITY DISTRIBUTION

The world's available freshwater supply is not distributed evenly around the globe, either seasonally or from year to year.
- About three-quarters of annual rainfall occurs in areas containing less than a third of the world's population.
- Two-thirds of the world's population live in the areas receiving only a quarter of the world's annual rainfall.

CHECK YOUR UNDERSTANDING
17. Distinguish between physical water scarcity and economic water scarcity.
18. Define the term "drought".

Environmental impacts of agricultural activities

IRRIGATION

Irrigation is the addition of water to areas where there is insufficient water for adequate crop growth. Water can be taken from surface stores, such as lakes, dams, reservoirs and rivers, or from groundwater. Types of irrigation range from total flooding, as in the case of paddy fields, to drip irrigation, where precise amounts are measured out to each individual plant

CONSEQUENCES OF IRRIGATION

- In Texas, irrigation has reduced the water table by as much as 50 m. By contrast, in the Indus Plain in Pakistan, irrigation has raised the water table by as much as 6 m since 1922, and caused widespread salinization.
- Irrigation can reduce the earth's albedo (reflectivity) by as much as 10%. This is because a reflective sandy surface may be replaced by one with dark green crops.
- Irrigation can also cause changes in precipitation. Large-scale irrigation in semi-arid areas, such as the High Plains of Texas, have been linked with increased rainfall, hailstorms and tornadoes. Under natural conditions semi-arid areas have sparse vegetation and dry soils in summer. However, when irrigated these areas have moist soils in summer and complete vegetation cover. Evapotranspiration rates increase and there have been increases in the amount of summer rainfall.

SALINIZATION

- Irrigation frequently leads to an increase in the amount of salt in the soil. This occurs when groundwater levels are close to the surface. Capillary forces bring water to the surface where it may be evaporated leaving behind any soluble salts that it is carrying. This is known as **salinization**.
- Some irrigation, especially paddy rice, requires huge amounts of water. As water evaporates in the hot sun, the salinity levels of the remaining water increase. This also occurs behind large dams.

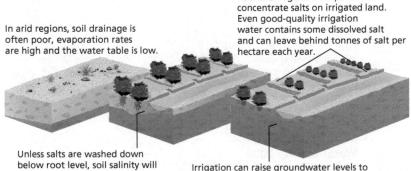

In arid regions, soil drainage is often poor, evaporation rates are high and the water table is low.

Poor drainage and evaporation concentrate salts on irrigated land. Even good-quality irrigation water contains some dissolved salt and can leave behind tonnes of salt per hectare each year.

Unless salts are washed down below root level, soil salinity will stunt growth and eventually kill off all but the most resistant plants.

Irrigation can raise groundwater levels to within a metre of the surface, bringing up more dissolved salts from the aquifer, subsoil and root zone.

EUTROPHICATION

Eutrophication, or nutrient enrichment, of water bodies has led to algal blooms, oxygen starvation and a decline in species diversity. While there is a strong body of evidence to link increased eutrophication with increased use of nitrogen fertilizers, some scientists argue that increased phosphates from farm sewage are the cause.

There are a number of stakeholders involved in the eutrophication process:

- chemical fertilizer companies that wish to sell fertilizers to farmers
- governments that wish to produce more food
- farmers who want to grow more food and make a greater profit
- health organizations and environmental organizations that wish to have safe water
- consumers who want cheaper food, but who may also end up paying the cost of making drinking water clean.

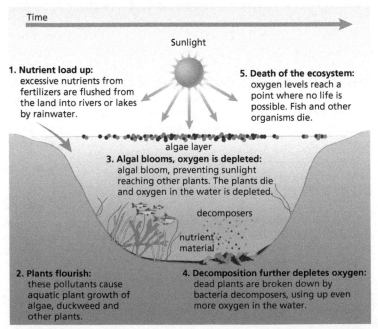

Time

Sunlight

1. **Nutrient load up:** excessive nutrients from fertilizers are flushed from the land into rivers or lakes by rainwater.

5. **Death of the ecosystem:** oxygen levels reach a point where no life is possible. Fish and other organisms die.

algae layer

3. **Algal blooms, oxygen is depleted:** algal bloom, preventing sunlight reaching other plants. The plants die and oxygen in the water is depleted.

decomposers

nutrient material

2. **Plants flourish:** these pollutants cause aquatic plant growth of algae, duckweed and other plants.

4. **Decomposition further depletes oxygen:** dead plants are broken down by bacteria decomposers, using up even more oxygen in the water.

CHECK YOUR UNDERSTANDING
19. Briefly explain the process of salinization.

20. Briefly explain the changes in a stream that result from eutrophication.

Human pressures on lakes and aquifers

POPULATION GROWTH

As populations grow, pressure on water resources increases. Population growth is uneven so there is increased pressure in certain locations. Urban areas experiencing rapid economic growth are likely to face the greatest increases in water stress.

The Middle East and North Africa (MENA) region contains over 6% of the world's population but only 1.4% of its freshwater. As the population increases, pressure on the water resources is likely to intensify. Twelve of the world's 15 water-scarce countries are in the MENA region. The Nubian Sandstone Aquifer System under the Sahara Desert is over 2 million km^2 in area and offers great potential for the region. Careful use of the aquifer and irrigation channels such as the Great Man-made River Project (from the Sahara to Libya) would enable the MENA region to access more freshwater resources.

POLLUTION

Lakes may be polluted by run-off from chemical fertilizers, phosphates, sewage, oil, acidification and industrial effluent. The largest sources of sulphur and nitrogen are China and India. Over recent decades there has been an increase in the acidification of lakes there, and also transboundary impacts on lakes in South Korea and Japan. The results of a UNEP survey are shown below.

Global variations in eutrophication

Region	Percentage of lakes and reservoirs suffering from eutrophication
Asia-Pacific	54
Europe	53
Africa	28
North America	48
South America	41

CASE STUDY

GROUNDWATER POLLUTION IN BANGLADESH

There has been an increase in the incidence of cancers in Bangladesh. It has been caused by naturally occurring arsenic in groundwater pumped up through the tube wells. Estimates by the World Health Organization suggest that as many as 85 million of its 125 million population will be affected by arsenic-contaminated drinking water.

For 30 years, following the lead of Unicef, Bangladesh has sunk millions of tube wells, providing a convenient supply of drinking water free from the bacterial contamination of surface water that was killing one-quarter of a million children a year. But the water from the wells was never tested for arsenic contamination, which occurs naturally in the groundwater. One in 10 people who drinks the water containing arsenic will ultimately die of lung, bladder or skin cancer.

The first cases of arsenic-induced skin lesions were identified across the border in West Bengal, India, in 1983. Arsenic poisoning is a slow disease. Skin cancer typically occurs 20 years after people start ingesting the poison. The real danger is internal cancers, especially of the bladder and lungs, which are usually fatal. Bangladeshi doctors were warned to expect an epidemic of cancers by 2010. The victims will be people in their thirties and forties who have been drinking the water all their lives – people in their most productive years.

One solution to the problem is to use concrete water butts to collect water from gutters. Other possible solutions include a filter system. Neither is as convenient as the tube wells they are designed to replace. Tube wells are easy to sink in the delta's soft alluvial soil, and for tens of millions of peasants the wells have revolutionized access to water.

COMMON MISTAKE

✗ *All groundwater is renewable.*

✓ Some groundwater can be considered a non-renewable resource as the water that helped fill the aquifer fell thousands of years ago in a wetter climate. If the annual use of groundwater exceeds its regeneration, the aquifer will decrease in size.

CHECK YOUR UNDERSTANDING

21. Explain how lakes may become polluted.
22. State the term used to identify a water-bearing rock.

Internationally shared water resources

A SOURCE OF CONFLICT

As populations grow, greater demands are made on water resources. Water resources are now becoming a limiting factor in many societies, and the availability of water for drinking, industry and agriculture needs to be considered. Many societies are now dependant primarily on groundwater, which is non-renewable. As societies develop, water needs increase. The increased demand for freshwater can lead to inequity of usage and political consequences. When water supplies fail, populations will be forced to take drastic steps, such as mass migration. Water shortages may also lead to civil unrest and wars.

The Grand Ethiopian Renaissance Dam

Ethiopia is building Africa's largest dam, the Grand Ethiopian Renaissance Dam, on the Blue Nile. It is designed to produce 6,000 megawatts of electricity, more than double Ethiopia's current output.

This opportunity for Ethiopia could spell disaster for Egypt.

The Nile provides nearly all of Egypt's water.

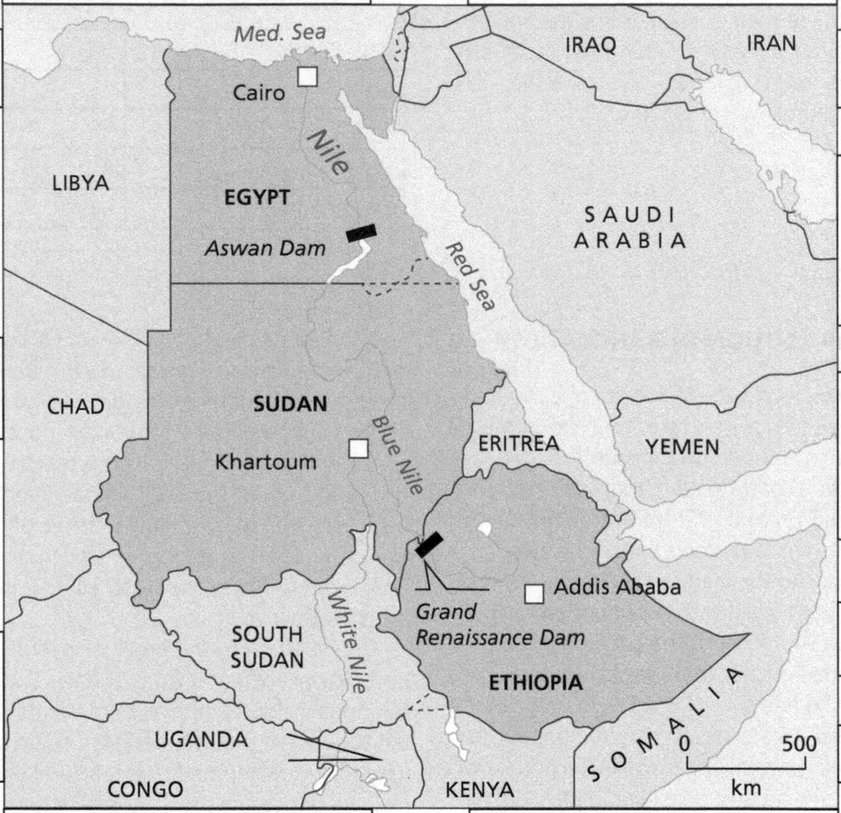

Sudan will receive some of the power produced by the dam.

By stabilizing the Nile's flow, it will also allow Sudan to prevent flooding, consume more water and increase agricultural output.

Egypt claims two-thirds of the flow based on a treaty it signed with Sudan in 1959.

The stakeholders include the governments of Egypt, Ethiopia and Sudan, as well as the people who will make use of the water.

The project will cost approximately US$4.8 bn.

This is no longer enough water to satisfy the growing population (1.8% growth in 2015) and agricultural sector.

The dam is just 20 km from the Sudan border.

In March 2015 the leaders of Egypt, Ethiopia and Sudan signed a declaration that approved construction of the dam as long as there is no "significant harm" to downstream countries.

There is uncertainty over the dam's ultimate use. Ethiopia insists that it will produce only power. Egyptians fear it will also be used for irrigation, reducing downstream supply.

Map labels: Med. Sea, IRAQ, IRAN, Cairo, LIBYA, EGYPT, Aswan Dam, Nile, Red Sea, SAUDI ARABIA, CHAD, SUDAN, Khartoum, Blue Nile, ERITREA, YEMEN, Addis Ababa, Grand Renaissance Dam, ETHIOPIA, SOUTH SUDAN, White Nile, SOMALIA, UGANDA, CONGO, KENYA, 0 500 km

EXAM TIP

Consider different ways that you could use the information about the Renaissance Dam. For example, the impact of dams, the management of international drainage basins, water security, water supply, renewable energy, sustainability, etc.

CHECK YOUR UNDERSTANDING

23. Outline the benefits to Sudan of the GERD.
24. Suggest reasons why Egypt does not approve of the development of the GERD.

Increased dam building

THE BUILDING OF LARGE DAMS

The number of large dams (more than 15 m high) that are being built is increasingly rapidly and is reaching a level of almost two completions every day. Famous dams include the Akosombo (Ghana), Tucurui (Brazil), Hoover (USA), and Kariba (Zimbabwe).

CASE STUDY

THE THREE GORGES DAM

The Three Gorges Dam on China's Yangtze River is the world's largest dam, at over 2 kilometres long and 100 metres high. The lake behind the dam is over 600 kilometres long. The dam was built to help meet China's ever-increasing need for electricity and water storage, as the population moves from a sustainable existence to a more western-style urban culture. Over a million people were moved out of the valley to make way for the dam and the lake. The Yangtze river basin provides 66% of China's rice and contains 400 million people. The river drains 1.8 million square kilometres and discharges 700 cubic kilometres of water annually.

Advantages and disadvantages of the Three Gorges Dam

Advantages	Disadvantages
The dam can generate up to 18,000 megawatts, eight times more than the Aswan Dam and 50% more than the world's next largest HEP dam, the Itaipu in Paraguay.	Up to 1.2 million people had to be moved to make way for the dam. Dozens of towns, for example, Wanxian and Fuling with 140,000 and 80,000 people respectively, had to be flooded. Much of the land available for resettlement is over 800 metres above sea level, therefore it is colder with infertile thin soils and on relatively steep slopes.
It has reduced China's dependency on coal.	To reduce the silt load, afforestation is needed but resettlement of people is causing greater pressure on the slopes above the dam.
It will supply energy to Shanghai and Chongqing, an area earmarked for economic development.	Up to 530 million tonnes of silt are carried through the Gorge annually. The port at the head of the lake may become silted up as a result of increased deposition and the development of a delta at the head of the lake. The mouth of the river may be starved of silt, and erosion of the coastline may result.
It protects 10 million people from flooding (over 300,000 people in China died as a result of flooding in the 20th century).	Most floods in recent years have come from rivers which join the Yangtze below the Three Gorges Dam.
It allows shipping above the dam: the dams have raised water levels by 90 metres, and turned the rapids in the gorge into a lake.	The region is seismically active and landslides are frequent. The weight of the water behind the lake may contribute to seismic instability.
It has generated thousands of jobs, both in its construction and the industrial development associated with the availability of cheap energy.	Archaeological treasures were drowned, including the Zhang Fei temple. The dam has interfered with aquatic life: the Yangtze River Dolphin is now believed to be extinct.
	It cost as much as $70 billion.

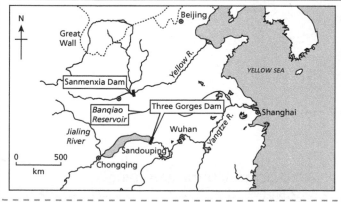

CHECK YOUR UNDERSTANDING

25. Describe the distribution of large dams built between 1945 and 2005 (as shown on page 17).

26. Outline the main advantages of large dams.

Integrated drainage basin management (IDBM)

IDBM PLANS

Integrated drainage basin management (IDBM) plans aim to deliver sustainable use of the world's limited freshwater resources. It uses a basin-wide framework for water management that is economically, socially and environmentally sustainable. It is not always possible, especially when neighbouring countries are in conflict with each other.

Examples of poor drainage basin management are frequent. For example, post-1940s development on the Missouri and Mississippi rivers in the USA focused on dam building, which benefitted downstream states at the expense of upstream states. Indigenous people were not consulted over their needs, and ecosystem needs were often ignored compared with economic needs. The 1993 floods cost an estimated $16 billion, due to a combination of floodplain construction and the failure of artificial levees.

In contrast, management of the Danube, Europe's second longest river, is seen as a success. The Danube has high biodiversity, and is well known for its lakes, wetlands, floodplain forests, and meadows and it provides important spawning grounds for fish. It is also a source of drinking water for 20 million people, and is economically important for industry, mining, farming and energy production. It is an important transport route. Over 80% of the original floodplain was lost due to construction projects, and other problems included river erosion, declining biodiversity,

draining of wetlands for biodiversity, pollution from industry and agriculture, toxic spills and flooding.

The basin is occupied by 17 countries, some of whom were divided by the Cold War, which made inert-basin management impossible. With political change in the late 1980s and 1990s, a new era of political and economic development occurred in much of Eastern Europe.

An International Commission for the Protection of the Danube River (ICPDR) was established, and it focuses on water quality, risks from accidents, monitoring and information management, river basin management and implementation of the EU Water Framework directive.

In addition, the lower basin countries (Bulgaria, Moldova, Romania and Ukraine) established the Lower Danube Green Corridor. This consists of a 400,000 ha network of protected wetland areas, 100,000 ha of newly protected wetlands, and the restoration of 200,000 other sites. This has increased the capacity of the Danube to reduce pollution, purify water, retain floodwaters, support fisheries and tourism, and provide new habitats for wildlife.

Another example comes from Costa Rica, where a tax on fossil-fuel use and payments from private hydroelectric companies is used to pay forest owners for maintaining their forests in upland areas. This benefits users in lowland areas through regulation of water quality and quantity. The power companies benefit through reduced sedimentation behind dams. However, some forest owners have complained that the payments made are not enough to be economically viable.

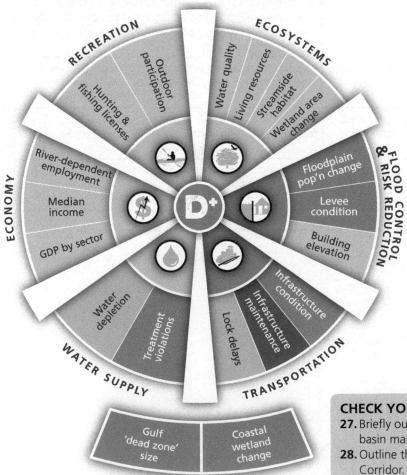

Watershed-wide indicators

CHECK YOUR UNDERSTANDING
27. Briefly outline the main aims of integrated drainage basin management plans.
28. Outline the successes of the Lower Danube Green Corridor.

Wetlands

RAMSAR CONVENTION

The Ramsar Convention, an international treaty to conserve wetlands, defines wetlands as "areas of marsh, fen, peatland or water, whether natural or artificial, permanent or temporary, with water that is static or flowing, fresh, brackish or salt". Thus, according to the Ramsar classification, there are marine, coastal and inland types, subdivided into 30 categories of natural wetland and nine human-made ones, such as reservoirs, barrages and gravel pits. Wetlands now represent only 6% of the earth's surface, of which 30% are bogs, 26% are fens, 20% are swamps, 15% are floodplains and 2% are lakes.

THE VALUE OF WETLANDS

Wetlands provide many important social, economic and environmental benefits.

Functions	Products	Attributes
Flood control	Fisheries	Biological diversity
Sediment accretion and deposition	Game	Culture and heritage
Groundwater recharge	Forage	
Groundwater discharge	Timber	
Water purification	Water	
Storage of organic matter		
Food-chain support/cycling		
Water transport		
Tourism/recreation		

LOSS AND DEGRADATION

The loss and degradation of wetlands is caused by several factors, including:
- increased demand for agricultural land
- population growth
- infrastructure development
- river flow regulation
- invasion of non-native species and pollution
- natural causes such as drought and hurricanes.

CASE STUDY

CHANGING RIVER MANAGEMENT – THE KISSIMMEE RIVER

Between 1962 and 1971 engineering changes were made to deepen, straighten and widen the Kissimmee River, which was transformed into a 90-kilometre, 10-metre deep drainage canal. The river was **channelized** to provide flood protection for land adjacent to the river.

The channelization of the Kissimee River had several unintended impacts:
- the loss of 2,000 to 14,000 hectares of wetlands
- a 90% reduction in wading bird and waterfowl usage
- a continuing long-term decline in game fish populations.

Concerns about the **sustainability** of existing ecosystems led to a state and federally supported restoration study. The result was a massive restoration project, on a scale unmatched elsewhere.

Between 1999 and 2015 over 100 square kilometres of river and associated wetlands were restored. The project created over 11,000 hectares of wetlands.

The costs of restoration
- The project cost over $410 million.
- Restoration of the river's floodplain may result in higher losses of water due to evapotranspiration during wet periods.

Benefits of restoration
- Higher water levels should ultimately support a natural river ecosystem again.
- Reestablishment of floodplain wetlands and the associated nutrient filtration function should result in decreased nutrient loads to Lake Okeechobee.
- Populations of key bird species such as wading birds and waterfowl have returned to the restored area, and in some cases numbers have more than tripled.
- Dissolved oxygen levels have doubled, which is critical for the survival of fish and other aquatic species.

Potential revenue associated with increased recreational usage (such as hunting and fishing) and ecotourism on the restored river could significantly enhance local and regional economies.

CHECK YOUR UNDERSTANDING
29. Outline the main advantages of wetlands.
30. Distinguish between natural and human-induced loss of wetlands.

Community participation

WATER SUPPLY

Water resources can be managed sustainably if individuals and communities make changes locally and this is supported by national government. Use can be reduced by self-imposed restraint. For example, using water only when it is essential, minimizing waste, and reusing supplies such as bath water. Education campaigns can increase local awareness of issues and encourage water conservation. There are many opportunities to increase freshwater supplies:

- retain water in reservoirs for use in dry seasons
- redistribute water from wetter areas to drier areas
- water conservation (for example, recycle grey water – water that has already been used so is not fit for drinking but could be used for other purposes).

Water harvesting includes:

- extraction from rivers and lakes (for example, by primitive forms of irrigation such as the shaduf and Archimedes screw), aided by gravity
- trapping behind dams and banks (bunds)
- pumping from aquifers (water-bearing rocks).

These can be achieved with either high-technology or low-technology methods.

Water can be collected from rivers and lakes, although this places a burden on those collecting it, especially women and children. In other locations, groundwater may be tapped by using pumps, but it is important that the rate of use does not exceed the rate of recharge.

SUSTAINABLE USE OF WATER

Water law principles of South Africa

All water is a resource common to all, the use of which should be subject to national control.
There shall be no ownership of water but only a right to its use.
The objective of managing the nation's water resources is to achieve optimum long-term social and economic benefit for our society from their use, recognizing that water allocations may have to change over time.
The water required to meet peoples' basic domestic needs should be reserved.
The development, apportionment and management of water resources should be carried out using the criteria of public interest, sustainability, equity and efficiency of use in a manner which reflects the value of water to society while ensuring that basic domestic needs, the requirements of the environment and international obligations are met.
Responsibility should, where possible, be *delegated* to a catchment or regional level in such a manner as to enable interested parties to participate and reach consensus.
The right of all citizens to have access to basic water services (the provision of potable water supply and the removal and disposal of human *excreta* and waste water) necessary to afford them a healthy environment on an equitable and economically and environmentally sustainable basis should be supported.

Source: Adapted from Department of Water Affairs and Forestry, South Africa, Water law principles

Check dam, Eastern Cape, South Africa

EXAM TIP
It is always useful to back up your answers with evidence. Some examples are given on this page. You could find out what is done in your home area to provide water, and what is being done to make sure that it is used (more) sustainably.

Stone plinth for water collection, Antigua

CHECK YOUR UNDERSTANDING
31. Explain the meaning of the term "water harvesting".
32. Outline the main objective of South Africa's water law principles.

Exam practice

The maps below show the number of large dams around the world in 1945 and 2005.

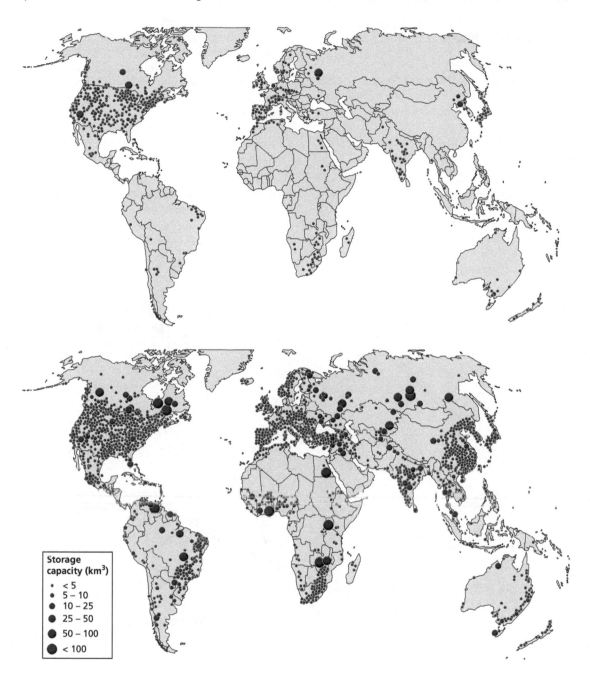

Storage
capacity (km³)

· < 5
· 5 – 10
● 10 – 25
● 25 – 50
● 50 – 100
● < 100

(a) (i) Describe the distribution of large dams in 1945. (2 marks)

(ii) Outline the main changes to the distribution of large dams by 2005. (2 marks)

(b) Explain one advantage and two disadvantages of the development of large dams. (2 + 2 + 2 marks)

(c) Either

Discuss the view that agricultural activities inevitably lead to a decline in water quality. (10 marks)

Or

Evaluate the efforts to protect wetlands from increasing human pressures. (10 marks)

Ocean currents

SURFACE OCEAN CURRENTS

Surface ocean currents are caused by the influence of prevailing winds blowing steadily across the sea. The main pattern of surface ocean currents is roughly circular flow, clockwise in the northern hemisphere and anticlockwise in the southern hemisphere.

The effect of ocean currents on temperatures depends on whether the current is cold or warm.

- **Warm currents** from equatorial regions raise the temperatures of polar areas (with the aid of prevailing westerly winds). However, the effect is only noticeable in winter. For example, the North Atlantic Drift raises the winter temperatures of north-west Europe.
- **Cold currents** such as the Labrador Current off the north-east coast of North America may reduce summer temperature, but only if the wind blows from the sea to the land.

UPWELLING CURRENTS

Many eastern oceans experience **upwelling currents**, where the ocean currents move cold water, rich in nutrients, from the ocean floor to the surface. Such upwelling currents are found off the coast of Peru, California and south-west Africa. These nutrient-rich waters support important fisheries. The best-known upwelling current, off the coast of Peru, disappears periodically during El Niño events.

Ocean conveyor belts

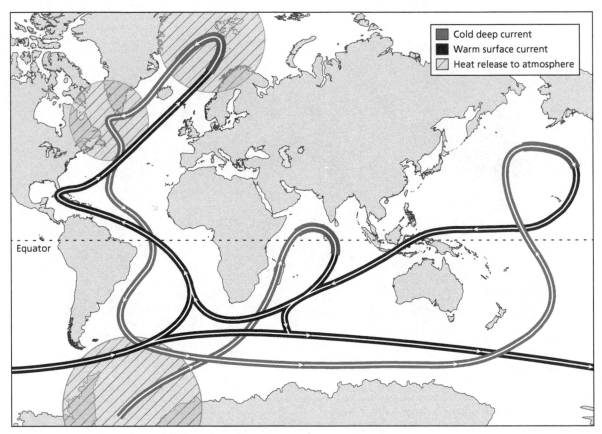

- Cold deep current
- Warm surface current
- Heat release to atmosphere

Equator

Driven by changes in temperature and salinity, large ocean currents are in constant motion, moving heat from the equator to the upper latitudes and then moving cold back toward the equator.

Known as "thermohaline circulation", this phenomenon includes the Gulf Stream, which moderates northern Europe's climate.

Ocean conveyor belts are important for the transfer of energy and nutrients around the world.

Some scientists speculate that global warming could weaken this circulation and leave some regions relatively cooler.

CHECK YOUR UNDERSTANDING

1. Outline the impact of ocean currents on the climates of ocean margins.
2. Identify two ways in which ocean currents may benefit human activities.

El Niño and La Niña

EL NIÑO

El Niño – the "Christ Child" – is a reversal of the normal atmospheric circulation in the southern Pacific Ocean. It involves a warming of the eastern Pacific that occurs at intervals between two and ten years, and lasts for up to two years.

LA NIÑA

La Niña is an intermittent cold current that flows from the east across the equatorial Pacific Ocean. It is an intensification of normal conditions whereby strong easterly winds push cold upwelling water off the coast of South America into the western Pacific. Its impact extends beyond the Pacific and has been linked with unusual rainfall patterns in the Sahel and in India, and with unusual temperature patterns in Canada.

NORMAL CONDITIONS IN THE PACIFIC OCEAN

The Walker circulation is the normal east–west circulation that occurs in low latitudes. Near South America winds blow offshore, causing upwelling of the cold, nutrient-rich waters. In contrast, warm surface water is pushed into the western Pacific. Under normal conditions, sea surface temperatures (SSTs) in the western Pacific are over 28°C, causing an area of low pressure and producing high rainfall. Towards South America SSTs are lower, high pressure exists, and conditions are dry.

High-altitude flow

Rising air

Descending air

Equator

Low pressure ← Surface flow ← High pressure

SOUTH AMERICA

AUSTRALIA

Pacific Ocean

Several degrees warmer and up to 1m higher than in Eastern Pacific

120°   180°   120°   60°

Upwelling cold water brings nutrients and encourages growth of plankton and fish stocks

EL NIÑO CONDITIONS IN THE PACIFIC OCEAN

During El Niño southern oscillation episodes, the normal pattern of air circulation is reversed. Water temperatures in the eastern Pacific rise as warm water from the western Pacific flows into the east Pacific. Low pressure develops over the eastern Pacific, whereas high pressure develops over the western Pacific. Consequently, heavy rainfall occurs over coastal South America whereas the western Pacific experiences warm, dry conditions.

The 2015–16 El Niño event is believed to have been the second strongest on record. It had many impacts. For example:
* around 100 million people experienced water and food shortages
* southern Africa experienced its driest year for 35 years

* in Zimbabwe food prices increased by over 50% compared with 2015
* the heavy rains in South America enabled the spread of the Zika virus (there was more stagnant water for breeding).

High-altitude flow

Descending air

Rising air

Fish catches down 20%

High pressure → Surface flow → Low pressure

Equator

SOUTH AMERICA

AUSTRALIA

Pacific Ocean

Drought in Australia's green belt Increase in bush fires

Floods in Peru and Chile

120°   180°   120°   60°

Warm currents force fish to move offshore to colder water – beyond range

MANAGING THE IMPACTS OF EL NIÑO AND LA NIÑA

It is difficult to manage the impact of these events. In the past El Niño events could not be predicted accurately. Now there are sensors across the Pacific that predict El Niño months in advance: one was predicted so far in advance that Peru was supplied with food and people moved from vulnerable areas. Some problems remain, however:
* they affect large parts of the globe, not just the Pacific
* some of the countries affected do not have the resources to cope
* there are indirect impacts on other parts of the world though trade and aid (teleconnections).

CHECK YOUR UNDERSTANDING
3. Compare the conditions in the south Pacific Ocean under normal conditions and during El Niño conditions.
4. Outline the impacts of the 2015 El Niño event.

Hurricanes

Hurricanes (known as "cyclones" in the South Pacific and Indian Ocean and as "typhoons" in Japan) are intense low-pressure systems that bring heavy rainfall, strong winds, and high waves and cause other hazards such as flooding and mudslides. Hurricanes are also characterised by enormous quantities of water. This is due to their origin over moist tropical seas. High intensity rainfall of up to 500 mm in 24 hours invariably causes flooding. Their path is erratic so it is not always possible to give more than 12 hours' warning. This is insufficient for proper evacuation measures.

A hurricane has a calm central area known as the eye. It also has very strong winds that cause most of the damage in a belt up to 300 km wide. The whole hurricane may be up to 800 km wide. In a fully developed hurricane, air pressure may drop by over 80 millibars.

Hurricanes move excess heat from low latitudes to higher latitudes. They start as small-scale tropical depressions, that is, localized areas of low pressure that cause warm air to rise. Some of these produce thunderstorms and may develop into tropical storms. However, only about 10% of tropical disturbances ever become true hurricanes, that is, tropical storms with wind speeds above 118 kph (74 mph). Hurricanes generally only form under set conditions.

- Sea temperatures must be over 27°C (since warm water provides the heat that drives the hurricane).
- Warm water needs to extend to a depth of 60 m.
- The low-pressure area has to be far enough away from the equator so that the Coriolis force (the force caused by the rotation of the earth) creates rotation in the rising air mass.

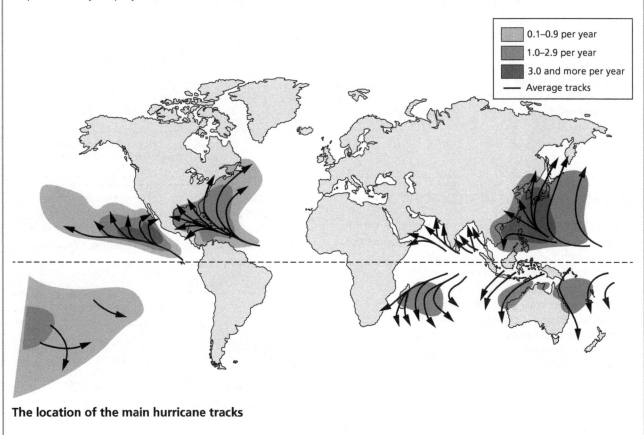

Legend:
- 0.1–0.9 per year
- 1.0–2.9 per year
- 3.0 and more per year
- Average tracks

The location of the main hurricane tracks

TYPHOON HAIYAN

The Philippines experiences about 20 typhoons every year. In November 2013, Typhoon Haiyan struck the Philippines. It was the most violent tropical cyclone to make landfall, with winds up to 315 kph (195 mph). Over 10,000 people were killed in the province of Leyte.

About 70–80% of the buildings in the area in the path of Haiyan were destroyed. The storm surge caused sea waters to rise by over 6 m. Power was knocked out and there was no mobile phone signal, making communication possible only by radio. Those living in the hardest-hit areas, such as the eastern Visayas, are among the poorest in the Philippines. Many had little or no savings so the typhoon caused an already vulnerable population even greater risk of future food and job insecurity.

The estimated economic cost is about $15 billion.

CHECK YOUR UNDERSTANDING
5. Describe the distribution of hurricanes as shown in the map.
6. Outline the main impacts of Typhoon Haiyan.

Oceans as a source and store of carbon dioxide

THE CARBON CYCLE

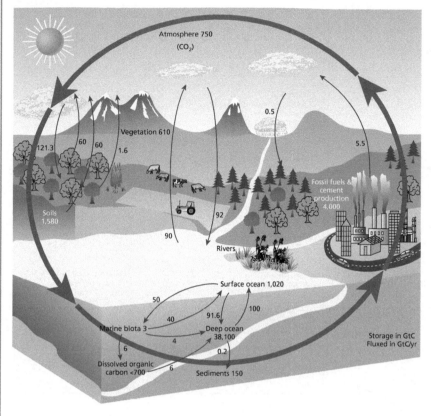

- Oceans are the largest CO_2 sink on Earth. Over geological time, over 90% of the world's carbon has settled in the ocean.
- Photosynthesis turns carbon dioxide into organic material. Over time, organic carbon settles into the deep ocean. In a thermohaline circulation, carbon on the ocean floor could be lifted to the surface, so the ocean could become a source of CO_2 rather than a sink.
- Oceans play a key role in the carbon cycle. The major reservoirs of carbon dioxide are the oceans ($38,000 \times 10^{12}$ kg of carbon), fossil fuels ($10,000 \times 10^{12}$ kg of carbon) and the atmosphere (750×10^{12} kg of carbon).
- Photosynthesis by plankton generates organic compounds of carbon dioxide. Some of this material passes through the food chain and sinks to the ocean floor where it is decomposed into sediments. Eventually it is destroyed at subduction zones.
- Carbon dioxide may be released during volcanic activity.
- The transfer of carbon dioxide from ocean to atmosphere involves a very long time-scale.

OCEAN ACIDIFICATION

Ocean acidification is caused by anthropogenic (man-made) sources – such as carbon emissions from industrial plants, power stations, vehicles and planes.

More acidic oceans are beginning to kill off coral reefs and shellfish beds and threaten stocks of fish. Increasing acidification reduces calcification in coral, resulting in slower growth and weaker skeletons. Oceans absorb around a million tonnes of carbon dioxide every hour and are now 30% more acidic than they were last century. On the Great Barrier Reef the growth rate of some coral pecies has declined by 14% since 1990 due to a combination of ocean acidification and temperature stress.

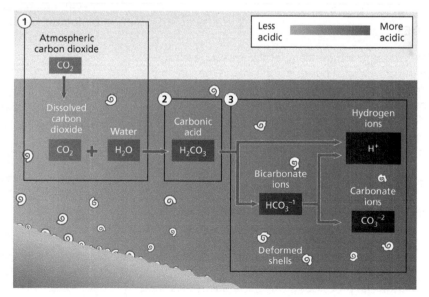

CHECK YOUR UNDERSTANDING

7. Outline the reasons for the acidification of the oceans.
8. Explain the process of ocean acidification.

Physical influences on coastal landscapes

WAVES

Constructive waves:
- Depositional waves
- Also called "spilling" or "swell" waves
- Long wavelength, low height
- Low frequency (6–8 per minute)
- High period (one every 8–10 seconds)
- Swash greater than backwash

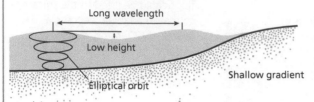

Destructive waves:
- Erosional waves
- Also called "surging", "storm" or "plunging waves"
- Short wavelength, high height
- High frequency (10–12 per minute)
- Low period (one every 5–6 seconds)
- Backwash greater than swash

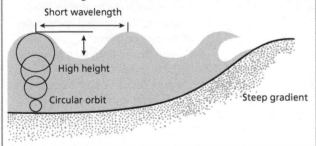

TIDES

Tides are regular movements of the sea's surface caused by the gravitational pull of the moon and the sun on the oceans. Tides are greatest in bays and funnel-shaped coastlines. The tidal range (difference between high tide and low tide) controls the vertical range of erosion, deposition, weathering and biological activity. It can also have a scouring effect and remove debris.

SEDIMENT

There are many sources of sediment supply.
- **Mass movements** provide large amounts of material which may bury beaches and protect cliffs.
- **Rivers** mostly carry sediment (fine-grained silts, clays and sands) to the coast.
- **Periglacial processes** provide frost-shattered shingle for beaches.
- **Erosion of cliffs** by the sea produces large amounts of material for beach building. This may protect the cliff from further erosion.
- The **sea** may transport sediments shorewards forming offshore bars and beaches.
- **Wind erosion** and transport carries a lot of fine sand.
- **Volcanic activity** may produce dust and ash for beaches.

SUBAERIAL AND WAVE PROCESSES

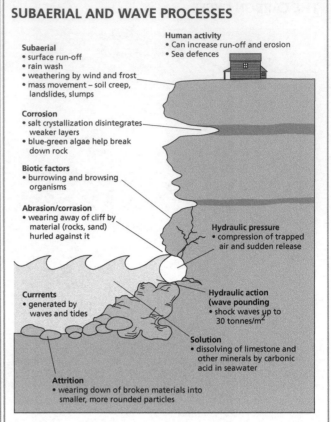

Sub-aerial, or cliff-face, processes include:
- **salt weathering:** the process by which sodium and magnesium compounds expand in joints and cracks, thereby weakening rock structures
- **freeze–thaw weathering:** the process whereby water freezes, expands and degrades jointed rocks
- **biological weathering:** carried out by molluscs, sponges and urchins. It is very important in low-energy coasts.

LITTORAL DRIFT

Littoral or longshore drift is the movement of sediment along the coastline.

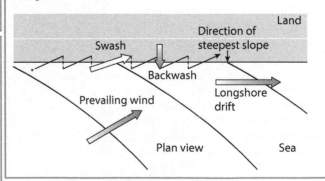

CHECK YOUR UNDERSTANDING
9. Contrast the movement of swash and backwash on a beach.
10. Outline two forms of hydraulic action.

Coastal landforms

CLIFFS

The profile of a cliff depends upon a number of factors including:
- geological structure
- subaerial and marine processes
- amount of undercutting
- rates of removal
- stage of development.

Rocks of low resistance are easily eroded and are unable to support an overhang. Jointing may determine the location of weaknesses in the rock, just as the angle of dip may control the shape of the cliff.

WAVE-CUT PLATFORM

Wave-cut platforms (also called shore platforms) include **intertidal platforms** (wave-cut platforms), **high-tide platforms** and **low-tide platforms**. Wave-cut platforms are most frequently found in high-energy environments and are typically less than 500 m wide with an angle of about 1°. A model of cliff- and shore-platform evolution shows how steep cliffs (1) are replaced by a lengthening platform and lower angle cliffs (5), subjected to subaerial processes rather than marine forces. Alternatively, platforms might have been formed by frost action, salt weathering, or biological action during lower sea levels and different climates.

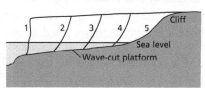

BEACHES

Beaches are an accumulation of sand or shingle in areas which are sheltered and/or have a large supply of sediment

Storm beach – a noticeable, semi-permanent ridge, found at the level of the highest spring tides

Berms – small-scale beach ridges built up by successive levels of tides or storms

Cusps – semi-circular embayments found in the shingle or at the shingle–sand interface

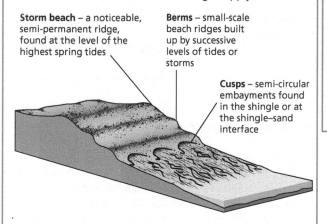

Beach form is affected by the size, shape and composition of materials, tidal range, and wave characteristics. Sediment size affects beach profile through its percolation rate.

STACKS

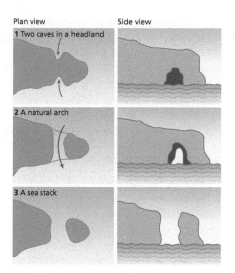

1 Wave refraction concentrates erosion on the sides of headlands. Weaknesses such as joints or cracks in the rock are exploited, forming caves.
2 Caves enlarge and are eroded further back into the headland until eventually the caves from each side meet and an arch is formed.
3 Continued erosion, weathering and mass movements enlarge the arch and cause the roof of the arch to collapse, forming a high standing stack.

Wave refraction concentrates wave energy on the sides of headlands. If there are lines of weakness, these may be eroded and widened. Over time these may be enlarged to form caves, and if the caves on either side of a headland merge, an arch is formed. Further erosion and weathering of the arch may cause the roof to collapse, leaving an upstanding **stack**. The eventual erosion of the stack produces a stump.

SPITS

A spit is a beach of sand or shingle linked at one end to land. It is found on indented coastlines or at river mouths. For example, along a coast where headlands and bays are common and near mouths (estuaries and rias), wave energy is reduced. They generally have a thin, attached end, the **proximal** end, and a larger, **distal** end with smaller recurves. Spits often become curved as waves undergo refraction. Cross-currents or occasional storm waves may assist this hooked formation.

EXAM TIP
If you are using a diagram to help explain the formation of a physical feature, make sure that you add labels that explain the processes that have formed it.

CHECK YOUR UNDERSTANDING
11. Briefly explain the formation of wave-cut platforms.
12. Explain the formation of a spit.

Eustatic and isostatic change

CHANGING SEA LEVELS

Sea levels change in connection with the growth and decay of ice sheets. **Eustatic** change refers to a global change in sea level. At the height of glacial advance, 18,000 years ago, sea level was 100–150 m below current sea level.

The level of the land also varies in relation to the sea. Land may rise as a result of tectonic uplift or following the removal of an ice sheet. The localized change in the level of the land relative to the level of the sea is known as **isostatic adjustment** or **isostasy**. Parts of Scandinavia and Canada are continuing to rise at rates of up to 20 mm/year.

A simple sequence of sea level change can be described:

- Temperatures decrease, glaciers and ice sheets advance and sea levels fall eustatically.
- Ice thickness increases and the land is lowered isostatically.
- Temperatures rise, ice melts, and sea levels rise eustatically.
- Continued melting releases pressure on the land and the land rises isostatically.

Although eustatic changes are global changes in sea level, not all places will show evidence of the rise or fall of sea level, and not to the same degree. This is because, in addition to eustatic changes, there are isostatic changes – local changes – meaning that different places will have different relative rises and falls in sea level.

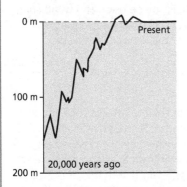

Sea level change on the south coast of Britain

As a result of global warming (the enhanced greenhouse effect) sea levels are rising, impacting especially on low-lying communities.

❓

CHECK YOUR UNDERSTANDING

13. Describe the changes in sea level on the south coast of Britain over the last 20,000 years.
14. Identify two conditions which can lead to advancing and retreating coasts.

ADVANCING AND RETREATING COASTLINES

According to Valentin's classification:

- **retreating coasts** include submerged coasts and coasts where the rate of erosion exceeds the rate of emergence/deposition
- **advancing coasts** include emerged coastlines and coasts where deposition is rapid.

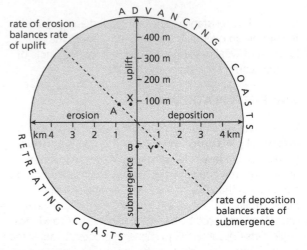

Advancing and retreating coasts

Advancing and retreating coasts result from sea-level changes and from erosion and deposition. Points A and X both experienced uplift of around 100 metres, but erosion has taken place at A, reducing the effect of uplift. Points B and Y have experienced about 100 metres of submergence, but Y has experienced deposition, so the overall change in its position relative to sea level is reduced.

FEATURES OF EMERGED COASTLINES

These include:
- raised beaches, such as the Portland Raised Beach UK
- coastal plains
- relict cliffs, such as those along the Fall Line in eastern USA
- raised mudflats, for example, the Carselands of the River Forth in Scotland.

SUBMERGED COASTLINES

These include:
- rias, such as the River Fal, drowned river valleys caused by rising sea levels or due to a sinking of the land
- fjords such as Milford Sound, New Zealand, and the Oslo Fjord, Norway, caused by drowning of U-shaped valleys.
- fjards or "drowned glacial lowlands".

Sand dune development

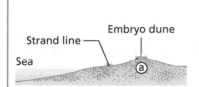

As the tide goes out, the sand dries out and is blown up the beach. At the top of the beach is a line of seaweed and litter called the strand line. A small embryo dune forms in the shelter behind the strand line. This dune can be easily destroyed unless colonized by plants.

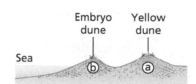

Sea couch grass colonizes and helps bind the sand. Once the dune grows to over 1 m high, marram grass replaces the sea couch. A yellow dune forms at 10–20 m high with the long-rooted marram forming a good sand trap.

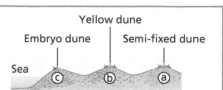

Once the yellow dune is over 10 m high, less sand builds up behind it and marram grass dies to form a thin humus layer. As soil begins to form, other plants are able to grow on the dune including dandelions. This kind of dune is called a semi-fixed dune. As the original dune (a) has developed, new embryo and yellow dunes have formed.

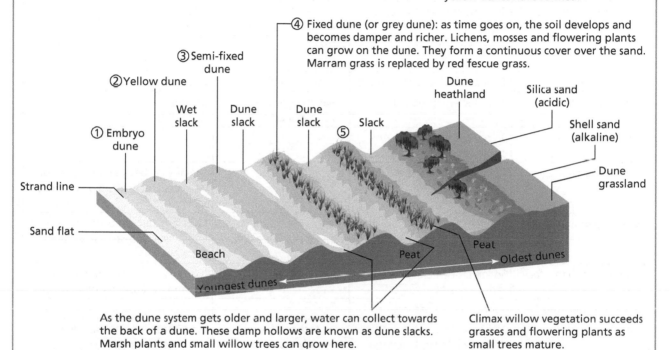

④ Fixed dune (or grey dune): as time goes on, the soil develops and becomes damper and richer. Lichens, mosses and flowering plants can grow on the dune. They form a continuous cover over the sand. Marram grass is replaced by red fescue grass.

As the dune system gets older and larger, water can collect towards the back of a dune. These damp hollows are known as dune slacks. Marsh plants and small willow trees can grow here.

Climax willow vegetation succeeds grasses and flowering plants as small trees mature.

Marram trapping sand on the yellow dune

Heather and gorse on the older dunes

COMMON MISTAKE

✗ *Vegetation development on sand dunes will always lead to the same species being present.*

✓ The type of vegetation that eventually develops depends on many factors such as soil type, rock type, climate, moisture content and human activities such as burning and the introduction of exotic species.

CHECK YOUR UNDERSTANDING

15. Suggest how wind speed and moisture availability vary between the beach and the fixed dunes.
16. Suggest why the dunes nearer the beach are called yellow dunes whereas those furthest away from the sea are called grey dunes.

Coastal management

COSTS AND BENEFITS OF COASTAL PROTECTION STRATEGIES

Type of management	Aims/methods	Strengths	Weaknesses
Hard engineering	To control natural processes		
Cliff-base management	*To stop cliff or beach erosion*		
Sea walls	Large-scale concrete curved walls designed to reflect wave energy	Easily made; good in areas of high density	Expensive. Life span about 30–40 years. Foundations may be undermined
Revetments	Porous design to absorb wave energy	Easily made; cheaper than sea walls	Life span limited
Gabions	Rocks held in wire cages absorb wave energy	Cheaper than sea walls and revetments	Small scale
Groynes	To prevent longshore drift	Relatively low cost; easily repaired	Cause erosion on downdrift side; interrupts sediment flow
Rock armour	Large rocks at base of cliff to absorb wave energy	Cheap	Unattractive; small-scale; may be removed in heavy storms
Offshore breakwaters	Reduce wave power offshore	Cheap to build	Disrupts local ecology
Rock strongpoints	To reduce longshore drift	Relatively low cost; easily repaired	Disrupts longshore drift; erosion downdrift
Cliff-face strategies	*To reduce the impacts of subaerial processes*		
Cliff drainage	Removal of water from rocks in the cliff	Cost effective	Drains may become new lines of weakness; dry cliffs may produce rockfalls
Cliff regrading	Lowering of slope angle to make cliff safer	Useful on clay (most other measures are not)	Uses large amounts of land – impractical in heavily populated areas
Vegetating the surface	To increase interception and reduce overland run-off	Relatively low cost	May increase moisture content of soil and increase risk of landslides
Soft engineering	Working with nature		
Off shore reefs	Waste materials, e.g. old tyres weighted down, to reduce speed of incoming wave	Low technology and relatively cost effective	Long-term impacts unknown
Beach nourishment	Sand pumped from seabed to replace eroded sand	Looks natural	Expensive; short-term solution
Managed retreat	Coastline allowed to retreat in certain places	Cost effective; maintains a natural coastline	Unpopular; political implications
"Do nothing"	Accept that nature will win	Cost effective!	Unpopular; political implications
Red-lining	Planning permission withdrawn; new line of defences set back from existing coastline	Cost-effective	Unpopular; political implications

COMMON MISTAKE

✗ *Coastal management schemes always protect the coastline.*

✓ Some forms of coastal protection, notably groynes, may actually damage stretches of coastline downdrift, as they trap sediment on the updrift side, and may starve stretches downdrift of beach material. Sea walls will only protect up to a certain height – intense storms may have waves that exceed that height.

CHECK YOUR UNDERSTANDING

17. Distinguish between gabions and rock armour.
18. Outline two cliff-face methods of coastal protection.

Conflicting pressures on coastlines due to land use

SOUFRIÈRE, ST LUCIA

Soufrière is located on the Caribbean island of St Lucia. Traditionally, agriculture and fishing have provided the main sources of employment and income. However, with the development of the tourism industry, fishermen and farmers are competing for resources with a variety of tourism-related users.

In the past, attempts to manage resources generally adopted a top-down approach in which regulations were established with little or no public consultation. The creation of the Soufrière Marine Management Area (SMMA) in 1995, introduced a **stewardship** approach to the sustainable management of resources within the area.

The SMMA agreement created five different types of zone within the area:
- marine reserve areas (MRA): designed to protect natural resources
- fishing priority areas: designed for maintaining and sustaining fishing activities
- recreational areas: beaches and marine (swimming and snorkelling) areas
- yachting areas: designated for pleasure boats and yachts
- multiple-use areas: for fishing, diving, snorkelling and other recreational activities.

Research has shown that coastal stewardship is successful when there is a clear benefit to be derived by those involved. If there are no perceivable benefits, as has been the case with beach-monitoring activities, such actions are often given low priority. Recent surveys suggest that the SMMA has seen an increase in the amount of fish caught and in fish biodiversity. Moreover, there has been less damage to coral reefs resulting from human activity.

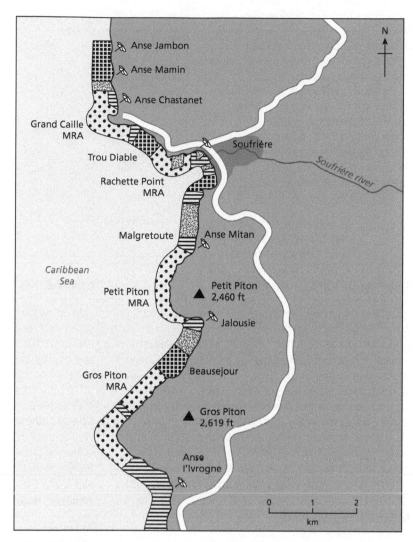

Marine Reserves
Allows fish stocks to regenerate and protects marine flora and fauna.
Access to the area is by permit and can be enjoyed by divers and snorkellers.

Fishing Priority Areas
Commercial fishing has precedence over all other activities in these areas.

Yachting areas
Yachting is not allowed in the SMMA. Moorings are provided in these areas only.

Multiple-use areas
Fishing, diving, snorkelling and other legitimate uses are allowed.

Recreational areas
Areas for public recreation – sunbathing, swimming.

CHECK YOUR UNDERSTANDING
19. Outline the function of the marine reserve areas.
20. Identify the varied stakeholders covered by the Soufrière Marine Management Area.

Coral reefs

Coral reefs are often described as the "rainforests of the sea" on account of their rich biodiversity. Some **coral** is believed to be 2 million years old, although most is less than 10,000 years old. Coral reefs contain nearly a million species of plants and animals, and about 25% of the world's sea fish breed, grow, spawn and evade predators in coral reefs. Some of the world's best coral reefs include Australia's Great Barrier Reef and many of the reefs around the Philippines and Indonesia.

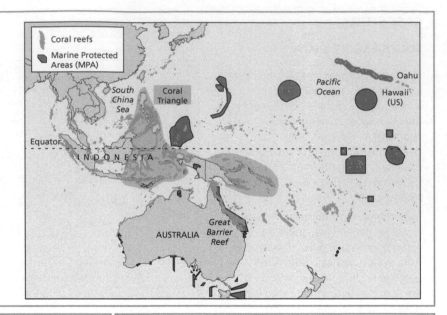

Coral reefs and Marine Protected Areas in the western Pacific Ocean

THE VALUE OF CORAL REEFS

Coral reefs have major biological and economic importance. About 4,000 species of fish and 800 species of reef-building corals have been identified. They protect coasts against erosion. Coral reefs generate large profits for some users. Tourism generated by the Great Barrier Reef is worth about A\$4.6 billion annually to Queensland alone. The global value of coral reefs in terms of fisheries, tourism and coastal protection is estimated to be US\$375 billion.

There are many stakeholders with an interest in coral reefs including fishermen, tourists and people involved in the tourism sector, conservationists and industrialists (coral can be used as a building material or for cement).

MANGROVE SWAMPS

Mangroves are salt-tolerant forests of trees and shrubs that grow in the tidal estuaries and coastal zones of tropical areas. The muddy waters, rich in nutrients from decaying leaves and wood, are home to a great variety of sponges, worms, crustaceans, molluscs and algae. Mangroves cover about 25% of the tropical coastline, the largest being the 570,000 ha of the Sundarbans in Bangladesh.

Examples of mangrove restoration

Geographic location	Notes
Australia – Brisbane airport	11 ha; approximately 50,000 plants along a drainage canal, partial offset for losses during runway construction
Bangladesh	148,500 ha
Pakistan	19,000 ha; 16,000 ha of plantation and 3,000 ha of assisted natural regeneration
Thailand	Mangrove restoration in abandoned shrimp ponds

MANAGING CORAL REEFS

One approach is to give ecologically sensitive areas special status such as a marine protected areas (MPAs). In theory, activities that are deemed harmful, such as fishing, drilling and mining, can then be restricted or banned, with penalties for rule-breakers. The Aichi Targets, agreed in 2010 under the UN Convention on Biological Diversity, seek to have at least 17% of inland water and 10% of coastal and marine areas under conservation by 2020.

The most urgent action relates to fishing vessels. Their remoteness makes it hard to stop vessels entering illegally. Satellites are sometimes used to police MPAs.

MANAGING MANGROVE SWAMPS

Mangroves provide humans with many ecological services, including fuelwood, charcoal, timber, thatching materials, dyes, poisons and food such as shellfish. Many fish species use mangrove swamps as nurseries. In addition, mangrove trees provide protection from tropical storms, and act as sediment traps.

Owing to the large range of benefits that mangroves provide, there are many stakeholders interested in mangrove swamps including fishermen, farmers, conservationists, local residents and politicians.

Management strategies include:
- a 30-year cycle of planting and harvesting
- restoration and afforestation
- managed realignment – allowing mangroves to migrate inland
- generic protection of mangrove ecosystems.

CHECK YOUR UNDERSTANDING
21. Briefly explain the value of coral reefs.
22. Outline the ecological services provided by mangrove swamps.

Sovereignty rights and exclusive economic zones

An **economic exclusive zone (EEZ)** is an area in which a coastal nation has sovereign rights over all the economic resources of the sea, seabed and subsoil, extending up to 200 nautical miles from the coast (the international nautical mile is 1.852 km). The term **"sovereign"** means having independent authority over a territory.

Exclusive economic zones have a major impact on the management and conservation of ocean resources since they recognize the right of coastal states to "exploit, develop and manage and conserve all resources … extending almost 200 nautical miles from its shore." Almost 90% of all known oil reserves under the sea and up to 98% of the world's fishing regions fall within an EEZ.

COMPETING CLAIMS IN THE ARCTIC

That Arctic sea ice has been melting at an accelerating rate for many years. This is opening up potential trade routes, and also providing access to valuable oil and gas reserves believed to lie beneath the Arctic.

The Arctic could hold up to a quarter of the world's undiscovered gas and oil reserves, that is, some 90 billion barrels of oil and vast amounts of natural gas. Nearly 85% of these deposits are offshore. Five countries are racing to establish the limits of their territory, stretching far beyond their land borders. Canada, Denmark, Norway, Russia and the USA are competing to gain better access to the Arctic's resource base.

There is also increased potential for fishing. Planktonic animals such as small crustaceans (copepods) and krill are now abundant, and feeding on the smaller plankton that survive the Arctic winter. The increase in primary productivity could support a larger cod population. As water temperatures rise, cod populations have migrated as far north as 80°N.

Environmental groups have called for a treaty similar to that regulating the Antarctic, which bans military activity and mineral mining.

Countries around the Arctic Ocean are rushing to stake claims on the Polar Basin seabed and its oil and gas reserves, made more tempting by rising energy prices. Resolving territorial disputes in the Arctic has gained urgency because scientists believe rising temperatures could leave most of the Arctic ice-free in summer months in a few decades' time. This would improve drilling access and open up the North-West Passage, a route through the Arctic Ocean linking the Atlantic and Pacific that would reduce the sea journey from New York to Singapore by thousands of miles.

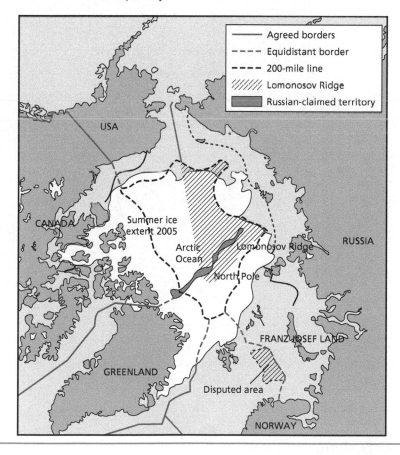

CHECK YOUR UNDERSTANDING

23. Briefly explain the meaning of the term "exclusive economic zone".
24. Outline the potential advantages of an ice-free Arctic Ocean.

Developing abiotic resources

The continental shelf contains sediments such as gravel, sand and mud. These come from the erosion of rocks and transport by rivers to the sea. Diamonds can be found in the continental shelf areas off Africa and Indonesia. Gold and manganese are found on the ocean floor.

The economics of mining has changed. Industrial commodity prices are much higher than in the 1970s, and technology has advanced. That means it may now become profitable to exploit the manganese crusts and other minerals recently discovered. A number of countries have been awarded licences by the International Seabed Authority to explore mining possibilities on the deep-ocean seabed. The Canadian company, Nautilus Minerals, plans to bring up ore containing copper and gold from the bottom of the Bismarck Sea north of Papua New Guinea, using technology developed by the offshore-oil industry.

HYDRATES

Hydrates are compounds that usually consist of methane molecules trapped in water. In the 1970s they were found on the slopes of continental shelves deep beneath the ocean surface. Some scientists believe these hydrates contain more energy than all the known deposits of fossil fuels. This makes them highly attractive to countries such as Japan and India, with little or no oil or gas. Some scientists argue that methane hydrates contain less energy than the energy needed to release and secure them.

OIL

Oil and gas deposits are found in the continental shelf. The Persian Gulf accounts for 66% of the world's proven oil reserves and 33% of the world's proven gas reserves. The continental shelf area of the Gulf of Mexico has been explored and developed since the 1940s.

Oil spills regularly contaminate coasts. Oil exploration is a major activity in such regions as the Gulf of Mexico and the South China Sea. The threats vary. There is evidence that acoustic prospecting for hydrocarbons may deter or disorientate some marine mammals. There is also growing evidence of widespread toxic effects on benthic (deep sea) communities in the vicinity of the 500-plus oil production platforms in the North Sea.

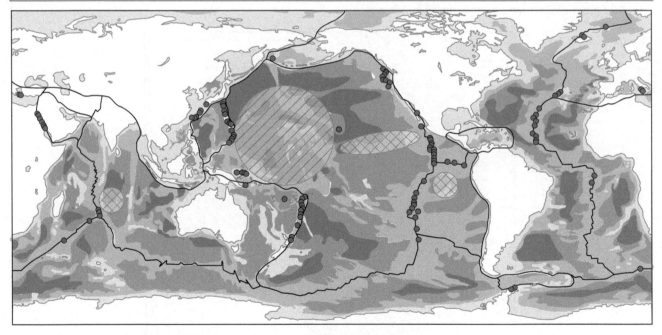

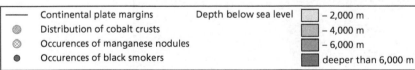

		Depth below sea level	
——	Continental plate margins		– 2,000 m
◍	Distribution of cobalt crusts		– 4,000 m
◍	Occurences of manganese nodules		– 6,000 m
●	Occurences of black smokers		deeper than 6,000 m

The location of deep ocean resources

❓

CHECK YOUR UNDERSTANDING

25. Suggest why the development of ocean abiotic resources is difficult.
26. Outline the problems associated with the development of oceanic oil resources.

Trends in biotic resource use

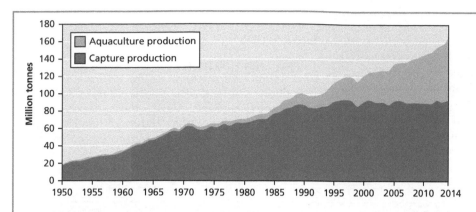

World capture fisheries and aquaculture production

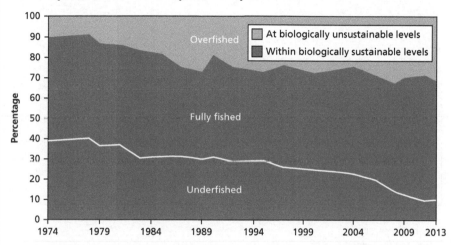

Notes: Dark shading = within biologically sustainable levels; light shading = at biologically unsustainable levels. The light line divides the stocks within biologically sustainable levels into two subcategories: fully fished (above the line) and underfished (below the line).

Global trends in the state of world marine fish stocks since 1974

World fisheries and aquaculture contributed almost 160 million tonnes of fish in 2012, valued at over $215 billion. Over 136 million tonnes were used as food for people.

The world's supply of fish as food has grown dramatically since 1961, with an average growth rate of 3.2% per year compared with a growth rate of 1.6% per year for the world's population. World food fish supply increased from an average of 9.9 kg per capita in the 1960s to 19 kg in 2012. Fish consumption was lowest in Africa, while Asia accounted for two-thirds of total consumption. Consumption in Asia reached 85.4 million tonnes, of which 42.8 million tonnes was consumed outside China. Globally, fish provides about 3.0 billion people with almost 20 per cent of their intake of animal protein.

Many argue that measures such as quotas, bans and the closing of fishing areas still fail to address the real problems of the fishing industry: too many fishermen are chasing too few fish and too many immature fish are being caught. For the fisheries to be protected, the number of boats and the number of men employed in fishing must be reduced. At the same time, the efficiencies which come from improved technology must be embraced.

CHECK YOUR UNDERSTANDING
27. Describe the changes in the size of world capture fisheries and aquaculture production.
28. Comment on the sustainability of the world's main fish stocks.

Managing ocean pollution

SOURCES OF POLLUTION

Marine-based activities that lead to pollution include the fishing industry, shipping (including the use of boats for transport, tourism and recreational fishing), offshore mining and extraction, illegal dumping at sea and discarded fishing gear. The main sources of marine litter include land-based activities such as discharge from stormwater drains, industrial outfalls, untreated municipal sewerage, littering of beaches and rivers and landfills.

CASE STUDY

DEEPWATER HORIZON

In April 2010 the Deepwater Horizon oil rig exploded and collapsed, releasing up to 4.9 million barrels of oil into the Gulf of Mexico. Over 160 km of coastline was affected, degrading marine and land-based ecosystems including oyster beds and shrimp farms.

RADIOACTIVE WASTE

Radioactive waste is also dumped in the oceans and usually comes from the nuclear power process, medical and research use of radioisotopes and industrial uses. Following the explosion of the nuclear power station at Fukushima Daiichi in Japan in 2011, radioactive waste was carried across the northern Pacific Ocean towards Canada and the USA. Between 1958 and 1992, the former Soviet Union dumped 18 unwanted nuclear reactors, several still containing their nuclear fuel, in the Arctic Ocean. Nuclear waste remains radioactive for decades.

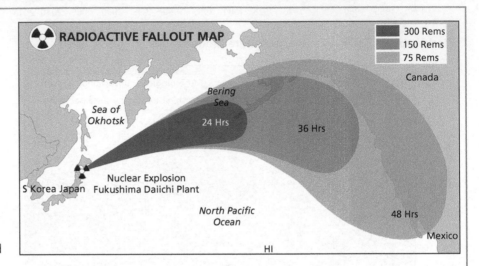

RADIOACTIVE FALLOUT MAP

300 Rems	
150 Rems	
75 Rems	

Canada

Sea of Okhotsk

Bering Sea

24 Hrs

36 Hrs

48 Hrs

Mexico

S Korea Japan

Nuclear Explosion
Fukushima Daiichi Plant

North Pacific Ocean

HI

PLASTIC

According to UNEP, in 2006 every square kilometre of sea held nearly 18,000 pieces of floating plastic. Much of it was concentrated in the Great Pacific Garbage Patch, a region containing as much as 100 million tonnes of plastic suspended in two separate gyres of garbage in an area twice the size of Texas. It can take centuries for plastic to decompose. Turtles, seals and birds inadvertently eat it, and not just in the Pacific. A Dutch study of 560 fulmars picked up dead in countries around the North Sea found 95% had plastic in their stomachs.

OIL

Shipping is a major cause of ocean pollution. Ships burn bunker oil, which may be responsible for about 60,000 deaths each year from chest and lung diseases, including cancer. Most of these occur near coastlines in Europe, East and South Asia. Oil spills have become rarer since 2010, when all single-hulled ships were banned.

ACTIONS

There have been some successes in the international handling of the marine environment. The United Nations Convention on the Law of the Sea, signed in 1982 but only entering into force in 1994, established a framework of law for the oceans, including rules for deep-sea mining and economic exclusion zones extending 200 nautical miles around nation states.

A series of international laws have effectively eliminated the discharge of toxic materials, from drums of radioactive waste to sewage sludge and air pollution from incinerator ships, into the waters around Europe. International public pressure in the mid-1990s forced the reversal by a major oil company of plans to scuttle the Brent Spar, a large structure from the North Sea offshore oil industry, into deep water west of Scotland. European agreements since then have indicated that all production platforms and other structures should be removed from the oil fields at the end of their lives wherever possible.

CHECK YOUR UNDERSTANDING

29. Using a named and located example, state how the oceans contribute to transboundary pollution.
30. Identify major sources of plastic pollution in the Pacific Ocean, and state the estimated size of the area affected.

Oceans as a source of conflict – geopolitics of oceans

GEOPOLITICS OF THE SOUTH CHINA SEA

The South China Sea is a vital trade artery, accounting for about 30% of the world's trade passes, worth over £3 trillion in value.

In 2015–16 China is alleged to have installed two launch batteries for surface-to-air missiles on Woody Island in the Paracel archipelago. The Paracels are also claimed by Vietnam and Taiwan. China insists that virtually all the sea belongs to it. The region is also rich in oil and gas.

China has also built over 4 km² of artificial land mass in the Spratly Islands to create land with facilities that could be used for military use, including an air strip with a 3000 m runway.

In 2016 China delegates visited Itu Aba, the biggest natural island in the Spratly archipelago in the South China Sea. Under UN law, Itu Aba is a rock that cannot sustain human life, so it is entitled to 12 nautical miles of territorial waters, but not the 200-mile exclusive economic zone accorded to habitable islands.

Three approaches are being tried to moderate China's behaviour – legal, diplomatic and military.

The legal case that the Philippines brought to the UNCLOS is to show that China's historic claim has no legal basis.

As for military deterrence, despite a marked increase in defence spending across the region in recent years, the USA is the only power capable of standing up to China.

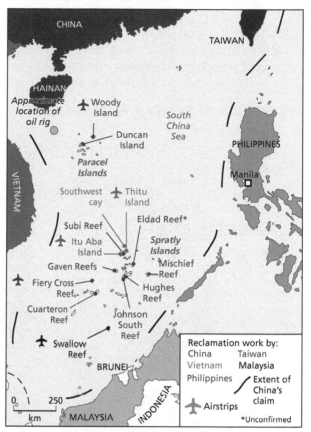

Geopolitical conflict in the South China Sea

CHECK YOUR UNDERSTANDING

31. Identify the countries involved in the conflict over the South China Sea.

32. Suggest reasons why there is a conflict over the South China Sea.

Exam practice

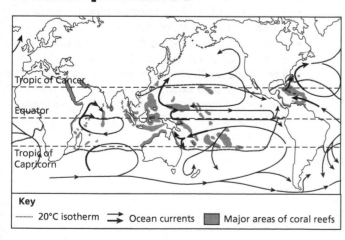

Key

------ 20°C isotherm → Ocean currents ▓ Major areas of coral reefs

(a) (i) Describe the distribution of the world's major coral reefs. (2 marks)

(ii) Explain two reasons for the distribution of coral reefs. (2 + 2 marks)

(b) Outline two advantages of coral reefs. (4 marks)

(c) **Either**

Discuss the view that oceanic pollution is an inevitable result of human activities, and cannot be managed. (10 marks)

Or

Evaluate the role of coastal processes in the development of coastal landforms. (10 marks)

1 THE CHARACTERISTICS OF EXTREME ENVIRONMENTS

The global distribution of extreme environments

Extreme environments include, among others:
- cold and high-altitude environments (polar, glacial areas, periglacial areas, high mountains in non-tropical areas)
- hot, arid environments (hot deserts and semi-arid areas).

These areas are relatively inaccessible and tend to be viewed as inhospitable to human habitation. Despite this, they provide numerous opportunities for settlement and economic activity.

DISTRIBUTION OF EXTREME ENVIRONMENTS

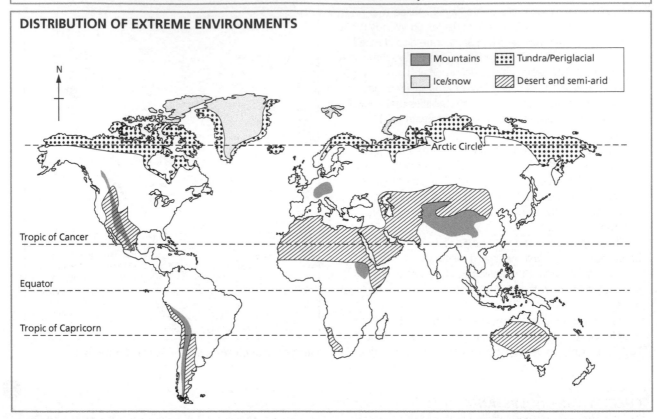

COLD AND HIGH-ALTITUDE ENVIRONMENTS

Cold environments are found in high latitudes and at high altitudes. Polar environments are located towards the North Pole and the South Pole, where levels of solar radiation are very low. In the northern hemisphere, there is a belt of **periglacial** (periglacial means "on the edge of glacial") environments. This zone is generally not found in the southern hemisphere, given the relative lack of land mass at around 60°–65°S.

Other cold environments are associated with high mountains. There are extensive areas of high ground in Asia, associated with the Himalayas. Other high-altitude areas include the Andes and the Rockies, located in the Americas.

DESERT AND SEMI-ARID ENVIRONMENTS

Desert and semi-desert areas cover as much as a third of the Earth's surface. They are generally located around the tropics and are associated with the subtropical high-pressure belt which limits the potential for rain formation. Four main factors influence the location of the world's main deserts. They include:
- the location of stable, subtropical **high-pressure conditions**, for example, the Sahara
- large distance from the sea (known as **continentality**), such as the central parts of the Sahara
- **rain-shadow effects**, as in Patagonia (South America)
- proximity to **cold upwelling currents**, which limit the amount of moisture held in the air, for example, off the west coast of southern Africa, helping to form the Namib desert.

CHECK YOUR UNDERSTANDING
1. Compare the distribution of periglacial areas with that of deserts and semi-arid environments.
2. Briefly explain the location of high mountain ranges.

Conditions in extreme environments

COLD AND HIGH-ALTITUDE ENVIRONMENTS

Cold environments are very varied. Mountain environments may have warm days and very cold nights. On average, temperatures decrease by about 10°C per 1000 m, so high-altitude areas will be cooler than surrounding low-altitude areas. Some high-altitude areas receive large amounts of relief rainfall. In contrast, other mountain areas receive low rainfall because they are in a rain shadow. Polar areas generally receive low rainfall.

Owing to their steep nature, mountains are difficult areas to build on, and they limit transport. Soils are often thin and infertile, and experience high rates of **erosion**. In contrast, in periglacial areas the low temperatures produce low rates of evaporation, and soils are frequently waterlogged. The growing season is relatively short as temperatures are above 6°C for only a few months of the year.

Climate associated with a periglacial area

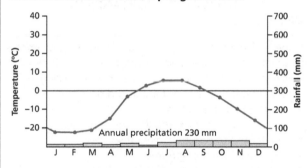

DESERT AND SEMI-ARID ENVIRONMENTS

In desert areas, the lack of water is a major barrier to development. Temperatures are hot throughout the year but, in the absence of freshwater, farming is difficult. In semi-arid areas, annual rainfall varies between 250 mm and 500 mm, so there is a greater possibility for farming, especially where water conservation methods are used. On the other hand, the guarantee of warm, dry conditions is excellent for tourism developments, especially in coastal areas. Desert areas may be at risk from flash floods.

Hot, desert climate

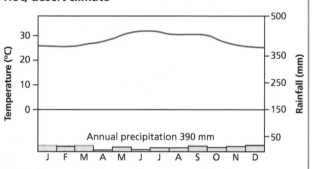

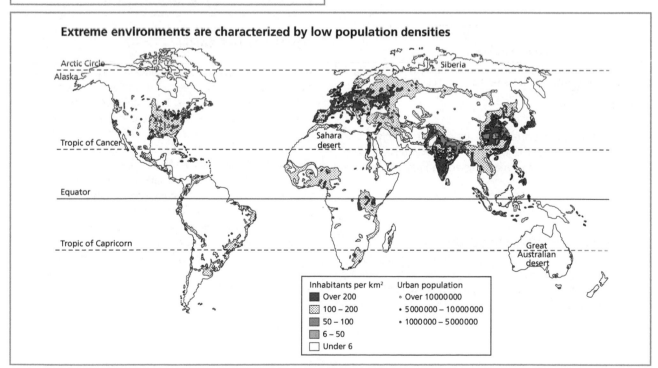

Extreme environments are characterized by low population densities

❓

CHECK YOUR UNDERSTANDING

3. Describe the main characteristics of the climate of a periglacial environment.
4. Describe the main characteristics of the climate of a hot desert climate.

People in extreme environments

POPULATION DENSITIES IN EXTREME ENVIRONMENTS

Extreme environments are characterized by low population densities. Examples include central Australia, Iceland, northern Canada, Namibia and the western Sahara. This is due to the extremes of climate, insufficient heat in Iceland and Canada, and insufficient water in the Sahara, for example. Each of these environments is a long way outside the comfort zones for human habitation. Nevertheless, comfort zones may be culturally biased. For example, Inuit populations are better able to deal with the cold than other populations.

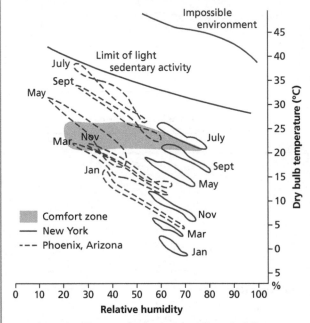

Note how the climates of New York and Phoenix fall outside the comfort zone in most of the six months plotted.

Comfort zones

Other factors are important, too. Iceland is relatively remote and isolated. This makes communications costly. It also increases the cost of materials and goods, which have to be imported.

❓

CHECK YOUR UNDERSTANDING
5. Describe the main characteristics of the comfort zone shown in the diagram.
6. How do indigenous people adapt to life in cold environments?

TRADITIONAL COPING MECHANISMS IN EXTREME ENVIRONMENTS

Traditionally, periglacial pastures have been used by Inuits for herding or hunting caribou. They tend to be migratory, moving north during the summer and heading south to the forest margins in winter. Many **indigenous people** have turned to rivers and the oceans. Fishing is an extremely important activity for many indigenous people in periglacial areas, for example, the Nenets of the Yamal peninsula.

Desert inhabitants are also migratory. The Bedouin and the Fulani are excellent examples. To cope with the extreme temperatures in the daytime they avoid direct sunlight and take a rest. They tend to travel in early morning and late afternoon. Their clothing – loose-fitting garments – also helps them to cope with high temperatures. It reduces sweating and allows them to remain reasonably comfortable.

COPING WITH WATER SHORTAGES IN DRY AREAS

Some coping mechanisms are "natural" and require farmers to adapt to the natural environment. Methods include:
- seasonal mobility (the traditional way of dealing with insufficient amounts of rainfall and pasture)
- reduction of size of herds
- exchange of livestock and livestock products
- greater use of drought-tolerant species
- use of wild species and tree crops
- windbreaks to reduce wind erosion of bare soil.

COPING IN THE SAHEL

The indigenous people of the Sahel in North Africa have adapted to their environmental conditions through a combination of strategies. They make use of the limited resources of the Sahel, and combat overgrazing by migrating to areas of seasonal growth. They leave vegetation around more permanent water sources for drier times. The livestock herds are diversified – cattle are kept for income in the meat market, sheep and goats for milk and meat for subsistence consumption. Herd diversification also allows pastoralists to make use of a greater variety of the available vegetation resources because the animals have different grazing patterns. The diet of the indigenous people varies with conditions. More milk is consumed in the wetter periods, whereas meat is more common in drier periods.

The changing distribution of extreme environments

THE ADVANCE AND RETREAT OF GLACIERS

A glacial system is the balance between inputs, storage and outputs. Inputs include **accumulation** of snow, avalanches, debris, heat and meltwater. The main store is that of ice, but the glacier also carries debris, **moraine**, and meltwater. The outputs are the losses due to **ablation**, the melting of snow and ice, and sublimation of ice to vapour, as well as sediment.

The **regime** of the glacier refers to whether the glacier is advancing or retreating:

- if accumulation > ablation, the glacier advances
- if accumulation < ablation, the glacier retreats
- if accumulation = ablation, the glacier is steady.

The past and present distribution of ice sheets and ice caps in the northern hemisphere

The glacier has a positive regime when accumulation exceeds ablation (melting, evaporation, calving, wind erosion, avalanche, and so on), so the glacier thickens and advances. A negative regime occurs when ablation exceeds accumulation, thus the glacier thins and retreats. Any glacier can be divided into two sections, namely an area of accumulation at high altitudes, and an area of ablation at lower altitude.

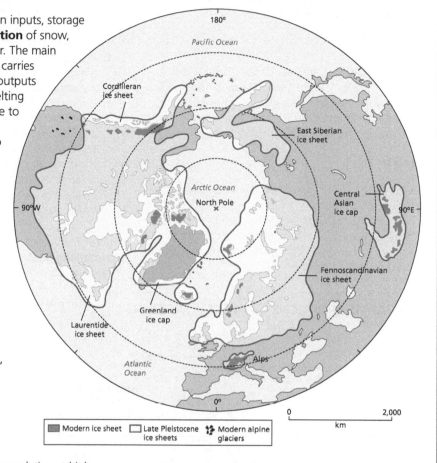

NATURAL DESERTIFICATION

Deserts also change in their distribution. This is partly due to long-term changes in climate, for example, increasing aridity in today's desert areas compared with wetter "pluvials" (rainy periods) which occurred around the same time as glacial advances in temperate areas.

The evidence for climate change in the Sahara

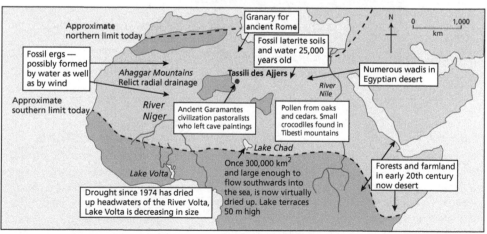

CHECK YOUR UNDERSTANDING
7. Compare the distribution of glacial environments during the Pleistocene with that of today.
8. Outline the evidence for climate change in the Sahara Desert.

Glacial erosion

METHODS OF GLACIAL EROSION INCLUDE PLUCKING AND ABRASION

Plucking

This occurs mostly at the base of the glacier and to an extent at the side. It is most effective in jointed rocks or those weakened by freeze-thaw. As the ice moves, meltwater seeps into the joints, freezes onto the rock and is then ripped out by the moving glacier.

Abrasion

The debris carried by the glacier scrapes and scratches the rock leaving **striations**.

The amount and rate of erosion depends on: (a) the local geology, (b) the velocity of the glacier, (c) the weight and thickness of the ice, and (d) the amount and character of the load carried.

In the northern hemisphere, **cirques** are generally found on north- or east-facing slopes of mountains where accumulation is highest and ablation is lowest. They are formed in stages.
1. A preglacial hollow is enlarged by freeze-thaw and removal by snow melt (this is also known as **nivation**).
2. Ice accumulates in the hollow.
3. Having reached a critical weight and depth, the ice moves out in a rotational manner, eroding the floor by plucking and abrasion.
4. Meltwater trickling down the bergschrund allows the cirque to grow by freeze-thaw. After glaciation an armchair-shaped hollow remains, frequently filled with a lake, for example, Blue Lake Cirque, New South Wales, Australia.

Other features of glacial erosion include **arêtes** and **pyramidal peaks** (horns) caused by the headward recession (cutting back) of two or more cirques. Glacial **troughs** (or U-shaped valleys) have steep sides and flat floors. In plan view they are straight since they have **truncated** the interlocking spurs of the preglacial valley. The ice may also carve deep **rock basins** frequently filled with **ribbon lakes**. **Hanging valleys** are formed by tributary glaciers which, unlike rivers, do not cut down to the level of the main valley, but are left suspended above, for example, Stickle Beck in the Lake District, UK. They are usually marked by waterfalls.

LANDFORMS PRODUCED BY GLACIAL EROSION

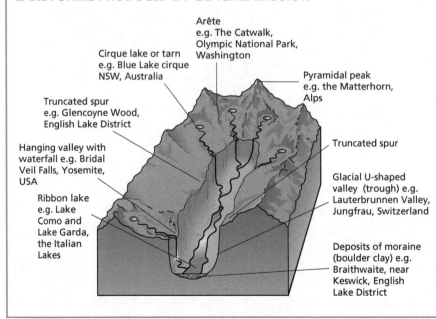

Arête
e.g. The Catwalk, Olympic National Park, Washington

Cirque lake or tarn e.g. Blue Lake cirque NSW, Australia

Pyramidal peak e.g. the Matterhorn, Alps

Truncated spur e.g. Glencoyne Wood, English Lake District

Hanging valley with waterfall e.g. Bridal Veil Falls, Yosemite, USA

Ribbon lake e.g. Lake Como and Lake Garda, the Italian Lakes

Truncated spur

Glacial U-shaped valley (trough) e.g. Lauterbrunnen Valley, Jungfrau, Switzerland

Deposits of moraine (boulder clay) e.g. Braithwaite, near Keswick, English Lake District

EXAM TIP ✓
Glacial erosion is complex and involves mechanisms other than plucking and abrasion. If you are writing about glacial erosion, be sure to mention that other mechanisms that aid glacial erosion include meltwater, freeze-thaw weathering and pressure release. Although not strictly glacial nor erosional, these processes are crucial in the development of glacial scenery.

COMMON MISTAKE ❗
✗ *All glacial troughs are U-shaped.*
✓ Many relict troughs have been acted upon by other processes (mass movements, weathering, human impacts) over thousands of years, and so may have had their shape altered.

CHECK YOUR UNDERSTANDING ❓
9. Briefly explain the formation of a cirque.
10. Outline the link between plucking and abrasion.

Glacial deposition

The term **drift** refers to all glacial and fluvioglacial (meltwater) deposits left after the ice has melted. Glacial deposits or **till** are angular and unsorted, and include erratics, drumlins and moraines. Till is often subdivided into **lodgement till**, material dropped by actively moving glaciers, and **ablation till**, deposits dropped by stagnant or retreating ice.

THE CHARACTERISTICS OF TILL

- Poor sorting – till contains a large range of grain sizes, for example, boulders, pebbles, clay
- Poor stratification – no regular sorting by size
- Mixture of rock types – from a variety of sources
- Many particles have striations (scratch marks)
- Long axis orientated in the direction of glacier flow
- Some compaction of deposits
- Mostly subangular particles

ERRATICS

Erratics are large boulders foreign to the local geology, for example, the Madison Boulder in New Hampshire, which is estimated to weigh over 4,600 tonnes!

THE CHARACTERISTICS OF MORAINE

Moraines are lines of loose rocks, weathered from the valley sides and carried by the glaciers. At the snout of the glacier is a crescent-shaped mound of **terminal moraine**. Its character is determined by the amount of load the glacier was carrying, the speed of movement and the rate of retreat. The **ice-contact slope** (up-valley) is always steeper than the down-valley slope. Cape Cod in Massachusetts, USA is a fine example of a terminal moraine.

Lateral moraines are ridges of materials found on the edge of a glacier. The lateral moraines on the Gorner Glacier in Switzerland are good examples. Where two glaciers merge and the two touching lateral moraines flow in the middle of the enlarged glacier they are known as medial moraines. Again, the Gorner Glacier contains many examples of medial moraines.

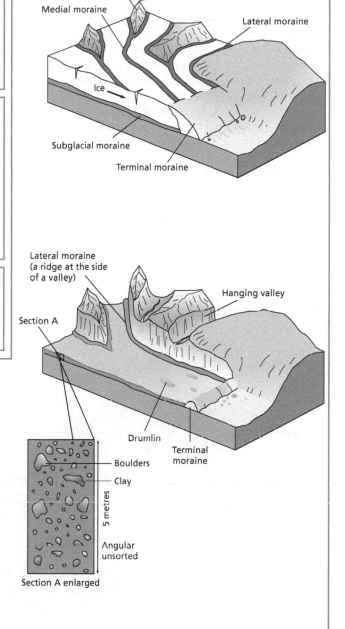

COMMON MISTAKE

✗ *All glacial deposits are typically angular and unsorted.*

✓ Meltwater from melting glaciers may cause a degree of rounding and sorting to occur. Many glacial deposits are influenced both by glacial action and fluvioglacial (meltwater) action.

CHECK YOUR UNDERSTANDING

11. Distinguish between drift and till.
12. Compare the location of terminal moraine with that of lateral moraine.

Periglacial environments

PERIGLACIAL PROCESSES

Freeze–thaw weathering is a process in which water freezes at 0°C and expands by 10%. It puts great pressure on jointed rocks. The more cycles of freeze–thaw per year, the greater its impact. **Frost heave** is a process whereby water that freezes in a soil lifts individual particles of soil or stones towards the surface. The stones in turn protect the ice underneath from melting initially, so when the ice under stones finally melts, other particles have fallen back to their original levels, and the stones do not fall back as much.

Solifluction literally means flowing soil. In winter water freezes in the soil causing expansion of the soil and segregation of individual soil particles. In spring the ice melts and water flows downhill. It cannot infiltrate the soil because of the impermeable permafrost. As it moves over the permafrost it carries segregated soil particles (peds) and deposits them further downslope as a solifluction lobe or terracette.

Periglacial landscape features

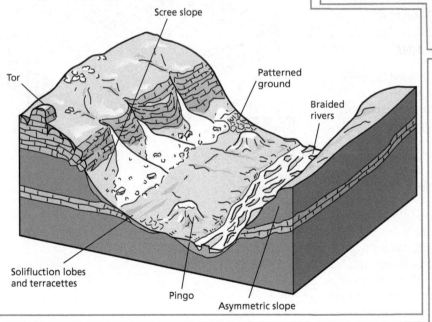

Scree slope
Tor
Patterned ground
Braided rivers
Solifluction lobes and terracettes
Pingo
Asymmetric slope

PERMAFROST

Periglacial areas are also associated with permafrost, impermeable permanently frozen ground. Approximately 20% of the world's surface is underlain by permafrost, in places up to 700 m deep. Three types of permafrost exist:
* continuous – where mean annual temperatures are −5 to −50°C
* discontinuous – where mean annual temperatures are −1.5 to −5°C
* sporadic – where mean annual temperatures are 0–−1.5°C.

The active layer is found at the surface. This layer thaws out seasonally and is characterized by intense mass movements. The depth of the active layer in Siberia varies from less than 1.6 m at 70°N to up to 4 m at 50°N.

PATTERNED GROUND

Patterned ground is a general term describing the stone circles, polygons and stripes that are found in soils subjected to intense frost action, for example, on the slopes of Kerid crater, southern Iceland. On steeper slopes stone stripes replace stone circles and polygons. Their exact mode of formation is unclear although ice sorting, differential frost heave, solifluction and the effect of vegetation are widely regarded to be responsible.

PINGOS

A **pingo** is an isolated, conical hill up to 100 m high and 1000 m wide. They are only found in periglacial areas. They are formed by the movement and freezing of water under pressure. Two types are generally identified: **open system** and **closed system** pingos. Open system pingos are formed where the source of the water is from a distant elevated source. In contrast, closed system pingos are formed when the supply of water is local, and the pingo is formed as a result of the expansion of permafrost. There are nearly 1,500 pingos in the Mackenzie Delta of Canada.

THERMOKARST

Thermokarst refers to depressions caused by subsidence due to the melting of permafrost. This may be because of broad climatic changes or local environment changes.

Local environmental changes include:
* cyclical changes in vegetation, which may affect the albedo or reflectivity of the surface
* shifting of stream channels, which may affect the amount of heat affecting permafrost
* fire, which rapidly destroys permafrost
* disruption of vegetation by human activity, which may remove surface layers and so open the permafrost to raised air temperatures in summer.

CHECK YOUR UNDERSTANDING
13. Distinguish between solifluction and frost heave.
14. Explain how human activity may lead to the formation of thermokarst.

Hot arid environments

WEATHERING IN DESERTS

- **Salt crystallization** causes the breakdown of rock by solutions of salt. There are two main types of **salt crystal growth**. First, in areas where temperatures fluctuate around 26–28°C sodium sulphate and sodium carbonate expand by about 300%. Second, when water evaporates, salt crystals may be left behind. As the temperature rises, the salts expand and exert pressure on rock.
- **Disintegration** is found in hot desert areas where there is a large diurnal (daytime and night-time) temperature range. Daytime temperatures may exceed 40°C whereas night-time temperatures are near freezing. Rocks heat up by day and contract by night. As rock is a poor conductor of heat, all of the stresses occur only in the outer layers. This causes peeling or **exfoliation** to occur. Moisture is essential for this to happen.

WIND ACTION IN DESERTS

Wind action is important in areas where winds are strong (over 20 km/h), turbulent, come largely from a constant direction and blow for a long time.

Near the surface wind speed is reduced by friction (the rougher the ground the more the wind speed is reduced). Sediment is more likely to be moved if there is a lack of vegetation, and it is dry, loose and small.

THE WORK OF WATER

Water is vital for the development of many desert landforms. It is important for the operation of mechanical and chemical weathering, and it is important for erosion, too. There are a number of sources of water in deserts.
- Rainfall may be low and irregular but it does occur, mostly as low intensity events although there are occasional flash floods.
- Deflation may expose the water table to produce an oasis.
- Rivers flow through deserts. These can be classified as **exotic** (come from a different climate zone), **endoreic** (flow to an inland basin), and **ephemeral** (seasonal).

Winds deposit the sand they carry as dunes. Their shape and size depends on the supply of sand, direction of wind, nature of the ground surface, and presence of vegetation. There are two types of wind erosion.
- **Deflation** is the progressive removal of small material leaving behind larger materials. This forms a stony desert or reg.
- **Abrasion** is the erosion carried out by wind-borne particles. They act like sandpaper, smoothing surfaces and exploiting weaker rocks. Most abrasion occurs within a metre of the surface, since this is where the largest, heaviest, most erosive particles are carried.

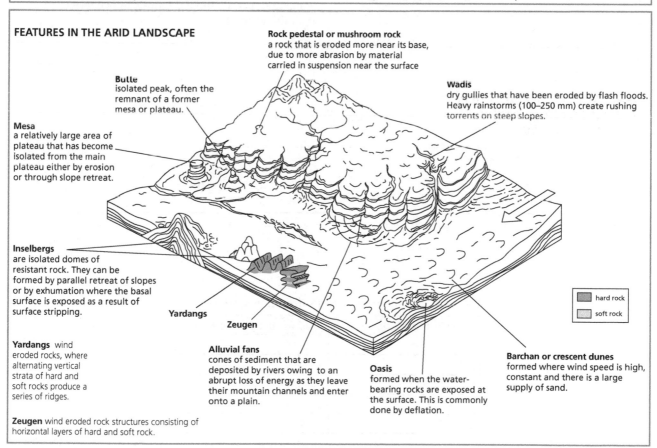

FEATURES IN THE ARID LANDSCAPE

Rock pedestal or mushroom rock
a rock that is eroded more near its base, due to more abrasion by material carried in suspension near the surface

Butte
isolated peak, often the remnant of a former mesa or plateau.

Mesa
a relatively large area of plateau that has become isolated from the main plateau either by erosion or through slope retreat.

Wadis
dry gullies that have been eroded by flash floods. Heavy rainstorms (100–250 mm) create rushing torrents on steep slopes.

Inselbergs
are isolated domes of resistant rock. They can be formed by parallel retreat of slopes or by exhumation where the basal surface is exposed as a result of surface stripping.

Yardangs

Zeugen

Yardangs wind eroded rocks, where alternating vertical strata of hard and soft rocks produce a series of ridges.

Alluvial fans
cones of sediment that are deposited by rivers owing to an abrupt loss of energy as they leave their mountain channels and enter onto a plain.

Oasis
formed when the water-bearing rocks are exposed at the surface. This is commonly done by deflation.

Barchan or crescent dunes
formed where wind speed is high, constant and there is a large supply of sand.

hard rock
soft rock

Zeugen wind eroded rock structures consisting of horizontal layers of hard and soft rock.

CHECK YOUR UNDERSTANDING
15. Explain briefly how rocks may be weathered by temperature changes in deserts.
16. Briefly explain two forms of wind erosion in hot, arid areas.

Agriculture in arid areas

HOT ARID AREAS

Hot arid areas offer a number of opportunities for agriculture – the abundance of heat and sunlight favour a lengthy growing season. In areas where water is available, the potential for farming is good. The main type of farming in most hot arid areas is **nomadism**. True nomads wander with their herds of camels, goats, sheep and/or cattle in search of water and new pasture. Oases, desalination plants and exotic rivers provide opportunities for settled agriculture. For example, the oasis at Douz, Tunisia produces dates, figs and oranges. Date palm is a particularly useful crop – the bark can be used for thatching or fencing, the leaves can be made into ropes and mats, and the dates can be eaten. Dams can control the supply of water so that there is enough in the dry season. In a number of oil-rich countries, such as Saudi Arabia and UAE, desalination plants provide a plentiful supply of water.

CHALLENGES – ARIDITY AND INFERTILITY

Agriculture in most arid and semi-arid areas is dominated by lack of freshwater and infertile soils. Poor transport can also make agriculture challenging. All arid and semi-arid areas have a negative water balance. That means the outputs from evapotranspiration and stores of water exceed the input from precipitation. Desert soils are arid (dry) due to low rainfall and high evapotranspiration. They are infertile due to:
- a low organic content resulting from limited biomass
- being thin

SUSTAINABILITY

Agriculture in hot arid areas could be made more sustainable by:
- planting drought-resistant and/or salt-tolerant vegetation to provide fodder for animals
- reducing herd size and pressure on limited amounts of vegetation
- using animal dung as fertilizer
- using solar panels to produce energy rather than using animal dung
- using more efficient types of irrigation
- building small-scale dams.

- lacking in clay
- the presence of soluble salts due to the lack of leaching.

Salinization may occur in areas where annual precipitation is less than 250 millimetres. In poorly drained locations, water evaporates and leaves behind large amounts of salts (see also p.10). The saline soils affect the growth of most crop plants by reducing the rate of water uptake by roots, and plants die due to wilting. Some crops, such as date palm and cotton, are more salt tolerant (halophytic).

However, for those without access to irrigation water and security of land tenure, agriculture is a high-risk venture.

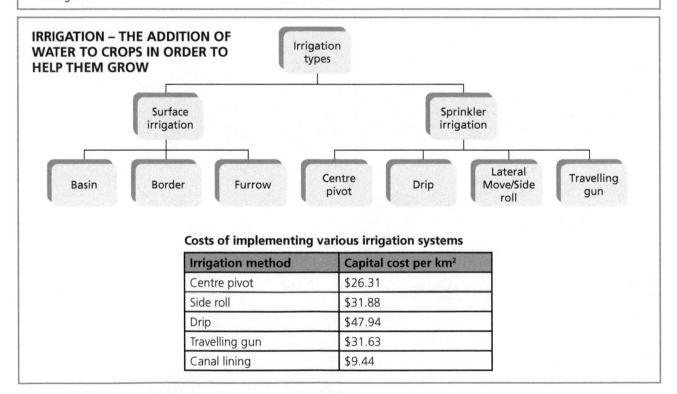

IRRIGATION – THE ADDITION OF WATER TO CROPS IN ORDER TO HELP THEM GROW

Irrigation types
- Surface irrigation
  - Basin
  - Border
  - Furrow
- Sprinkler irrigation
  - Centre pivot
  - Drip
  - Lateral Move/Side roll
  - Travelling gun

Costs of implementing various irrigation systems

Irrigation method	Capital cost per km²
Centre pivot	$26.31
Side roll	$31.88
Drip	$47.94
Travelling gun	$31.63
Canal lining	$9.44

CHECK YOUR UNDERSTANDING
17. Distinguish between aridity and infertility.
18. Outline the main sources of water in hot arid areas.

Human and physical opportunities for mineral extraction in hot arid environments

MINERAL EXTRACTION IN HOT ARID AREAS

There is enormous potential for the development of mineral resources in hot arid areas. Oil and gas in the Middle East, copper in Chile and uranium mining in Australia illustrate the potential riches that occur in hot arid areas.

However, exploitation may be difficult because many of these areas are remote, and transport is expensive. In addition, housing and basic utilities such as water, energy and waste disposal need to be provided.

The shortage of water may require desalination plants to be built or groundwater reserves to be tapped. Moreover, since many hot arid areas are remote and inaccessible, transporting the resources to be exported is difficult and expensive.

Resource exploitation in hot arid areas can have negative environmental impacts. The movement of vehicles and people can introduce exotic species. The mines and their waste can be a major source of dust that is linked to an increase in respiratory problems.

CASE STUDY

ROSEMONT COPPER, ARIZONA

Rosemont Copper is a copper mining project in Pima County, Arizona, an undeveloped area approximately 50 km south-east of Tucson.

The arguments for and against Rosemont Copper, Arizona

Arguments in favour of the proposal	Arguments against the proposal
• Copper is needed for a clean-energy economy; for example, hybrid cars contain twice as much copper as conventional cars. • The mine would import 105% of the water needed. • The used land will be reclaimed and restored. • The mine would employ 400 people directly for at least 19 years and support 1,700 indirect jobs. • The project would support linked industries – contractors providing goods and services to the mine operation during the nearly two decades of operation.	• Open-pit copper mines pollute the air and nearby water supplies. • The mine structures would be visible from Arizona State Route 83, a designated scenic route. • Destruction of habitat and individuals representing nine endangered and threatened species. • Complete loss of 85 historic properties that include Native American remains and prehistoric sites. • Degradation of air quality throughout the region. • Significant traffic increases along Highways 83 and 82, as the mine would generate 55–88 round-trip truck shipments daily.

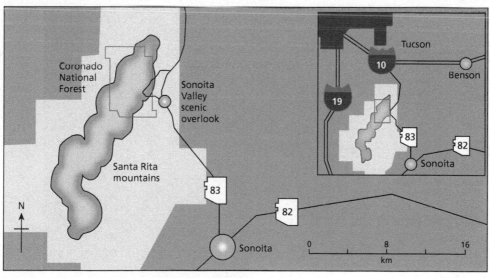

Proposed mine at Rosemont Ranch, Arizona

CHECK YOUR UNDERSTANDING

19. Identify the advantages of mineral extraction in hot arid areas.

20. Outline the environmental damage that may be caused by resource exploitation in hot arid environments.

Opportunities and challenges for mineral extraction in cold environments

The exploitation of cold environments for their mineral resources creates opportunities and challenges. Resource development can improve the economies of these regions, but it can also put pressure on the fragile environment and create conflict between different stakeholders.

FRAGILITY OF PERIGLACIAL AREAS

Periglacial areas are fragile for three reasons:
- The ecosystem is highly susceptible to human impact due to the limited number of species involved.
- The extremely low temperatures reduce decomposition, and pollution has a very long-lasting effect on ecosystems.
- Permafrost is easily disrupted – heat from buildings and pipelines, and changes in the vegetation cover, cause thawing of the permafrost, which can lead to subsidence.
- Close to rivers, owing to an abundant supply of water, frost heave is very significant and can lift piles and structures out of the ground. Piles for carrying oil pipelines therefore need to be embedded deep in the permafrost to overcome mass movement in the active layer. In Prudhoe Bay, Alaska, they are 11m deep. However, this is extremely expensive: each cost more than $3000 in the early 1970s.

The human impact on periglacial areas

The hazards associated with the use of cold environments are diverse and may be intensified by human impact. Problems include mass movements, flooding, thermokarst subsidence, low temperatures, poor soils, a short growing season and a lack of light.

Problems with pipelines

Unstable permafrost pipeline above ground

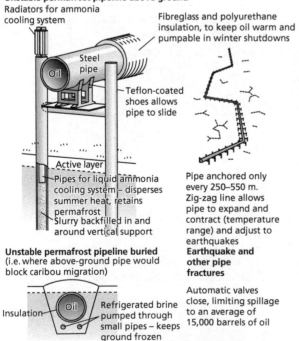

Radiators for ammonia cooling system

Fibreglass and polyurethane insulation, to keep oil warm and pumpable in winter shutdowns

Steel pipe

Oil

Teflon-coated shoes allows pipe to slide

Active layer

Pipes for liquid ammonia cooling system – disperses summer heat, retains permafrost

Slurry backfilled in and around vertical support

Pipe anchored only every 250–550 m. Zig-zag line allows pipe to expand and contract (temperature range) and adjust to earthquakes
Earthquake and other pipe fractures

Automatic valves close, limiting spillage to an average of 15,000 barrels of oil

Unstable permafrost pipeline buried (i.e. where above-ground pipe would block caribou migration)

Insulation

Oil

Refrigerated brine pumped through small pipes – keeps ground frozen

CHECK YOUR UNDERSTANDING
21. Explain why permafrost can be a problem in cold environments.
22. Briefly explain the term "resource nationalism".

CASE STUDY

RESOURCE NATIONALISM IN GREENLAND

Greenland has major physical and political challenges to development. The country is geographically isolated and lacks both the physical and human infrastructure to support rapid development. Some 80% of the land mass is covered in thick ice. Nuuk, the capital, has a population of just 16,000. Much of the country is inaccessible. There are no roads connecting the scattered communities and a single ship serves the settlements on the western side of the island.

There has been a long history of mining in Greenland. Cryolite was mined for 130 years between 1857 and 1987.

In 2009, Nuuk achieved full home rule, including control of natural resources. Denmark remained in charge of security and foreign affairs. Greenland's resource nationalism was expected to bring great benefits to Greenland. However, the oil price collapse since 2013 has made Arctic oil exploration too expensive and not worthwhile.

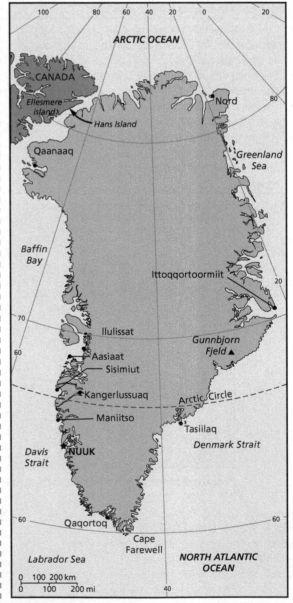

Settlements in Greenland

Tourism

CASE STUDY

ZUNI PUEBLO, NEW MEXICO, USA

Zuni Pueblo is the largest and most traditional of the 19 New Mexican Pueblos. It has a unique language, culture, and history resulting in part from its geographic isolation in a remote area of one of the most sparsely populated regions of the USA. Tourism is an attractive option because of the relatively low capital investment and the potentially high economic returns.

- **Sociocultural concerns:** Zuni needs to control the development of tourism in order to safeguard against the negative consequences that could affect the social and cultural life there. There have been proposals for motel complexes, casinos and golf courses, but none have been implemented.
- **Environmental concerns:** Water, air, soil and biodiversity are resources that can be easily affected by tourism. Some developments are allowed such as big-game hunting by non-Indians as long as they have a Zuni guide. Overall, the impact of tourism has been limited largely because it has been controlled.

Hot, arid environments: tourism
Tourism in hot arid areas needs to be controlled to limit the potential damage it could do.

Responsible tourism in Zuni

Check in with the Visitor Center before starting your visit to Zuni Pueblo. Remember, you are visiting an active community of residents' daily lives and homes – not a museum or theme park.

Consider capturing visual memories instead of photographs! Assume that ALL "cultural" activities within the Pueblo are off-limits to photograph, video or audio record or sketch unless specifically informed otherwise. Always inquire first and ask permission before photographing any activity involving people. NO photography is permitted of images inside the Old Mission.

Observe with quiet respect any traditional dances and events that you may encounter. Applause is as inappropriate as in a church setting.

Exercise common sense by not climbing around fragile archaeological structures or adobe walls. Removal of artifacts or objects from these areas is a Federal offence.

Respect our community by not using alcohol or drugs and not bringing weapons.

Hike only in designated areas (check at Visitor Center) and not around archaeological ruin sites.

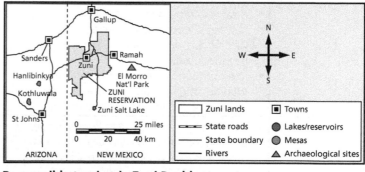

Responsible tourism in Zuni Pueblo

Causes, acceleration, consequences and management of desertification

WHAT IS DESERTIFICATION?

Desertification is the spread of desert-like conditions into previously productive areas. Currently, about 25% of the global land territory and nearly 16% of the world's population are threatened by desertification.

Desertification occurs when already fragile land in arid and semi-arid areas is over-exploited. This overuse can be caused by overgrazing and deforestation.

CONSEQUENCES OF DESERTIFICATION

Environmental	Economic	Social and cultural
• Loss of soil nutrients through wind and water erosion • Changes in composition of vegetation and loss of biodiversity as vegetation is removed • Reduction in land available for cropping and pasture • Increased sedimentation of streams because of soil erosion and sediment accumulations in reservoirs • Expansion of area under sand dunes	• Reduced income from traditional economy (pastoralism and cultivation of food crops) • Decreased availability of fuelwood, necessitating purchase of oil/ kerosene • Increased dependence on food aid • Increased rural poverty	• Loss of traditional knowledge and skills • Forced migration due to food scarcity • Social tensions in reception areas for migrants

STRATEGIES FOR PREVENTING DESERTIFICATION

Cause of desertification	Strategies for prevention	Problems and drawbacks
Overgrazing	• Improved stock quality: through vaccination programmes and the introduction of better breeds, yields of meat, wool, and milk can be increased without increasing the herd size. • Better management: reducing herd sizes and grazing over wider areas would both reduce soil damage.	• Vaccination programmes improve survival rates, leading to bigger herds. • Population pressure often prevents these measures.
Overcultivation	• Use of fertilizers: these can double yields of grain crops, reducing the need to open up new land for farming. • New or improved crops: many new crops or new varieties of traditional crops with high-yielding and drought-resistant qualities could be introduced. • Improved farming methods: use of crop rotation, irrigation and grain storage can all be increased to reduce pressure on land.	• Cost to farmers. • Artificial fertilizers may damage the soil. • Some crops need expensive fertilizer. • Risk of crop failure. • Some methods require expensive technology and special skills.
Deforestation	• Agroforestry: combines agriculture with forestry, allowing the farmer to continue cropping while using trees for fodder, fuel and building timber. Trees, protect, shade and fertilize the soil. • Social forestry: village-based tree-planting schemes involve all members of a community. • Alternative fuels: oil, gas, and kerosene can be substituted for wood as sources of fuel.	• Long growth time before benefits of trees are realized. • Expensive irrigation and maintenance may be needed. • Expensive: special equipment may be needed.

CHECK YOUR UNDERSTANDING
25. Define the term "desertification".
26. Explain how farming practices can cause desertification.

Increasing international competition for access to resources in extreme environments

As ice caps are melting, a military race is also building in the region. The USA and Russia are competing for extremely valuable resources in the Arctic. The region is opening up two major shipping lanes, and oil and gas reserves are worth trillions of dollars. The potential for economic competition is fierce, especially among the eight members of the Arctic Council: Canada, Denmark, Norway, Iceland, Finland, Sweden, Russia and the USA.

CASE STUDY

THE YAMAL MEGAPROJECT

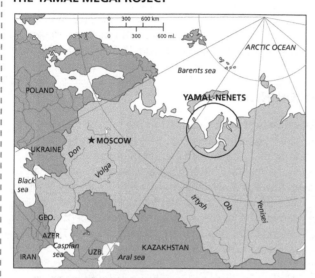

The location of the Yamal Peninsula

The Yamal Peninsula is located in Northern Siberia. Yamal is a remote, wind-blown region underlain by permafrost. The Nenets are indigenous nomadic reindeer herders who have used the Yamal Peninsula for over 1,000 years. They graze their reindeer in the Yamal during the summer and move south during the winter. The Nenets are now facing threats from climate change and also from oil and gas exploration.

Yamal has the world's largest natural gas reserves, with an estimated 55 trillion cubic metres (tcm). Russia's largest energy project in history puts the future of nomadic herding at considerable risk. Roads and pipelines are difficult for reindeer to cross, and oil pollution is threatening the quality of pasture and freshwater. The project was initiated in the 1990s, and the first of its gas supplies was produced in 2012.

Environmental impacts
- Oil and gas operations in the Yamal Peninsula destroyed over 64,000 km² of tundra in just ten years of exploration.
- Fish yields on the River Ob have decreased, fish spawning grounds have been polluted.

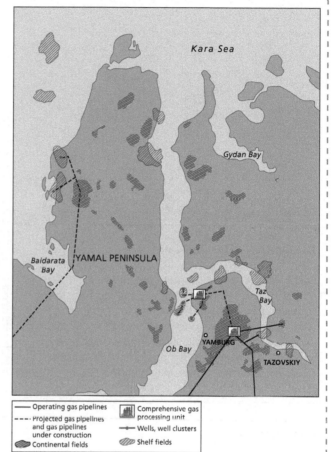

Yamal oil and gas fields

- The River Ob used to provide 60% of the former Soviet Union's fish catch; nearly 30 fisheries on tributaries of the Ob have been destroyed.
- Reindeers' migratory routes have been bisected by railroads, and some reindeer have been shot.

According to a Spokesperson from Survival International, "The Nenets people have lived on and stewarded the tundra's fragile ecology for hundreds of years. No developments should take place on their land without their consent, and they need to receive fair compensation for any damages caused".

CHECK YOUR UNDERSTANDING
27. Briefly explain how the exploitation of oil has negatively affected the Nenets' lifestyle.
28. State how much natural gas is believed to exist in the Yamal Peninsula.

New technology and sustainable development in extreme environments

SUSTAINABLE AGRICULTURE: BUSTAN, EGYPT

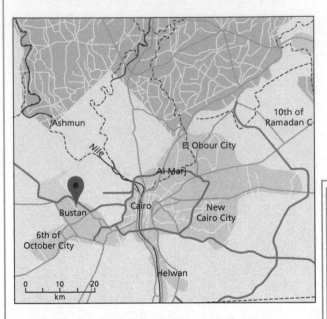

The location of Bustan, Cairo

Bustan (Arabic for "orchard") is the first commercial aquaponics farm in Egypt. Water circulates from tanks containing fish through hydroponic trays that grow vegetables. Each tank contains about a thousand tilapia fish, which are native to Egypt. Water from the pond is then used to irrigate the olive trees that produce a high quality olive oil.

- This organic and closed system mimics natural processes and enables waste to be efficiently reused.

- The fish tanks provide 90% of the nutrients plants need to grow.
- The ammonia that results from the fish breathing is converted into nitrogen and absorbed by the plants before being sent back to the fish tanks, ammonia-free and healthy.
- Bustan is a labour-intensive farm and uses sustainable biological pest control methods, such as ladybirds to kill aphids, in order to avoid chemical inputs.
- Bustan uses 90% less water than traditional farming methods in Egypt. It produces 6–8 tonnes of fish per year and could potentially yield 45,000 heads of lettuce if it were to grow just a single type of vegetable.

DESALINIZATION

Desalination or **desalinization** is a process that removes salt from seawater. Seawater is desalinated to produce fresh water fit for human consumption (potable water) and for irrigation. The costs of desalinating seawater are generally high due to the large amount of energy required.

According to Global Water Intelligence:

- Around 1% of the world's population depends on desalinated water to meet their daily needs. This will rise to 14% by 2025.
- The world's largest single desalination project is Ras Al-Khair in Saudi Arabia, which produced over one million cubic metres per day in 2014.
- In Israel over 40% of domestic water comes from seawater desalination, the largest contribution to any one country.

CASE STUDY

SOLAR POWER: DESERTEC

In 2005 a researcher from Germany, Nadine May, suggested that only a small amount of the Sahara would be needed to produce enough solar energy for the world. Her data was used to produce a map suggesting how much of the Sahara would be needed to power the world, to power Europe, and just Germany. It claimed that in just six hours, the world's deserts receive more energy from the sun than humans consume in a year.

As it happened, the fall in the price of solar panels and wind turbines in the EU led Desertec to conclude in 2013 that Europe could produce most of its renewable energy from within its own borders.

Nevertheless, solar energy in the Northern Sahara is still going ahead, with projects in Tunisia, Morocco and Algeria. The Tu Nur project in Tunisia will send power to Europe in 2018. This seems inappropriate as Tunisia depends on neighbouring Algeria for some of its own energy. The Moroccan government has attracted funding from overseas lenders to develop the world's largest concentrating solar power (CSP) plant at Ourzazate. Originally it was planned as an export project to Spain, but it is now promoted as a project to increase Morocco's supply of renewable energy.

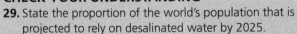

CHECK YOUR UNDERSTANDING
29. State the proportion of the world's population that is projected to rely on desalinated water by 2025.
30. Explain why hot arid areas have excellent potential for solar power.

The impacts and management of global climate change in extreme environments

CLIMATE CHANGE IN EXTREME ENVIRONMENTS

The primary **environmental effects** of climate change include:
- falling crop yields and rising food insecurity
- increased water stress
- flooding of low-lying areas
- the spread of diseases such as malaria.

The **sociopolitical** effects of climate change include:
- conflict over resources such as water and pasture
- loss of territory and border disputes
- forced migration resulting in urban overcrowding
- tension over energy supplies, for example, from deforestation.

COPING STRATEGIES

These include:
- improving soil fertility by the careful use of fertilizers
- using indigenous plant species
- improving the efficiency of irrigation systems and pest control
- adopting water and soil conservation techniques using diguettes (stone rows) to reduce run-off on slopes.

CLIMATE CHANGE IN COLD AREAS

Global climate change is likely to have the greatest impact in areas where the mean annual temperature is just below freezing. Up to 40% of permafrost areas are at risk of degradation and the development of thermokarst (subsidence). In contrast, very cold glacial areas (such as in central Antarctica) will remain well below freezing even if they experience a rise in temperature.

Arctic temperatures have increased at almost twice the global average rate in the last 100 years. Up to 70,000 km^2 of sea ice are disappearing annually.

Disadvantages of global climate change
A warming climate brings many problems for the Inuit. Unpredictable sea ice can be fatal. For many, the traditional hunter-gatherer existence is proving difficult as changes in the environment make their way of life harder. The effect of climate change – more shipping, mining, and oil and gas exploration – may threaten the environment.

Advantages of global climate change
Rising temperatures may have some advantages for economic activities. Farming may become more productive as net primary productivity increases and the length of the growing season increases. In some places, forestry, especially coniferous forestry, may be possible. There may also be increased possibilities for tourism. For example, Greenland had over 90,000 overseas tourists visit and stay in paid accommodation during 2015.

Factors enabling adaptation to climate change

Communications	The presence of diversified media and accessibility of information about weather in general and hazards in particular
Transport	A system which functions even during extreme events
Finance	Access to banking, credit and insurance products which spread risk before, during and after extreme events
Economic diversification	Access to a range of economic and livelihood options
Education	Basic language and other skills necessary to understand risks and shift livelihood strategies as necessary
Organization and representation	Right to organize and to have access to and voice concerns through diverse public, private and civil society organizations
Knowledge generation, planning and learning	The social and scientific basis to learn from experience, proactively identify hazards, analyse risk and develop response strategies that are tailored to local conditions

CHECK YOUR UNDERSTANDING
31. Outline the potential problems of rising temperatures for cold areas.
32. Outline the methods that could be used in hot arid areas to reduce the effect of rising temperatures.

Exam practice

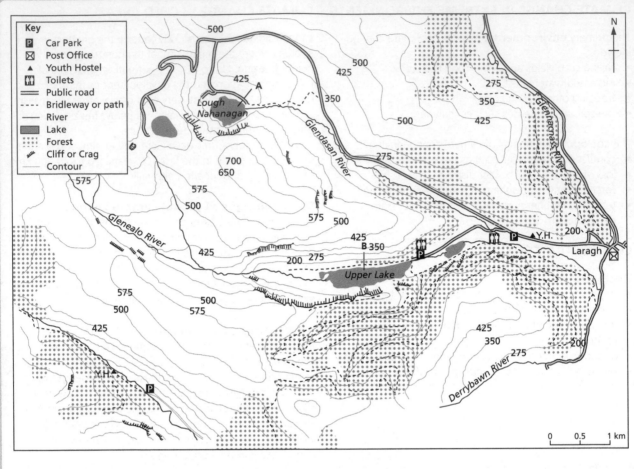

(a) (i) Identify the two types of lakes at A and B. (2 marks)

(ii) Briefly explain the formation of the type of lake found at A. (3 marks)

(b) Outline the opportunities for tourism in this extreme environment. (5 marks)

(c) Either

Discuss the view that desertification is both inevitable and unmanageable. (10 marks)

Or

Examine the view that new technology and sustainable development will transform the
economies of extreme environments. (10 marks)

1 GEOPHYSICAL SYSTEMS

Plate movement

MECHANISMS OF PLATE MOVEMENT

The theory of plate tectonics states that the Earth is made up of many layers. On the outside there is a very thin rigid crust, that is composed of thicker continental crust and thinner oceanic crust.

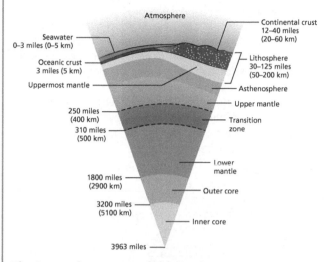

The internal structure of the Earth

The flow of heat from the Earth's interior to the surface comes from two main sources – radiogenic (that is, radioactive decay of materials in the mantle and the crust) and primordial heat (that is, the heat lost by the Earth as it continues to cool from its original formation).

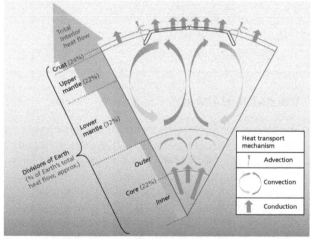

Global internal heat flow

Large-scale convection currents occur in the Earth's interior.

Subduction refers to the plunging of one plate beneath another. Subduction zones form where an oceanic lithospheric plate collides with another plate – whether continental or oceanic.

Rifting occurs at constructive plate boundaries, for example, the East Africa Rift Valley or the rift at Thingvellir, Iceland, where the North American Plate and the Eurasian Plate are moving away from each other. In each case, hotspot activity is believed to be the main cause of rifting. The rift valleys created consist of rock that is hotter and less dense than the older, colder plate. Hot material wells up beneath the ridges to fill the gaps created by the spreading plates.

(a) Upwelling convection in mantle causes the oceanic crust to form a ridge

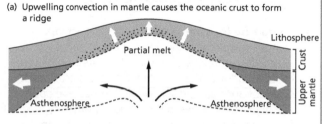

(b) Lateral tension develops, causing rift faulting and downward movement of the central block; magma intrudes along faults, giving surface lava

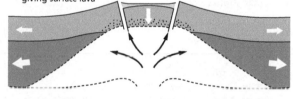

(c) Lateral movement continues with further intrusions parallel to original rift faults

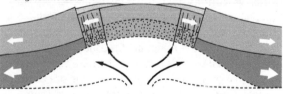

(d) Main rifting sequence is repeated periodically as upwelling continues

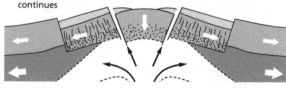

Rifting

CHECK YOUR UNDERSTANDING

1. Identify the main forms of internal heat flow within the Earth.
2. Describe the processes that occur at a subduction zone.

Volcanoes

CHARACTERISTICS OF VOLCANOES

There are many types of volcano formed by various kinds of eruption. Three of the most common are shield, composite and cinder volcanoes.

- **Shield volcanoes** build up when there is no explosive activity, therefore no ejected fragments. Shield volcanoes are formed from very hot, runny basaltic lava. Because it is so hot, the lava can flow great distances. It builds up shield volcanoes, which have gently sloping sides, a shallow crater and a large circumference.

Basic lava 'shield'

- **Composite (or strato) volcanoes** are the most common type of volcano and are formed by alternating eruptions of fragmental material followed by lava outflows. These volcanoes are characterized by slopes of 30° near the summit and 5° near the base. The highest volcanoes in the world are of this type, for example, Mount Etna and Vesuvius in Italy, and Chimborazo and Popacatepetl in Mexico.

Composite strato volcano

- **Cinder volcanoes** are formed by fragments of solid material which accumulate as a steep conical hill around the vent to form a cone. The shape depends on the nature of the material. It is usually concave as the material spreads out near the base and has a steep angle of 30°–40° depending on the size of the material. Cinder and ash cones are not usually very high (up to 300 m) with the exception Volcano Du Fuego in Guatemala which is 3,350 m and all ash.

Cinder Volcano

VOLCANIC ERUPTIONS

- Lava eruptions: The amount of silica in a lava eruption is what makes the difference between the volcanoes in Iceland and Hawaii (which erupt continuously) and those in Japan and the Philippines (where eruptions are infrequent but violent).
  Icelandic lava eruptions are characterized by persistent fissure eruption. Large quantities of basaltic lava may build up vast horizontal plains.
  Unlike the Icelandic lava eruptions, Hawaiian eruptions involve a central vent. Occasional pyroclastic activity occurs, but this is less important than the lava eruption. Runny basaltic lava flows down the sides of the volcano and gases escape easily.

- Pyroclastic eruptions: Strombolian eruptions are explosive eruptions that produce pyroclastic rock. Eruptions are commonly marked by a white cloud of steam emitted from the crater.

Strombolian

- Vulcanian eruptions: these are violent and occur when the pressure of trapped gases in viscous magma becomes sufficient to blow off the overlying crust of solidified lava.

Vulcanian

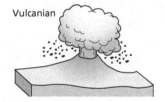

VOLCANIC HAZARDS

There are a number of primary and secondary hazards associated with volcanic eruptions. The primary hazards are the direct impacts of the eruption, for example, lava flows, ash fallout, pyroclastic flows and gas emissions. Secondary hazards may be due to the way that the ejected material reacts or changes form. For example, ash may join with rainwater to form mudflows (lahars). Lahars, or volcanic mudflows, are another hazard associated with volcanoes. A combination of heavy rain and unstable ash increase the hazard of lahars.

Landslides are also associated with volcanic activity. Two main types occur – debris avalanches and lahars. Debris avalanches commonly occur around the same time as an eruption, and they may also help the eruption to occur.

CHECK YOUR UNDERSTANDING
3. Describe the main characteristics of a shield volcano.
4. Outline the main hazards associated with volcanic eruptions.

Earthquakes

An earthquake is a series of seismic vibrations or shock waves which originate from the focus – the point at which the plates release their tension or compression suddenly. The epicentre marks the point on the surface of the Earth immediately above the focus of the earthquake. Shallow focus earthquakes occur relatively close to the ground surface, whereas deep focus earthquakes occur at considerable depth under the ground.

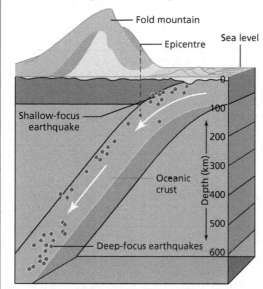

Shallow- and deep-focus earthquakes, and the epicentre

WAVE TYPES

- Primary (P) waves or pressure waves are the fastest wave type and can move through solids and liquids – they shake the earth backwards and forwards.
- Secondary (S) waves or shear waves move with a sideways motion and are unable to move through liquids – they make the ground move horizontally, causing a lot of damage. When P-waves and S-waves reach the surface, some of them are transformed into surface waves:
    - Love waves cause the ground to move sideways.
    - Rayleigh waves cause the ground to move up and down.

Love waves and Rayleigh waves travel slowly through the crust, but they cause the most damage.

❓

CHECK YOUR UNDERSTANDING
5. Distinguish between the terms "focus" and "epicentre".
6. Outline the potential impact of earthquake hazards.

EARTHQUAKES AND PLATE BOUNDARIES

The movement of oceanic crust into the subduction zone creates some of the deepest earthquakes recorded, up to 700 km below the ground. When the oceanic crust slides into the hotter fluid mantle it takes time to warm up. As the slab descends, it distorts and cracks and eventually creates earthquakes.

Along mid-ocean ridges brittle faults occur as magma cools, solidifies and then cracks due to upwelling magma from below. Earthquakes here are small because the brittle faults cannot extend more than a few kilometres. Nevertheless, many earthquakes occur a long way from any plate boundary. Some of these are related to human activity, such as the construction of large dams, mining, fracking and the testing of nuclear weapons.

Earthquake hazards and impacts

Primary hazard	Impacts
Ground shaking	Loss of life
Secondary hazard	Loss of livelihood
Ground failure and soil liquefaction	Total or partial destruction of buildings
Landslides and rockfalls	Interruption of water supplies
Debris flow and mudflow	Breakage of sewage disposal systems
Tsunamis	Loss of public utilities such as electricity and gas
	Floods due to collapsed dams
	Release of hazardous material
	Fires
	Spread of chronic illness due to lack of sanitary conditions

Earthquakes may cause other geomorphological hazards such as landslides, liquefaction and tsunamis.

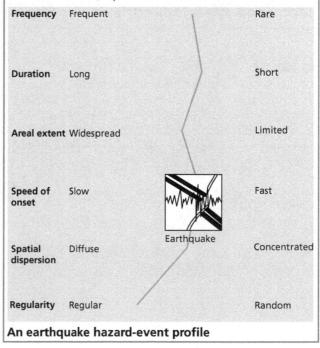

An earthquake hazard-event profile

Mass movements

Mass movements include any large-scale movement of the Earth's surface that are not accompanied by a moving agent such as a river, glacier or ocean wave.

Types of mass movement

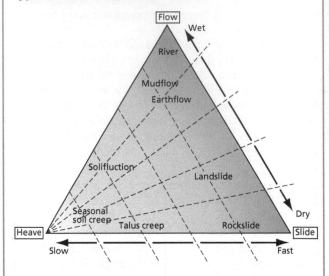

A classification of mass movements by speed

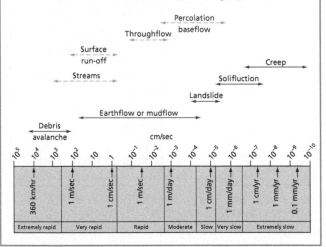

CAUSES OF MASS MOVEMENTS

Increasing stress and decreasing resistance

Factor	Examples
Factors contributing to increased shear stress	
Removal of lateral support through undercutting or slope steepening	Erosion by rivers and glaciers, wave action, faulting, previous rockfalls or slides
Removal of underlying support	Undercutting by rivers and waves, subsurface solution, loss of strength by exposure of sediments
Loading of slope	Weight of water, vegetation, accumulation of debris
Lateral pressure	Water in cracks, freezing in cracks, swelling, pressure release
Transient stresses	Earthquakes, movement of trees in wind
Factors contributing to reduced shear strength	
Weathering effects	Disintegration of granular rocks, hydration of clay minerals, solution of cementing minerals in rock or soil
Changes in pore water	Saturation, softening of material pressure
Changes of structure	Creation of fissures in clays, remoulding of sands and clays
Organic effects	Burrowing of animals, decay of roots

EXAM TIP ✔

Classifications of mass movement are simplifications. There is a lot of overlap between, for example, landslides, earthflows and mudflows. Sometimes the names are used interchangeably. Sometimes, the descriptions of the speed (extremely rapid, very rapid, extremely slow) are more helpful.

COMMON MISTAKE ❗

✗ *All mass movements are fast and destructive.*

✔ Some mass movements are large (and infrequent) and may kill people or destroy homes and/or livelihoods, but many are very small and may be relatively frequent, such as soil creep, but cause little impact to people and livelihoods.

CHECK YOUR UNDERSTANDING ❓

7. Compare the speed of a landslide with that of soil creep.
8. Identify (i) one dry and fast mass movement and (ii) a wet and slow mass movement.

Spatial distribution of geophysical hazard events

EARTHQUAKES

Tectonic hazards include earthquakes, volcanoes and tsunamis. Most of the world's earthquakes occur near plate boundaries, for example, along the centre of the Atlantic Ocean in association with the Mid-Atlantic Ridge. Similarly, there are many earthquakes around the edge of the Pacific Ocean. In some cases these chains are quite broad.

Broad belts of earthquakes are associated with subduction zones, whereas narrower belts of earthquakes are associated with constructive plate margins. Collision boundaries are also associated with broad belts of earthquakes, whereas conservative plate boundaries give a relatively narrow belt of earthquakes.

Many earthquakes are related to human activities, for example, the construction of large dams, mining and the testing of weapons. These may occur quite far from plate boundaries.

CHECK YOUR UNDERSTANDING

9. Compare the distribution of earthquakes with that of volcanoes.
10. Outline the main features of the distribution of fatal landslides.

VOLCANOES

Most volcanoes are found at plate boundaries although some occur over hotspots. About three-quarters of the Earth's 550 historically active volcanoes lie along the Pacific Ring of Fire.

Not all volcanoes are formed at plate boundaries. Those in Hawaii, for example, are found in the middle of the ocean and occur at a hotspot. A hotspot is a plume of hot material rising from the deep within the mantle – which is responsible for the volcanoes.

LANDSLIDES

Landslides occur all round the world. However, fatal landslides tend to be more common in areas that have:
- active tectonic processes that lead to high rates of uplift and occasional earthquakes
- high rainfall, including high short-term intensities
- a high population density.

Most fatal landslides occur in low-income countries where defensive structures are lacking. Those locations include:
- the southern edge of the Himalayas
- central China
- south-west India
- along the western boundary of the Philippine sea plate
- central Indonesia
- the Caribbean and central Mexico
- along the western edge of South America.

LOCATION OF NATURAL HAZARDS

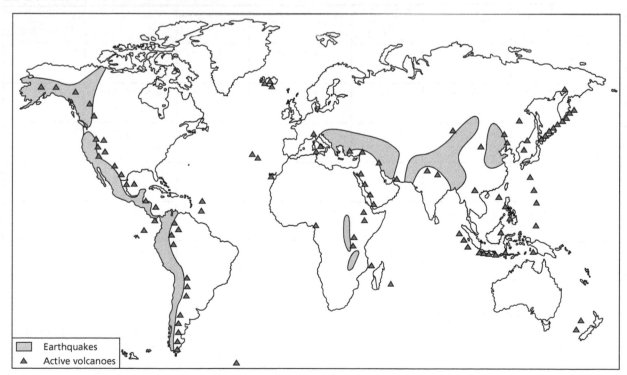

Earthquakes
Active volcanoes

The relevance of hazard magnitude and frequency/recurrence for risk management

RECURRENCE INTERVAL

The **recurrence interval** or **return period** is the expected frequency of occurrence measured in years for an event of a particular size. In general, small events have a high frequency/short return period whereas large events have a very low frequency/high return period. Thus there are fewer highly destructive earthquakes but many minor ones. These are generalized into high-frequency low-magnitude events versus low-frequency high-magnitude events. Low-frequency high-magnitude events cause the most destruction and require the greatest management.

EARTHQUAKE FREQUENCY AND MAGNITUDE

The Richter scale was developed in 1935 to measure the magnitude of earthquakes. The scale is logarithmic, so an earthquake of 5.0 on the Richter scale is 10 times more powerful than one of 4.0 and 100 times more powerful than one of 3.0. Scientists are increasingly using the Moment Magnitude Scale (M) which measures the amount of energy released and produces figures that are similar to the Richter scale. For every increase of 1.0 on the M scale the amount of energy released increases by over 30. Every increase of 0.2 represents a doubling of the energy released.

Annual frequency of occurrence of earthquakes of different magnitude based on observations since 1900

Descriptor	Magnitude	Annual average	Hazard potential
Great	≥ 8	1	Total destruction, high loss of life
Major	7–7.9	18	Serious building damage, major loss of life
Strong	6–6.9	120	Large losses, especially in urban areas
Moderate	5–5.9	800	Significant losses in populated areas
Light	4–4.9	6,200	Usually felt, some structural damage
Minor	3–3.9	49,000	Typically felt but usually little damage

MEASURING VOLCANOES

The strength of a volcano is measured by the Volcanic Explosive Index (VEI). This is based on the amount of material ejected in the explosion, the height of the cloud it creates and the amount of damage caused. Any explosion above level 5 is considered to be very large and violent. A VEI 8, or supervolcano, ejects more than 1,000 km³ of material, 10 times more than a VEI 7. The last eruption of a VEI 8 was about 74,000 years ago.

Magnitude and frequency of volcanic eruptions

VEI	Classification	Description	Height of eruption column	Volume of materials erupted	Frequency of eruption	Example	Occurrences in last 10,000 years
3	Vulcanian/ Pelean	Severe	3-15 km	>10,000,000 m^{-3}	Yearly	Nevado del Ruiz (1985)	868
4	Pelean/Plinian	Cataclysmic	10-25 km	>0.1 km^{-3}	$\geq$10 years	Soufriere Hills (1995)	278
5	Plinian	Paroxysmal	>25 km	>1km $^{-3}$	$\geq$50 years	Mt St Helens (1980)	84
6	Plinian/Ultra-Plinian	Colossal	>25 km	>10 km^{-3}	$\geq$100 years	Krakatoa (1883)	39
7	Plinian/Ultra-Plinian	Super-colossal	>25 km	>100 km^{-3}	$\geq$1000 years	Tambora	4
8	Plinian/Ultra-Plinian	Mega-colossal	>25 km	>1,000 km^{-3}	$\geq$10, 000 years	Toba (73,000 BP)	None

CHECK YOUR UNDERSTANDING

11. Describe the relationship between hazard magnitude and frequency.

12. Compare the frequency of a VEI 8 (supervolcano) with that of a VEI 5 (Plinian eruption), for example Mt St Helens.

Geophysical hazard risk

VULNERABILITY

The concept of **vulnerability** refers to the geographic conditions that increase the susceptibility of a community to a hazard. It includes not only the physical effects of a natural hazard but also the status of people and property in the affected area. A number of factors can increase people's vulnerability to natural hazards, especially catastrophic events.

Economic factors
- Levels of wealth and development: this influences the quality of housing that people live in.
- Building styles and building codes: in Haiti in 2010, many of the buildings collapsed because they had been poorly built).
- Access to technology: people with greater access to communications are better able to keep up to date with warnings and forecasts.
- Insurance cover: the poor cannot afford insurance cover.

Social factors
- Education: people with a better education generally have a higher income and can afford better-quality housing.
- Public education: educational programmes in Japan have helped people to deal with earthquakes.

- Awareness of hazards: the 2004 tsunami in South Asia alerted many people to the dangers that tsunamis present.
- Gender: many women are carers for their children and/ or their parents and they may feel responsible for them following an event.

Demographic factors
- Population density: large urban areas such as Port-au-Prince in Haiti are especially vulnerable to natural hazards.
- Age: elderly people may be less mobile than younger populations.
- Disability: the mortality rate of the disabled in the Japanese tsunami of 2011 was twice that of those without a disability.

Political factors
- The nature of society: the failure of the Burmese government to allow aid to the victims of Cyclone Nargis in 2008 increased the number of deaths from disease and malnutrition.
- Effectiveness of lines of communication: the earthquake in Sichuan (China) in 2008 brought a swift response from the government, which mobilized 100,000 troops and allowed overseas aid into the country.

THE PROGRESSION OF VULNERABILITY

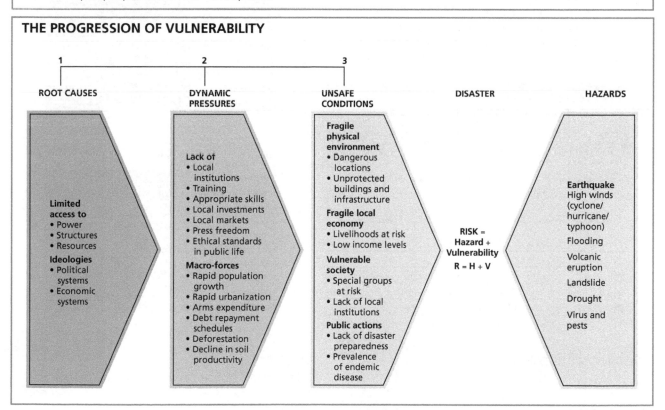

CHECK YOUR UNDERSTANDING
13. Suggest how political factors may affect hazard risk.
14. Briefly explain why some people are more vulnerable to hazards than others.

Geographic factors affecting the impacts of geophysical events

FACTORS THAT AFFECT THE IMPACTS OF GEOPHYSICAL EVENTS

The impacts of geophysical events depend on a number of interrelated factors.

- The magnitude and the frequency of events: for example, the stronger the earthquake, the more damage it can do. The more aftershocks there are the greater the damage that is done. Earthquakes that occur close to the surface are potentially more damaging than earthquakes deep underground since overlying rocks will absorb more of the energy of the deep-focus earthquakes.
- Population density: a geophysical event that hits an urban area of high population density could inflict far more damage than one that hits a rural area of low population density.
- Type of buildings: high-income countries generally have better quality buildings that have been built to be earthquake resistant.
- Time of day: an earthquake during a busy time, such as rush hour, may cause more deaths than one at a quiet time.
- Distance from the event: the impact of a volcano may decrease with distance from the volcano, whereas the effect of an earthquake may be greater further away from the epicentre.
- Types of rocks and sediments: loose materials may act like liquid when shaken, a process known as liquefaction. Solid rock is much safer, and buildings built on flat areas of solid rock are more earthquake resistant.

- Secondary hazards: these may cause more fatalities than the original event. For example, following the Pinatubo eruption and Nevado del Ruiz, more people died due to lahars than during the volcanic eruption.
- Economic development: high-income countries will usually be better prepared and have more effective emergency response services, better access to technology and better health services.

Country	WRI		
Australia	4.22 %	15.05 %	28.01 %
Brazil	4.09 %	9.53 %	42.92 %
Cambodia	16.58 %	27.65 %	59.96 %
Canada	3.01 %	10.25 %	29.42 %
Central Afr. Rep.	7.03 %	9.39 %	74.80 %
Chile	11.65 %	30.95 %	37.66 %
China	6.39 %	14.43 %	44.29 %
Costa Rica	17.00 %	42.61 %	39.89 %
Ecuador	7.53 %	16.15 %	46.63 %
Germany	2.95 %	11.41 %	25.87 %

Country	WRI		
Haiti	11.68 %	16.26 %	71.85 %
Indonesia	10.24 %	19.36 %	52.87 %
Jamaica	11.83 %	25.82 %	45.81 %
Japan	12.99 %	45.91 %	28.29 %
Liberia	7.84 %	10.96 %	71.54 %
Malawi	7.98 %	12.34 %	64.66 %
Myanmar	8.90 %	14.87 %	59.86 %
Nepal	5.12 %	9.16 %	55.91 %
Netherlands	8.24 %	30.57 %	26.94 %
Papua New Guinea	16.43 %	24.94 %	65.90 %

Country	WRI		
Philippines	26.70 %	52.46 %	50.90 %
Senegal	10.38 %	17.57 %	59.08 %
Serbia	7.12 %	18.05 %	39.46 %
Sri Lanka	7.32 %	14.79 %	49.52 %
Sudan	7.99 %	11.86 %	67.37 %
Syria	5.69 %	10.56 %	53.85 %
USA	3.76 %	12.25 %	30.68 %
Vanuatu	36.28 %	63.66 %	56.99 %
Viet Nam	12.53 %	25.35 %	49.43 %
Zimbabwe	10.06 %	14.96 %	67.24 %

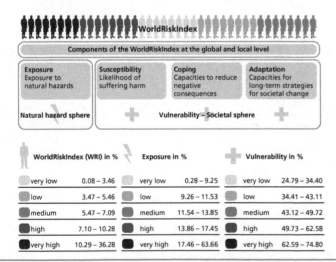

WorldRiskIndex

Components of the WorldRiskIndex at the global and local level

Exposure – Exposure to natural hazards	Susceptibility – Likelihood of suffering harm	Coping – Capacities to reduce negative consequences	Adaptation – Capacities for long-term strategies for societal change
Natural hazard sphere	Vulnerability – Societal sphere		

WorldRiskIndex (WRI) in %		Exposure in %		Vulnerability in %	
very low	0.08 – 3.46	very low	0.28 – 9.25	very low	24.79 – 34.40
low	3.47 – 5.46	low	9.26 – 11.53	low	34.41 – 43.11
medium	5.47 – 7.09	medium	11.54 – 13.85	medium	43.12 – 49.72
high	7.10 – 10.28	high	13.86 – 17.45	high	49.73 – 62.58
very high	10.29 – 36.28	very high	17.46 – 63.66	very high	62.59 – 74.80

CHECK YOUR UNDERSTANDING

15. Suggest how earthquakes could have a greater impact if they occurred on a weekday rather than at the weekend.
16. Briefly explain how the magnitude and frequency of hazard may affect its impact.

WHAT IS RISK?

Risk is expressed as the likelihood of loss of life, injury, or destruction and damage from a disaster in a given period of time. Disaster risk is widely recognized as the consequence of the interaction between a hazard and the characteristics that make people and places vulnerable and exposed.

FACTORS AFFECTING THE PERCEPTION OF RISK

At an individual level there are three important influences on an individual's response. For example:
- experience – the more experience of environmental hazards the greater the adjustment to the hazard
- material well-being – those who are better off have more choice
- personality – is the person a leader or a follower, a risk-taker or risk averse?

Ultimately there are three choices: do nothing and accept the hazard; adjust to the situation of living in a hazardous environment; leave the area.

HAZARD PROFILES

A hazard profile is a description and analysis of a specific type of local hazard. Primary hazards include earthquakes, floods and wildfires. Secondary hazards include landslides, tsunamis and fires. Primary hazards account for the greatest losses in terms of loss of life and economic damage.

It is possible to characterize hazards in many ways.
- **Magnitude:** the size of the event, for example, the size of an earthquake on the Richter scale, or M-scale.
- **Frequency (recurrence interval):** how often an event of a certain size occurs, for example, a flood of 1 m height may occur, on average, every year. By contrast, in the same stream a flood of 2 metres in height might occur only every ten years.
- **Duration:** the length of time that the environmental hazard exists. This varies from a matter of hours for urban smog, to possibly decades for a drought.
- **Areal extent:** the size of the area covered by the hazard. It can range from very small scale, such as an avalanche chute, to continental (for a drought, for example).
- **Spatial concentration:** the distribution of hazards over space, for example, they could be concentrated in certain areas, such as tectonic plate boundaries, coastal locations, valleys and so on.
- **Speed of onset:** this is rather like the time lag in a flood hydrograph. It is the time difference between the start of the event and the peak of the event. It varies from rapid events, such as the Kobe earthquake, to slower events such as drought in the Sahel of Africa.
- **Regularity:** some hazards are regular such as cyclones whereas others are much more random such as earthquakes and volcanoes.

EXAMPLES OF HAZARD EVENT PROFILES

California natural hazards profile

Frequency: How often is an event likely to occur?

1/100 years   Various times/1 year

Primary Hazards
● Earthquakes

Secondary Hazards
○ Volcanoes
○ Landslides

Magnitude: How extensive an area could the event affect?

Less than 10%   More than 50%

Duration: How long will the event last?

Minutes   Weeks

Speed of onset: How much warning time before is there an event occurs?

More than 12 hrs   Less than 1 hr

In California, earthquakes are relatively common, they affect large areas, occur without warning and are over quickly. In contrast, landslides are less common, they affect smaller areas, are shortlived and happen relatively quickly. Another hazard profile is shown for Nevada, below.

Nevada hazard profile

Profile Characteristics
in decreasing order of importance:

Intensity (damage to property values, in dollars)

$1000 $1 Billion

Frequency (annual probability of return)

.01   5.0

Spatial Extent (area affected)

5,000 sq. ft.   50,000 sq. ml.

Duration (time)

10 Seconds   90 days

Speed of Onset (time between initiation and exhibi.)

3 Seconds   1 Month

Natural Hazards of Nevada
in decreasing order of importance:
○ >6.0 Earthquakes
● Landslides
● Volcanic Ash

CHECK YOUR UNDERSTANDING

17. Describe the hazard risk profile for volcanoes in California.
18. Suggest reasons why the hazard profile for earthquakes in Nevada differs from the hazard profile for earthquakes in California.

Earthquakes

CASE STUDY

HAITI'S EARTHQUAKE, 2010

In 2010 an earthquake measuring 7.0 on the Richter scale occurred just 25 km west of Port-au-Prince, at a depth of only 13 km below the surface. Further aftershocks were as strong as 5.9, and occurred just 9 km below the surface, 56 km south-west of the city. A third of the population was affected: about 230,000 people were killed, 250,000 more were injured and around a million people were made homeless.

The city and the region around it are mainly shanty settlements of overcrowded, badly constructed buildings, hopelessly ill-suited to withstanding any shaking. Many of Port-au Prince's 2 million residents lived in tin-roofed shacks perched on unstable, steep ravines.

The Red Cross estimated that 3 million people – a third of Haiti's population – needed emergency aid. Haiti is one of the world's poorest countries, and it was overpopulated and vulnerable even before the disaster. The country had only two fire stations. The earthquake degraded an already feeble health service by destroying many hospitals and clinics.

A long-term strategy for rebuilding Haiti is vital. Even before the earthquake Haiti was poor, environmentally degraded, aid dependent and had few basic services.

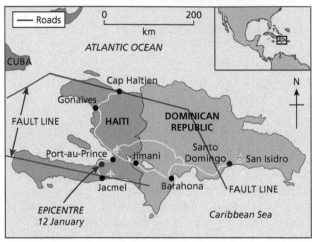

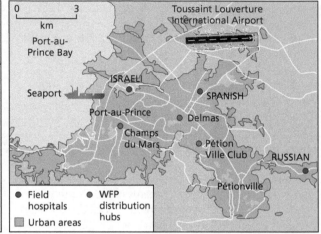

CASE STUDY

CHRISTCHURCH, NEW ZEALAND, EARTHQUAKES, 2010–2012

Christchurch is New Zealand's second-largest urban area with a population of 386,000. The 2010 Christchurch earthquake was a 7.1 magnitude earthquake which struck the South Island of New Zealand at 4.35 am. Aftershocks continued into 2012. The strongest aftershock was a 6.3 magnitude earthquake, which occurred in 2011.

There was a relative lack of casualties compared with the 2010 Haiti earthquake. The lack of casualties in New Zealand was partly because the earthquake happened in the early hours of a Saturday morning, when most people were asleep, many of them in timber-framed homes. Futhermore, building standards in New Zealand are high.

Another earthquake hit Christchurch in 2011. It was a powerful natural event that severely damaged the city, killing 185 people. The 6.3 magnitude earthquake struck the region at 12.51 pm local time on Tuesday 22 February 2011. The earthquake was centred 2 km west of the port town of Lyttleton and 10 km south-east of the centre of Christchurch. Although smaller in magnitude than the 2010 quake, the earthquake was more damaging and deadly for a number of reasons.

- The epicentre was closer to Christchurch.
- The earthquake was shallower at 5 km underground, whereas the September 2010 quake was measured at 10 km deep.
- The February 2011 earthquake occurred at lunchtime on a weekday when the CBD was busy.
- Many buildings had already been already weakened by previous earthquakes.
- Liquefaction was significantly greater than during the 2010 earthquake, causing the upwelling of more than 400,000 tonnes of silt.

CHECK YOUR UNDERSTANDING

19. Describe the main characteristics of the earthquake hazard profile (p.53).
20. Identify the main reasons why the 2011 Christchurch earthquake was more devastating than the 2010 Christchurch earthquake.

Volcanic eruptions

SOUFRIÈRE HILLS, MONTSERRAT

Montserrat is a small island in the Caribbean, and it has been affected by the Soufrière Hills volcano since 1995. The cause of the volcano is the plunging of the South American and North American Plates under the Caribbean Plate.

In July 1995 the Soufrière Hills volcano erupted after being dormant for nearly 400 years. At first the volcano gave off clouds of ash and steam. The biggest settlement and capital, Plymouth, with a population of just 4,000, was covered in ash and abandoned. This has had a severe impact on Montserrat as Plymouth contained all of the government offices and most of the shops and services. The southern third of the island had to be evacuated.

The hazard posed by the volcano was just one aspect of the risk experienced on Montserrat. For the displaced people there were other hazards.

All public services (government, health and education) had to be moved to the north of the island. Montserrat's population fell from 11,000 to 4,500. Many people fled to nearby Antigua. Some "refugees" stayed on in Montserrat living in tents.

The northern part of the island has been redeveloped with new homes, hospitals, upgraded roads and a new airport.

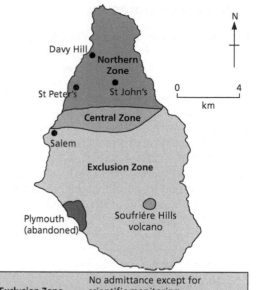

Exclusion Zone	No admittance except for scientific monitoring and national security matters.
Central Zone	Residential area only. All residents in state of alert. All have rapid means of exit 24 hours a day. All residents must have hard hats and dust masks.
Northern Zone	Significantly lower risk, suitable for residential and commercial occupation.

Volcanic hazard risk in Montserrat

MOUNT SINABUNG

Mount Sinabung is a volcano in northern Sumatra, Indonesia. Indonesia is located on the Pacific Ring of Fire and has some 120 active volcanoes. Sinabung is currently one of the most active volcanoes there. It erupted in 1600, and was then dormant until 2010. Since 2010 it has been active. The area surrounding the volcano is populated due to the fertile plateau that the volcano has helped to form.

Since the recent volcanic activity began in 2010, more than 30,000 villagers have been evacuated from their homes. The Department of Health provided medicines and doctors to the evacuees, and the National Disaster Management Agency provided food and face masks. Kitchens were set up; blankets, sleeping mats and tents were provided. Only one fatality was reported – a person who suffered from respiratory problems.

In 2013 crops failed due to the ash fall. In February 2014 eruptions caused the deaths of at least 16 people. Ironically, this eruption occurred just after residents living more than 5 km from Sinabung were allowed to return home. In 2016 Sinabung erupted again, killing seven people.

EXAM TIP
When using maps, make use of the scale to show, for example, the spatial extent of a hazard/protection scheme. For example, in the map showing the volcanic hazard risk in Montserrat, the exclusion zone is about 8 km × 8 km.

CHECK YOUR UNDERSTANDING
21. Describe the variations in the volcanic hazard risk, as shown on the map of volcanic hazard risk in Montserrat.
22. Outline one advantage of Mount Sinabung to the people of Indonesia.

Landslides

MANAGING URBAN LANDSLIDES

Kalimpong is a hill station located in West Bengal, India. It is located at a height of 1,250 m overlooking the Teesta river. Dumsi Pakha is an area of comparatively low wealth below the main town. This area shows how urban mismanagement in steep terrains can lead to landslide problems, with catastrophic outcomes for the residents.

The two key elements that cause landslide problems are poor water management and slope disruption. Water management on the slopes above Dumsi Pakha is a key factor. In Kalimpong town, water is discharged without any control into gullies than run through Dumsi Pakha. The channel is dry towards the end of the dry season, and is choked with garbage. There is extensive instability on the flanks of the channel. This instability is being exacerbated by the large flows that the channel transports during the monsoon.

Elsewhere in the settlement the problems are more local. Some houses are built by first creating a terrace by excavating into the slope. By creating a mini-terrace to build on, a steeper slope is made behind the terrace, and that increases instability.

Water management is a serious problem, with a lack of sewers and storm drains. Large volumes of water flow unregulated into natural channels during heavy rainfall, causing problems downstream. Close to the town, the dumping of construction waste in an uncontrolled manner is adding to the problem. Adding extra weight to the top of an active landslide will inevitably make the stability problem worse.

LANDSLIDES IN SRI LANKA, 2016

The death toll from three days of torrential rain and landslides in Sri Lanka, during May 2016, rose to 37, with more than 150 people missing. Rains forced more than 137,000 to leave their homes. The worst affected area was in the Central Hills.

The Red Cross said rescue operations had been made more difficult as roads leading to the mountainous area were treacherous.

The area had received more than 100 mm of rain between 15 and 16 May. Dozens of houses were destroyed when mountain slopes collapsed, forcing hundreds of villagers to evacuate.

A range of costs commonly associated with landslide problems	
Personal cost	Fatal accidents
	Injuries
	Psychiatric problems
Immediate costs	Evacuation and provision of temporary or replacement housing
	Mobilization of relief workers and emergency services
	Transport delays
	Cost of repair
Indirect costs	Compensation
	Increased insurance premiums
	Depreciated property or land values
Cost of prevention	Research into the nature and extent of landslide problems at universities
	Design and construction of preventative measures including drainage and regrading
	Costs of monitoring potentially unstable slopes

COMMON MISTAKE

✗ *All landslides are caused by natural processes.*

✓ Some landslides are caused by natural processes, but many are caused by human activities such as road building, and can be triggered by the vibrations caused by heavy traffic.

CHECK YOUR UNDERSTANDING

23. Briefly outline the two main causes of landslides at Kalimpong.
24. Explain why it is costly to manage landslides

Global geophysical hazard and disaster trends

TRENDS IN GEOPHYSICAL HAZARDS

The frequency of geophysical disasters remained broadly constant between 1994 and 2013. In contrast, population growth and economic development have varied considerably. Earthquakes (including tsunamis) have killed more people than all other natural disasters put together. They are rarer than floods but can cause very high numbers of casualties in a short period. Tsunamis were the deadliest natural disaster. Asia bore the brunt of natural disasters, notably Indonesia, Pakistan and China. Low-income countries bear a disproportionate burden of natural hazards, including a huge disparity in death rates.

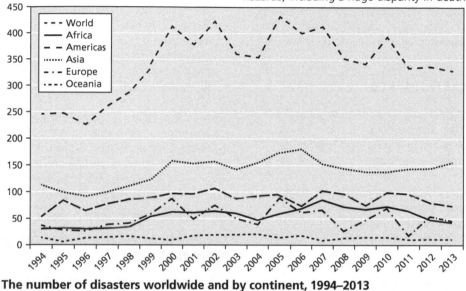

The number of disasters worldwide and by continent, 1994–2013

POPULATION GROWTH AND URBAN GROWTH

Urbanization within highly seismic zones has increased significantly over recent decades. Slums and squatter settlements frequently expand onto high-risk areas such as slopes and embankments. People are becoming more susceptible to natural disasters, largely due to population growth and globalization. There are increasingly more people in high-risk areas.

The world's population is expected to reach 8.5 billion by 2030 and 9.7 billion in 2050. Half of the world's population growth is expected to be concentrated in nine countries: India, Nigeria, Pakistan, D.R. Congo, Ethiopia, Tanzania, the USA, Indonesia and Uganda.

The concentration of population growth in the poorest countries presents its own set of challenges. These include eradicating poverty and inequality, combating hunger and malnutrition, expanding educational enrolment and health systems, and providing adequate housing, all of which are crucial if sustainable development is to be achieved.

Number of disasters and number of deaths per income group, 1994–2013

	Disasters (%)	Number of disasters	Deaths (%)	Number of deaths
High income	26	1,700	13	182,000
Upper middle-income	30	1,992	19	252,000
Lower middle-income	27	1,751	35	474,000
Low income	17	1,119	33	441,000

Source: Based on data in CRED, 2015

CHECK YOUR UNDERSTANDING

25. Describe the trend in natural disasters for the world and by continent.

26. Outline the main impact of natural hazards on income groups.

Geophysical hazard adaptation

PREDICTIONS, FORECASTS AND WARNINGS

Scientific predictions are used to provide precise statements on the time, place and size of a future event. Forecasts are more general statements about future events. They are commonly expressed as probabilities. Effective warning depends on science and technology, as well as communications systems and people's ability to interpret them. Earthquakes remain the most difficult natural hazard to predict and forecast, although areas that are "at risk" are well known.

PREDICTING EARTHQUAKES

The main ways of preparing for earthquakes include:
• better forecasting and warning
• improve building design and building location
• establish emergency procedures.

There are a number of ways of predicting and monitoring earthquakes, which involve the measurement of:
• small-scale ground surface changes
• small-scale uplift or subsidence
• ground tilt
• changes in rock stress
• micro-earthquake activity (clusters of small quakes)
• anomalies in the Earth's magnetic field
• changes in radon gas concentration
• changes in electrical resistivity of rocks.

Monitoring for earthquake prediction

Instrument	Purpose
Seismometers	To record micro-earthquakes
Magnetometer	To record changes in the Earth's magnetic field
Near-surface seismometer	To record larger shocks
Vibroseis truck	To create shear waves to probe the earthquake zone
Strain meters	To monitor surface deformation
Sensors in wells	To monitor changes in groundwater levels
Satellite relays	To relay data to the US Geological Survey
Laser survey equipment	To measure surface movement

PREDICTING VOLCANOES

Volcanic eruptions are easier to predict than earthquakes since there are certain signs. The main ways of predicting volcanic eruptions include:
• seismometers, to record swarms of tiny earthquakes that occur as the magma rises
• chemical sensors, to measure increased sulphur levels
• lasers/GPS, to detect the physical swelling of the volcano/crater
• ultrasound, to monitor low-frequency waves in the magma resulting from the surge of gas and molten rock
• observations, such as at Gunung Agung (Java).

However, it is not always possible to state exactly when a volcanic eruption will happen. It is also difficult to predict the timescale of an eruption. Some volcanoes may erupt for days, while others go on erupting for years. In general, volcanoes at destructive plate boundaries tend to produce more-explosive volcanoes, whereas those at hotspots, such as Hawaii, produce more frequent but less-explosive eruptions.

GEOPHYSICAL HAZARD ADAPTATION: PREPAREDNESS

• Land-use zoning is an important aspect of geophysical hazard adaptation. Different land uses may be prevented from locations in a zone that is known to be at risk of a hazard. For example, densely populated buildings, hospitals and fire services should not be built close to a fault line or in areas at risk of landslides. In some volcanic areas, residents are evacuated, for example, from the slopes of Mount Sinabung, and an exclusion zone may be formed, as in the case of Montserrat. Building codes can be enforced to ensure that buildings are of an adequate standard to survive a hazard event.
• Insurance: nations and the international community are generally not well prepared for rare events. One method of preparedness is to take out insurance cover against hazard events. However, some geophysical events are considered by the insurance industry as "acts of God", and so insurance cover is not available. Another factor is that most LIC residents are unable to afford insurance, even if it is available. In addition, it is always much harder to justify spending money on an event that might not occur. It is far easier to spend money after an event has happened.
• New technology can be used to record the swelling of volcanoes and changes in water chemistry, and mobile phones with GPS can be used to inform agencies about geophysical changes.

❓

CHECK YOUR UNDERSTANDING
27. Briefly explain how land-use zoning operates.
28. Suggest reasons why some people do not have insurance against natural hazards.

Pre-event management

MANAGING LANDSLIDES

There are many ways to reduce the risks associated with landslides, such as:

- terracing steep slopes
- drainage of water from slopes
- building restraining structures such as stone walls
- erosion control at the base of cliffs
- diversion of roads away from active areas.

MANAGING THE RISK OF EARTHQUAKES

Buildings can be designed to withstand the ground-shaking that occurs in an earthquake. Single-storey buildings are more suitable than multi-storey structures as the potential for swaying is reduced. Some tall buildings are built with a "soft storey" at the bottom, such as a car park raised on pillars. This collapses in an earthquake, so that the upper floors sink down onto it and this cushions the impact. Basement isolation – mounting the foundations of a building on rubber mounts which allow the ground to move under the building – is widely used in earthquake-prone areas. This isolates the building from the tremors. Building reinforcement strategies include building on foundations built deep into underlying bedrock, and the use of steel-constructed frames that can withstand shaking. Land-use planning is another important way of reducing earthquake risk.

	Pakistan	Haiti	Peru	Indonesia
Most destructive quake	8 October 2005	12 January 2010	31 May 1970	26 December 2004
Location	Northern Pakistan/Kashmir	Port-au-Prince area	Chimbote	Sumatra
Magnitude	7.6	7.0	7.9	9.1
Fatalities	75,000	222,500	70,000	227,900 (including the global tsumami deaths)

Light walls and gables
Lightweight structures are subject to smaller forces and are less likely to fall when the ground shakes.

Quake-resistant houses are being built in Pakistan – of straw. The compressed bales are held together by nylon netting and sandwiched between layers of plaster.

Light roofs
In Haiti heavy concrete roofs collapsed on many homes; sheet-metal roofs on wooden trusses are more resistant.

Small windows
Small, regularly spaced openings create fewer weak spots in walls. But the bigger problem in Haiti was that walls were not properly reinforced.

Reinforced walls
The reinforcing rods need not be made of metal. Natural materials such as eucalyptus or bamboo work well too.

In Peru the walls of some adobe houses have been reinforced with a plastic mesh to prevent collapse.

Bamboo

Mesh

Crown beam

Corner column

Confined masonry
In Indonesia and elsewhere, brick walls can be framed and connected to the roof by corner columns and a crown beam of reinforced concrete. In a quake the structure moves as a unit.

Shock absorbers
Tyres filled with stones or sand and fastened between floor and foundation can serve as cheap ground-motion absorbers for many types of building.

Safe houses

CONTROLLING VOLCANOES

It is possible to manage lava flows by diverting them using dry channels or explosives, or by pumping water onto the lava front to cool it. There is little that can be done to reduce the impacts from pyroclastic flows other than to evacuate the area. GPS can be used to monitor the swelling of volcanoes, which could indicate an imminent eruption.

MANAGING THE RISK OF TSUNAMIS

Tsunamis are generally managed using sea walls and early warning systems. However, cost constraints usually dictate the height of the wall that can be built. Walls can only provide a certain amount of protection and will not stop bigger waves.

CHECK YOUR UNDERSTANDING

29. Suggest how small windows and light roofs help in an earthquake.
30. Outline a cheap method of producing shock absorbers.

Post-event management strategies

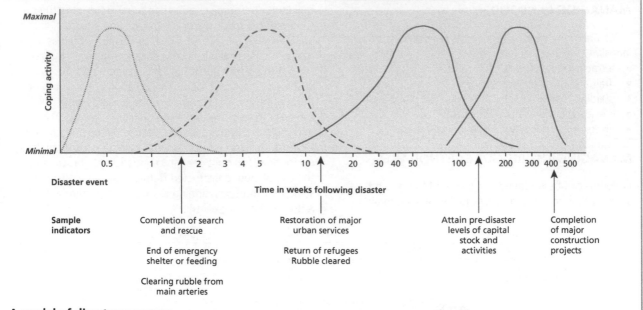

A model of disaster recovery

Rescue

In the immediate aftermath of a disaster the main priority is to rescue people. This may involve the use of search and rescue teams and sniffer dogs. Thermal sensors may be used to find people alive among the wreckage. The number of survivors decreases very quickly. Few survive after 72 hours.

Rehabilitation

Rehabilitation refers to people being able to make their homes safe and live in them again.

Reconstruction

For some residents, rehabilitation is not possible, and so reconstruction (rebuilding) is necessary. This can be a very long, drawn-out process, taking up to a decade for major construction projects.

The overall aim of rehabilitation and reconstruction is to get communities back to their pre-disaster level, and to promote continuing human development.

RESCUE, REHABILITATION AND RECONSTRUCTION IN PRACTICE

Following the 2004 South Asian tsunami, the Indonesian government produced a master plan for the rehabilitation and reconstruction of the region. The government:
- provided help for survivors of the disaster
- buried the bodies of the dead
- enhanced basic facilities and infrastructure to be able to provide adequate services for the victims.

Within eight years, Banda Aceh had been reconstructed. Some $7 billion of contributions helped the Indonesian government to rebuild homes and communities first, then to restore the infrastructure, and finally redevelop the economy.

USING PHONES FOR HAZARD MAPPING AND TO TRACK MISSING CHILDREN

The use of ICT allows information about the scale and location of a hazard to be distributed quickly. In a project in Rio de Janeiro, UNICEF has been training young people to map social and environmental risk.

RapidFTR was created by a New York student. It is an open-source app designed to reunite children with their families in rapidly developing disaster situations, and it is being actively developed by UNICEF.

Using phones and laptops, the app enables humanitarian workers to register information about missing children. The quicker children are found, the less likely they are to suffer violence, exploitation or trafficking.

COMMON MISTAKE

✗ *Mobile phones make the management of geophysical hazards much easier.*

✓ Although, in theory, mobile phones can make the management of geophysical hazards easier, not everyone has access to a mobile phone. Even those who do might not have the RapidFTR app. Just because the app exists does not guarantee its successful use.

CHECK YOUR UNDERSTANDING
31. State the length of search and rescue, as shown in the diagram.
32. Identify the main process taking place between 4 and 5 weeks after the disaster.

EXAM TIP
When writing about disaster recovery and rehabilitation, show that there is a difference between rehabilitation in HICs and LICs. In many LICs, rehabilitation and recovery takes much longer than in HICs, as the government does not have the funds available.

Exam practice

The diagram shows a hazard response curve.

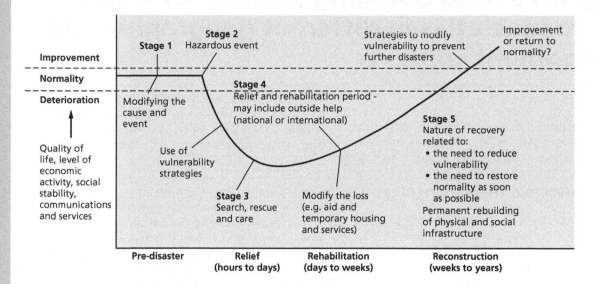

(a) **(i)** For one named geophysical hazard, describe how it could be modified and/or its impacts reduced (Stage 1 of the hazard response curve). (2 marks)

(ii) Suggest why the quality of life may decrease dramatically in Stage 2. (2 marks)

(b) Contrast the main characteristics of the relief/rehabilitation period with that of the reconstruction period.
(3 + 3 marks)

(c) **Either**

Examine the impact of social factors on geophysical hazard risk. (10 marks)

Or

Examine the role of hazard magnitude and frequency in the development of hazard management strategies.
(10 marks)

1 CHANGING LEISURE PATTERNS

The growth and changing purpose of leisure time for societies in different geographic and development contexts

DEFINITIONS

Leisure may be defined in terms of time, activities, or states of mind. In terms of time, leisure can be seen as free time. Leisure can also be defined as specific activities conventionally thought of as "leisurely". A more thorough definition may be based on what the majority of people would list as leisure activities, such as watching TV, participating in sports or exercise, reading, watching movies, and so on.

FACTORS AFFECTING THE GROWTH OF LEISURE

The growth of leisure has been facilitated by many factors:
- a reduction in the length of the working day
- a reduction in the length of the working week
- an increase in wages
- an increase in disposable income
- the growth of leisure activities
- an increase in people taking early retirement
- an increase in self-employment and flexitime
- developments in technology (such as washing machines and freezers) which enable people to spend less time on chores, and other developments in technology, such as TVs and the internet, which provide opportunities for leisure.

ECONOMIC DEVELOPMENT AND PARTICIPATION IN LEISURE

There is a strong link between income and leisure activity. Societies in LICs have less disposable income and therefore less chance of engaging in leisure pursuits, particularly if this involves purchasing expensive equipment.

For some indigenous populations, leisure activities include story-telling and music. For others, their quality of life and standard of living is so low that it limits the amount of leisure time and leisure activities.

As life expectancy increases, leisure activities in retirement become more important. The leisure industry has had to accommodate the needs of the over-65s who make up 25% of the population in many HICs.

Gender differences can also be noticed. Men generally have more leisure time than women. This is largely due to the household and parenting responsibilities of women, although this is changing. Women's status in society is closely linked to the amount of leisure time they have.

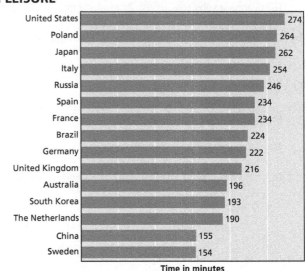

Average daily TV viewing time per person in selected countries worldwide in 2015 (in minutes)

CHECK YOUR UNDERSTANDING

1. Suggest why leisure activities increase with levels of development.
2. Outline the main characteristics of the countries where people watch most TV each day.

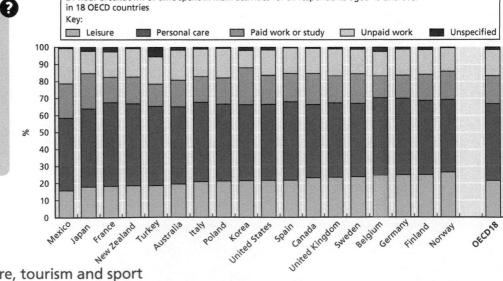

The categorization of tourist activities and sporting activities

CATEGORIZATION OF TOURIST ACTIVITIES

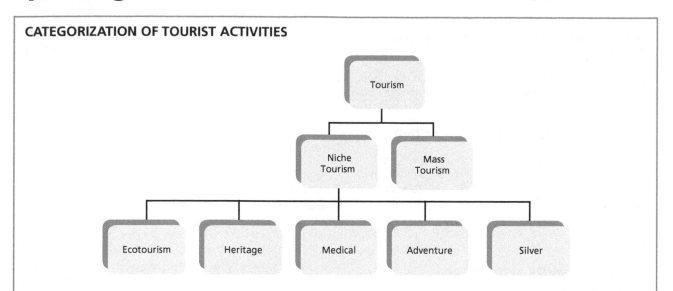

Types of tourism

There are many types of tourism. These vary greatly in terms of their cost, duration and destination. For example, skiing holidays are generally expensive due to the equipment needed. As they mainly occur in mountainous areas, much of the food and drink has to be imported, and therefore it costs more. Most tourists usually go skiing for one week.

In contrast, beach holidays may be cheaper because less specialist equipment is needed. Large-scale tourism is a high-density form of tourism and holidays often last for between 7 and 14 days. This type of tourism is most frequently found in coastal areas, and is usually provided as a package holiday. In contrast, ecotourism is generally a low-impact form of tourism and is often associated with nature.

CATEGORIZATION OF SPORTING ACTIVITIES

The table below classifies sporting activities. Some sports are very popular such as swimming, fishing and jogging. Some are very expensive such as skiing and yachting, whereas others are cheap, such as running or football.

Groupings of sports

Group 1 Athletics and rugby	American football, archery, curling, Gaelic sports, gymnastics, lacrosse, orienteering, rugby league, rugby union, track and field athletics, triathlon
Group 2 Dancing and yoga	Keep fit, aerobics, dance exercise, pilates, trampolining, yoga
Group 3 Outdoor sports	Angling or fishing, BMX, cyclo-cross, mountain biking, climbing/mountaineering, cycling (for health, recreation, training), cycling (to get to places, e.g. work, shops), hill trekking or backpacking, motor sports, rambling/walking for pleasure, shooting
Group 4 Swimming, cycling and gym	Health, fitness or conditioning activities, swimming or diving (indoor or outdoor), cycling for fitness or pleasure
Group 5 Racquet sports and running	Badminton, ice-skating, jogging, cross-country, road running, squash, table tennis, tennis
Group 6 Bowling	Bowls (lawn) (outdoor), bowls (indoor)
Group 7 Cricket, football, pub sports and tenpin bowling	Cricket, croquet, darts, football (indoors and outdoors), golf, pitch and putt, putting, skittles, snooker, pool, billiards (excluding bar billiards), tenpin bowling
Group 8 Boxing, martial arts and weightlifting	Boxing, judo, karate, other martial arts (including self-defence), taekwondo, weight training (including body building), weightlifting
Group 9 Minor team sports	Baseball/softball, basketball, hockey, netball, rounders, volleyball
Group 10 Water sports	Any other water sport, canoeing, rowing, skiing, waterskiing, windsurfing or boardsailing, yachting or dinghy sailing

EXAM TIP

The table is a just summary of sport – there are many sports not included, for example, Australian rules football and playing with conkers. If there is a sport you like, refer to it in your answer.

CHECK YOUR UNDERSTANDING

3. Distinguish between mass tourism and ecotourism.
4. Choose any two activities from two different groups, and comment on their likely cost, popularity and site.

The link between economic development and participation in leisure activities

CHANGES IN LEISURE TIME AND ACTIVITIES

As countries develop, there is a change in the leisure activities that people take part in. This occurs for many reasons. Very poor people may not be able to afford appliances such as televisions. However, as incomes improve, watching TV becomes an important leisure activity. It is not just about being able to afford the leisure activity, it is also about having the time to enjoy the activity or partake in it. For more wealthy countries/communities, the combination of more income and more leisure time leads to a greater range of leisure activities.

CASE STUDY

CHILDREN'S LEISURE ACTIVITIES IN SUDAN, AN LIC

When rural children are collecting water, running errands to the shop, weeding plots and/or herding sheep and goats, they introduce elements of play into their activities. For example, when boys are out herding, games such as shedduck (play fighting where participants have to hop with one leg behind them) are played. By combining games with overseeing the herds, boys can make the work more enjoyable and the time passes more quickly. Children use scrap metal to make dolls, tractors, houses and models of local shops. They then use these to act out domestic life and agricultural cycles. An awareness of how trade and wage labour operate is exemplified in their acting out of paying for crops using money made out of broken china.

LEISURE IN THE BRICs

Growth of the leisure industry in the BRIC countries (Brazil, Russia, India and China) has been accelerating due to rising disposable income, a growing middle class, rapid urbanization, greater online connectivity and an aging population.

Across the BRICs, visiting the cinema is a favourite pastime. India led the world with 2.9 billion cinema attendances in 2013. Although cinema attendances have been falling in India relative to TV, gaming and other online entertainment, cinema attendance in Brazil, Russia and China has been rising.

As the BRIC population ages, the demand for leisure activities such as health resorts, spas, theatres, concert halls, exhibitions and cruises is set to increase dramatically.

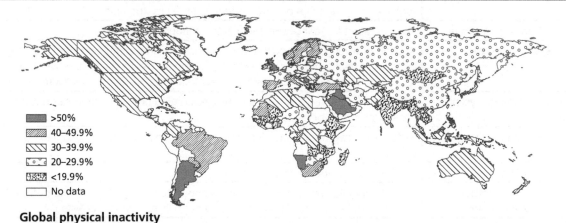

> >50%
> 40–49.9%
> 30–39.9%
> 20–29.9%
> <19.9%
> No data

Global physical inactivity

COMMON MISTAKE

✗ As the population ages, there will be increased demand for new facilities and activities in all countries.

✓ Demand for new facilities and activities among the elderly comes from those who are well off and can afford to participate – there are many who are too poor or infirm to participate in leisure activities.

CHECK YOUR UNDERSTANDING

5. Explain briefly why many people in NICs do not have much leisure time.
6. Explain why leisure activities in low income countries are limited.

Factors affecting participation in sports and tourism

PHYSICAL, DEMOGRAPHIC AND SOCIO-ECONOMIC FACTORS

The participation rate refers to the proportion of a population that takes part in a specific sporting activity. There are many factors that influence a person's decision to participate in sport and tourism. These include physical, demographic and socio-economic factors, including age, sex, economic circumstances, ethnicity, health, stage in life cycle, amount of free time and disposable income.

AFFLUENCE

Most tourists and many participants in sport are affluent. Most people who go on holiday travel a relatively large distance from their home, that is, they visit a place that they do not normally experience. Many sports take place in sporting venues such as tennis courts and swimming pools, which generally charge an entry fee. Most participants in golf are from HICs and NICs, although there is a growing number of golf courses in LICs to cater for tourists.

Socio-economic status	Walk	Cycle	Sport and active recreation
Managerial and professional	74.3	13.2	62.7
Intermediate occupations	67.4	10.0	51.7
Small employers and self-employed	64.2	9.6	55.9
Lower supervisory and technical occupations	66.7	9.9	49.4
Semi-routine and routine	62.0	10.3	53.6
Never worked and long-term unemployed	67.0	10.3	53.6

GENDER

There is a relatively low participation rate of Muslim women in athletics and swimming, for example. The convention for Muslim women to remain robed means that successful Muslim athletes, such as the Moroccan middle-distance runner, Hasna Benhassi, receive much criticism at home.

PERSONALITY

Some people prefer team sports while others may prefer more solitary sports such as running. Some people do not like to participate in sport. Likewise, tourist destinations very between those who prefer adventure, those who prefer cultural heritage, and those who prefer to seek out "sun, sand and sea".

PLACE OF RESIDENCE

Many people who live in mountainous areas ski regularly, just as those who live in a coastal area may go surfing. Good access to rivers and lakes may promote fishing, just as hilly areas may promote cycling. Many people prefer to spend their holidays in a different type of environment to that in which they live.

STAGE IN FAMILY LIFE CYCLE

The family life cycle refers to the responsibilities that a person may have towards their family. Some of the stages include:
* young, single people not living at home
* a young couple, without children
* a couple with children being supported at home/ university
* older couple with no dependent children living at home.

These stages may affect the amount of time and disposable income that is available to participate in sport or to take a holiday.

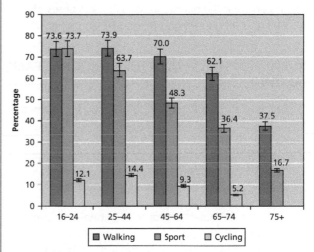

Sport and active recreation by age in the UK

In a survey of participation in sport in the UK, 16–24-year-olds were most likely to play football outdoors (over 20%), and football was the most popular sport among this age group. Health and fitness (going to the gym) was a top ten activity for all age groups and highest for those aged 16–24 (over 20%) and 25–44-year olds (18%). The 16–44-year-olds were the most likely to swim, while people aged 75 and over were the least likely to. Golf was a top ten activity for all age groups except 16–24-year-olds.

CHECK YOUR UNDERSTANDING
7. Suggest how changes in life cycle can affect participation in tourism.
8. Describe the variations in age and participation in the three sporting activities shown in the diagram.

Rural and urban tourism hotspots

DEFINITION

Hotspots are areas of intense sporting or leisure activity that attract above average numbers of visitors. Tourists are attracted to these hotspots because they have primary and secondary resources and are accessible.

Primary tourist/recreational resources are the pre-existing attractions for tourism or recreation (that is, those not built specifically for the purpose), including climate, scenery, wildlife, indigenous people, cultural and heritage sites. These are distinguished from **secondary tourist/ recreational resources**, which include accommodation, catering, entertainment and shopping.

CASE STUDY

TOURISM IN AN URBAN HOTSPOT: OXFORD, UK

Oxford is a world-famous historic university city. It is a small city (population c. 150,000) but it attracts around 5 million visitors each year.

The most popular attractions include the Ashmolean Museum and a number of Oxford University colleges, the Old Bodleian Library and the Sheldonian Theatre.

Tourism generates around £300 million in Oxford annually. The number of jobs sustained directly and indirectly by tourism is estimated at about 7,300 or 3–4% of the working population.

Problems related to tourism include traffic levels in the city, in particular the number of coaches and buses, the availability and cleanliness of toilets and overcrowding.

The Oxford Tourism Strategy aims to ensure visitor satisfaction, encourage an increase in tourist spending within the city, and minimize the environmental problems which result from tourism. In particular, the strategy attempts to:

- provide a larger coach park
- increase the use of public transport and park and ride
- encourage walking tours, registered sightseeing buses and cycles
- provide comprehensive on-street information to encourage visits to lesser-known attractions and places of interest
- increase the number of visitors in the off season in order to spread the tourist load.

RURAL HOTSPOT – KILLARNEY NATIONAL PARK, IRELAND

The scenery of the Killarney area, including the National Park, is world renowned. It is a major attraction and the area is one of the most visited tourist venues in Ireland. Over a million visitors travel to Kerry each year bringing an estimated £160 million to the area. Of these the majority visit Killarney, a town with a resident population of 14,000 and over 4,000 tourist rooms!

The five basic objectives for the National Park are to:
- conserve nature
- conserve other significant features and qualities
- encourage public appreciation of the heritage and the need for conservation
- develop a harmonious relationship between the park and the community

- enable the park to contribute to science through environmental monitoring and research.

The Killarney National Park Management Plan identifies four main zones:
- Natural zone: where conservation of nature is the main objective.
- Cultural zone: where the primary objective is the conservation of noteworthy features resulting from human activities including demesne landscapes, archaeological and historic sites, buildings and structures.
- Intensive management zone: where basic park objectives other than conservation are emphasized, provided park resources are not adversely affected.
- Resource restoration zone: removal of non-native conifer plantations.

COMMON MISTAKE

✗ A hotspot is an overcrowded tourist resort such as a national park.

✓ A hotspot is a relatively small area, not the whole of Oxford or Killarney National Park but a small part within it, such as the university area of Oxford, or Muckross House in Killarney National Park.

CHECK YOUR UNDERSTANDING

9. Suggest why Oxford has become an urban tourism hotspot.
10. Briefly explain the land-use zoning used in Killarney National Park.

Spheres of influence for sporting and touristic facilities

DEFINITIONS

The **sphere of influence** refers to the area from which a facility or an attraction draws its support. A small playground may have a low **threshold** population (the minimum number of people needed to support a facility or service) and a very small **range** (the maximum distance people will travel to visit a facility/service), hence it will have a small sphere of influence (catchment area). In contrast, a national sports stadium will derive its support from a much wider area.

SPORTS FACILITIES

An example of a sports site at the bottom of the scale could be a small play area, with a sphere of influence of about 1 km, providing facilities/activities for young children. At a higher level, there may be a sports facility with a variety of indoor and outdoor activities – a gym or a sports centre – providing for a range of up to 10 km.

At the top end of the hierarchy, there may be a top-level sports complex and athletics stadium/swimming pool, serving a larger population from a greater distance. In the UK, it has been suggested that the threshold population for an 18-hole golf course is about 30,000 people, and that about 3 ha of playing fields should be provided for every 1,000 people. In the USA, baseball pitches ("diamonds") and tennis courts have a threshold population of 6,000 and 1,000 respectively.

A simple hierarchy is outlined below.

A hierarchy of sporting facilities in different settlements

Sphere of influence	Community size	Recommended facilities	Activities offered
<5km	Village 500–1,500	Community hall Community open space Mobile library	Badminton, keep fit, yoga, football, cricket
<10km	Small country town 2,000–6,000	As above plus: Tennis courts, gyms Sports hall Swimming pool	As above plus tennis, netball, gym, hockey
<25km	Town	As above plus: Specialist sports venues Golf courses Skateboard parks Bowling	As above plus bowling, golf, skateboarding, judo, karate
<50km	City	Sports stadia Athletics grounds	As above plus home grounds of sports clubs (football, rugby, hockey, athletics grounds)
	Capital city	National sports centre for selected sports	As above, but for national teams

TOURIST FACILITIES

Tourist attractions have a national and/or international sphere of influence. The examples of Oxford and Killarney National Park both attract national and international tourists. Most national parks are located in environments that are attractive, rugged, of significant natural beauty or significance, and are usually located away from large urban areas. However, there are some natural parks that are located close to urban areas, and some within urban areas, such as Sangay Gandhi National Park in Mumbai and Nairobi National Park. Some tourist destinations have a truly global sphere of influence, such as New York, London and Paris, whereas others are more localized such as Centre Parcs in the UK and Camp America, USA.

In some countries, it is possible to recognize national/international tourist attractions in a country, for example, in South Africa Kruger National Park has an international sphere of influence whereas the Addo Elephant Park has a local/national sphere of influence.

EXAM TIP
When writing about the sphere of influence of tourism or sporting events, be sure to give named examples.

CHECK YOUR UNDERSTANDING
11. Define the term "sphere of influence".
12. Describe how the sphere of influence varies for facilities in a village and a city.

Factors affecting the geography of a national sports league

Most teams in the top league of their national sports derive support from a wide area. Some clubs have a global following, although many of the fans can only follow their team on television or the internet. In contrast, lower league teams tend to have a much smaller sphere of influence. They do not have the success to generate international (or even national) interest. Hence, their support is more localized. However, some fans move on to university or for a job, so the fan base may become broader.

RUGBY IN SOUTH AFRICA

Rugby is one of the most popular sports in South Africa, along with soccer and cricket. The country has competed at a high level in these sports. In apartheid South Africa, rugby was the "white person's game", whereas most black and coloured communities played soccer. In 2017, the Cheetahs received an invitation to play in the Rugby Union Pro12 league, which features team from Ireland, Wales, Scotland and Italy.

Currie Cup

The Currie Cup is the main provincial rugby competition in South Africa. It has been contested since 1892. Up to and including 2015, the most successful province was Western Province (Western Cape), followed by the Northern Transvaal/ Blue Bulls. Since rugby became a professional sport in the early 1990s, no single team has dominated the Currie Cup.

The Currie Cup takes place roughly between July and October. The format divides 14 teams into eight Premier Division and six First Division teams.

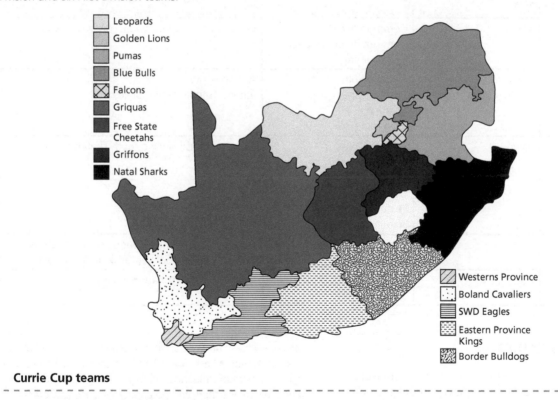

Currie Cup teams

GAELIC GAMES IN IRELAND

In Ireland, the Gaelic Games (hurling, Gaelic football and camogie) are arranged at a county level. Teams form the Republic of Ireland (26 counties) and Northern Ireland (6 counties) compete in a national league and in a knock-out competition. The league is based on merit, whereas the knock-out competition is based on a provincial knock-out competition (there are four provinces), followed by the All-Ireland semi-finals and final. Support for the counties are largely from within each county, although some may have moved away from their county of birth/ residence for jobs, university studies and so on.

CHECK YOUR UNDERSTANDING

13. Explain why some teams have an international fan base.
14. State the name of the provincial rugby competition in South Africa.

Costs and benefits of large-scale temporary leisure sites

Large-scale sporting, musical, cultural or religious festivals have many costs but also many benefits. Some events may take place in a location that is not normally used for leisure for the rest of the year, such as the Glastonbury Music Festival, UK. Others may be temporary events in areas that are associated with leisure and tourism, such as the triathlon in the grounds of Blenheim Palace, Woodstock, UK.

THE GLASTONBURY FESTIVAL

The Glastonbury Festival is the largest open-air music festival in Europe. Festival-goers are attracted by the opportunity to see a particular type of performance or performer and to be part of the festival culture.

Glastonbury began as the Pilton Pop Festival in 1970. The festival is held in a huge open-air arena which covers approximately 800 acres. This site now accommodates up to 250,000 people.

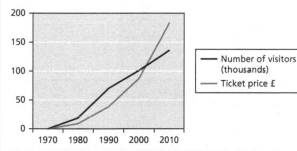

The growth of visitor numbers and the increase in the price of attending the Glastonbury Festival

ECONOMIC IMPACTS OF THE GLASTONBURY FESTIVAL

The local area benefits financially from the festival. On average, each visitor to the festival spends about £300. The income from the festival makes substantial contributions to charities such as Water Aid, Greenpeace and Oxfam and may also be used to support local projects such as the provision of sporting facilities. On the other hand, the Glastonbury Festival has negative impacts including dealing with large numbers of people. Problems include congestion, litter, pollution, drug-taking and theft.

The festival is not wholly welcomed by the local community. The activities of local residents are severely restricted and the festival is a burden for them. Nevertheless, it also offers local people an employment opportunity.

ENVIRONMENTAL IMPACTS OF THE GLASTONBURY FESTIVAL

- **Air pollution:** CO_2 emissions increase dramatically during the three-day festival. Car transport is still very popular because of its flexibility, but attempts have been made to encourage supporters to use public transport.
- **Waste disposal:** Rubbish is one of the greatest problems at the site. This consists mainly of human waste, empty plastic water bottles and tents. The tents are a big problem because many of the fans leave them behind at the end of the festival and their condition is not good enough for charity use.
- **Noise pollution:** Many of the performances continue through the night, which causes disturbance to local residents.
- **Provision of resources:** Energy, water and food need to be supplied and residue disposed of.

REDUCING THE IMPACT OF THE FESTIVAL

Waste management at Glastonbury Festival
- Over 175,000 people visit Glastonbury every year.
- It costs about £800,000 to dispose of the rubbish.
- Over one million plastic bottles are used at the Festival over five days. To reduce this number, organizers provide over 400 water taps around the site.
- Over 11 million litres of water are drunk at the Festival each year.
- Over 15,000 refuse bins are used at the Festival each year.
- Glastonbury aims to recycle 60% of the Festival waste each year.
- Over 1,300 volunteers help to clear up the waste.
- Over 10,000 native trees have been planted since 2000 in order to protect the natural habitat.
- All Festival programmes are placed in reusable, organic bags, not plastic ones.

CHECK YOUR UNDERSTANDING
15. Describe the trends in (a) visitor numbers and (b) ticket prices at the Glastonbury Festival, 1970–2010.
16. Briefly describe the problems associated with the Glastonbury Festival.

Niche national tourism strategies

DEFINITION

Niche tourism refers to special-interest tourism catering for relatively small numbers of tourists. There are many types of niche tourism, including adventure, movie location and heritage tourism.

ADVENTURE TOURISM

Adventure tourism is a form of niche tourism that involves travel to a remote area and some level of risk. It has increased in popularity in recent decades and attracts high-value customers. For example, to climb Mount Everest costs around $50,000 per person, however, not all adventure tourism is expensive. Adventure tourism also supports local economies more than mass tourism. About 66% of revenue spent in the adventure tourism sector remains in the destination. Adventure tourism also encourages sustainable practices – it involves local communities, supports local businesses, and promotes environmental protection for future use.

MOVIE LOCATION TOURISM

There are many links between places used in TV programmes and films and their attraction for fans. Films such as *Lord of the Rings* and *The Hobbit* have attracted many tourists to New Zealand. Some television series attract large numbers of visitors. The TV detective series *Inspector Morse* is said to have boosted tourism in Oxford. Other popular TV series include *Game of Thrones*, leading to an increase in the number of tourists to Northern Ireland.

HERITAGE TOURISM

Heritage tourism is travelling to experience the place, artefacts, historic sites and indigenous people of an area. It is sometimes referred to as cultural tourism.

Machu Picchu is a World Heritage Site and in 2007 it was selected as one of the new seven wonders of the world along with the Great Wall of China, the Coliseum and the Taj Mahal.

Machu Picchu: heritage tourism costs and benefits

Benefits	Disadvantages
Social benefits • Improvement of infrastructure (electricity, water supply, sewerage and communications) benefits local people. • Encourages the celebration of customs and cultural events.	**Social costs** • Cultural conflicts and abandonment of traditional customs and moral values. • Increase in local crime.
Economic benefits • It provides valuable foreign exchange, which can be invested in local services and projects, connected to development. • Extra tax revenue for the government comes from accommodation, restaurants, airports, and ticket sales. • It provides direct employment (e.g. accommodation and guides) and indirect employment (e.g. food production and housing construction). • Tourism can produce a "multiplier effect" whereby money generated in one sector of the economy benefits other sectors.	**Economic costs** • It inflates local prices of goods and services. • Jobs in tourism are mainly seasonal. • Tourism is volatile and subject to downturns due to economic recession or terrorism. • "Leakage" can easily occur. This means that money generated by tourism does not stay in the country but returns home.
Environmental benefits • Tourism has conserved natural and cultural resources that would have become derelict otherwise.	**Environmental costs** • Visitors are causing footpath erosion. • Heavy rainfall, steep slopes, deforestation and trampling of vegetation can lead to landslides. • The local infrastructure cannot cope with recent urbanization. The Urubamba river is overloaded with untreated sewage and its banks are covered with garbage.

CHECK YOUR UNDERSTANDING
17. Outline the economic benefits of heritage tourism.
18. Explain two environmental problems associated with heritage tourism.

The role of transnational corporations (TNCs) in expanding international tourism destinations

TNCS AND SECONDARY TOURISM RESOURCES

In many LICs tourism has the potential to stimulate economic growth and development. It generates over 6% of global GNP and 13% of consumer spending. Many LICs possess primary resources and some secondary resources such as hotels. LICs may not have all the infrastructure in place to develop their own tourist industry. So they have been forced to rely upon the TNCs concerned with tourism to organize and market these resources and products. The TNCs are usually based in HICs – the eight largest hotel chains in the world are all US-based companies.

There are five main areas of TNC involvement in the tourism industry – airlines, hotels, cruise lines, tour operators and travel agents. All TNCs have been increasingly integrating their businesses.

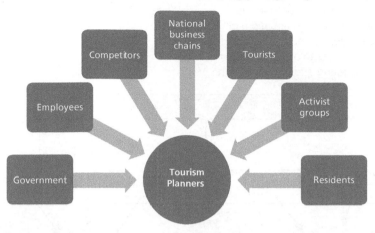

The stakeholders in tourism planning

The advantages and disadvantages of TNC involvement in expanding tourism destinations

Advantages	Disadvantages
1 The presence of the TNC hotel (with its numerous national flags on display) can boost tourist demand.	1 Labour exploitation – much of the employment is part-time, seasonal, unskilled and low paid.
2 Tourists from HICs (and increasingly those from LICs too) demand high standards. TNCs are often more able to provide these than the smaller independent operators.	2 Removal of capital – TNCs get to use resources which could have benefitted the local population or host country.
3 TNCs can introduce a diverse range of new technologies and skills into an economy including advanced management, environmental and financial systems.	3 Outflow of wealth (leakage) – much of the income generated by tourism (transport, accommodation, restaurants and bars) leaves the host country and passes to the countries that the TNCs are based in.
4 TNCs improve the productivity and sustainability of the sector and the economy.	4 Outside decision-making – decisions about where to invest, what type of facility to invest in and so on are made on the basis of how it will benefit the TNC and its shareholders rather than the local community in which it is to be based.
5 TNCs can help to provide a greater range of tourist activities in some destinations, which can lead to innovation by local firms.	5 The outflow of wealth from LICs is widening the global economic divide – tourism is helping reinforce global inequalities by providing competitively priced holidays in some low income and emerging economies.
6 TNCs generate employment. In some cases they generate proportionally more employment than local firms in that country.	
7 TNCs often pay higher wages and offer better packages for employees than local firms.	
8 TNCs may make an effort to establish linkages with local suppliers.	

EXAM TIP
Make sure that you can refer to a named TNC and give specific details of involvement in touristic destinations.

CHECK YOUR UNDERSTANDING
19. Outline two advantages of TNC involvement in tourism.
20. Suggest how the stakeholders presented in the diagram do not necessarily place the needs of local residents first.

Tourism and sport at the international scale   77

The costs and benefits of tourism as a national development strategy

THE BENEFITS OF TOURISM

Tourism as a development strategy has many attractions. It is a positive route towards economic development for poor countries, especially when they lack the raw materials for manufacturing. It can be an effective way for a country to overcome its problems of balance of payments.

Tourism is labour intensive and can overcome the problem of unemployment. It provides jobs directly (catering, transport, guiding and accommodation) or indirectly (construction, engineering and food production). Tourism also provides opportunities to acquire new skills, for example, in languages, catering and entertainment.

Tourist links

Tourism can create a multiplier effect, which means that income gained by local people is circulated through the economy. Tourism can redistribute wealth globally, nationally and locally provided leakage is not allowed to drain the economy. Tourism adds diversity to the export base of the country and thereby helps to stabilize its foreign exchange earnings.

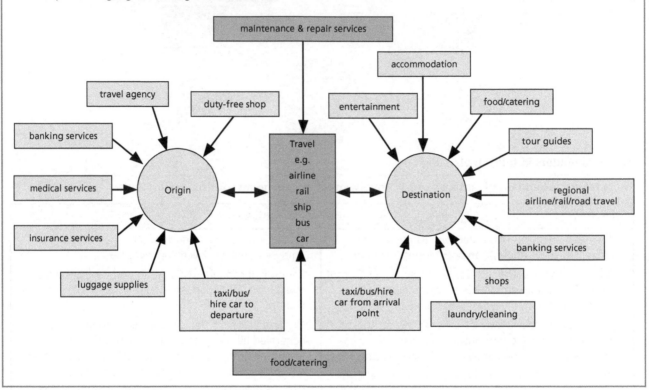

THE COSTS OF TOURISM

Many costs have arisen due to tourism as a national development strategy. A weak currency attracts tourists but makes imports (for tourists) costly and marketing expensive. Tourism puts undue pressure on natural ecosystems, leading to a decline in ecosystem services, increased soil erosion, litter, pollution, and the decline of biodiversity. Much tourist-related employment is unskilled, seasonal, part-time, poorly paid and lacking any rights for the workers. Important resources, especially water, are provided for tourists while local people may have to go without. A large proportion of profits goes to overseas companies (leakage), tour operators, hotel chains, airline companies and so on. Many of the tourist facilities are owned by TNCs and local decision-making making is minimized. Crime is increasingly directed at tourists; a lot of it is petty crime but there have been very serious incidents, involving terrorist attacks on tourists. Finally, tourism is by nature very unpredictable, varying with many factors such as the strength of the economy, cost, safety, alternative opportunities, and stage in the family life cycle.

CHECK YOUR UNDERSTANDING
21. Outline the economic advantages of tourism as a national development strategy.
22. Study the diagram of tourist links. Identify the links in the destination that may be controlled by external TNCs.

Hosting international sporting events

There are many political, economic and cultural factors that affect the hosting of an international event. Most countries that bid to host such an event need to be in a position where they can guarantee that they can run the event successfully. NICs that have hosted the events include China (2008 Olympic Games) and Brazil (2014 World Cup and 2016 Olympic Games). Hosting such an event requires stadia, transport infrastructure, hotels and other facilities for tourists/fans, and training facilities for the athletes/players taking part. Political factors involved in the allocation of events such as the Olympic Games and football's World Cup include lobbying by national governments and alleged corruption of organizations such as the IOC and FIFA.

The advantages and disadvantages of hosting an event such as the Olympic Games or World Cup

Advantages	Disadvantages
Prestige, it is considered an honour to host the event and if the event is a success the host city gains in reputation.	There may be financial problems; stadia and facilities used in the 2004 Athens Olympic Games have been abandoned and become derelict.
The event may make a profit through sales of radio and TV rights, tickets and merchandise, as well as spending in hotels, restaurants etc.	A large number of visitors puts a strain on hotels, transport, water supplies etc.
It gives a boost to sports facilities and other facilities. Cities build or improve their facilities to host events.	Large events are security risks; due to the international television coverage they are now potential terrorist targets.
Economic spin-offs, trade and tourism in particular.	If an event does not do well, the host country's image suffers. The host will have difficulty attracting other events if, indeed, it wants to.
It unites the country and gives a sense of pride.	

CASE STUDY

LONDON OLYMPIC AND PARALYMPIC GAMES 2012

The London 2012 Olympic Games and Paralympic Games helped to regenerate one of London's poorest areas. The event created 12,000 new jobs and up to £17 billion was spent upgrading infrastructure and building 2,800 new homes. More than 46,000 people worked in the Olympic Park and Olympic Village. The Games encouraged developments in the East End of London, such as Westfield shopping centre.

A new Olympic Park, the Queen Elizabeth Park, opened on 1 January 2013 along the River Lea. Around 2,800 housing units were created from the Athletes' Village and 50% of these were affordable housing for local people.

Nevertheless, there were many people who were not happy with the Games. Some people and businesses did not want to move but were forcibly relocated. London and the South East received the bulk of the funds, increasing inequalities with the rest of the country.

Financial costs and benefits of the 2012 London Olympic and Paralympic Games

Costs	Benefits
Running the Games: £1.5 bn	INCOME
Olympic stadium: £560 m	Lottery £1.5 bn
Athletes' village and park: £650 m	TV and marketing £560 m
Security: £200 m	Sponsorship and official suppliers: £450 m
Redevelopment: £800 m	Ticket revenues: £300 m
Transport and infrastructure: £7 bn	Licensing: £50 m
	London Development Agency: £250 m
	Council tax levy: £625m
Total £10.710 billion	Total £3.735 billion

COMMON MISTAKE

✗ Hosting large-scale sporting events always brings economic success.

✓ Although national "prestige" is said to be a benefit of hosting international sporting events, many people are not in favour of hosting large-scale events since they cost a great deal to set up, and they might not make a profit.

CHECK YOUR UNDERSTANDING

23. Outline the economic advantages of staging international sporting events.
24. Outline the economic disadvantages of staging international sporting events.

Unsustainable touristic growth

CARRYING CAPACITY

Carrying capacity can be thought of in three main ways:
- **physical carrying capacity**, which is the measure of absolute space, for example, the number of spaces within a car park
- **ecological capacity**, the level of use that an environment can sustain before environmental damage occurs
- **perceptual capacity**, the level of crowding that a tourist will tolerate before deciding the location is too full.

TOURISM IN VENICE – AN URBAN HOTSPOT

The historic centre of Venice comprises 700 ha with buildings protected from alterations by government legislation. The optimum carrying capacity for Venice is 9,780 tourists who use hotel accommodation, and 1,460 tourists staying in non-hotel accommodation. There are clear seasonal variations with an increase in visitor numbers in summer and at weekends. Research has estimated that an average of 37,500 day trippers visit Venice every day in August. A limit of 25,000 visitors a day has been suggested as the maximum carrying capacity for Venice.

The large volume of visitors creates a range of social and economic problems for planners, such as congestion and competition for scarce resources.

Day tripping is becoming increasingly important, while residential tourism is becoming less important. Day trippers contribute less to the local economy than visitors who stay.

To control the number of day trippers, the Venetian authorities have:
- denied access to the city by unauthorized tour coaches via the main coach terminal
- built gates around the city and charged visitors to enter.

Since 2000, Venice has seen an increasing number of cruise liners and cruise passengers. In 2015, more than 650 boats docked, bringing over 2.2 million passengers. However, tourism receipts fell by around €300 million. Cruise passengers do not stay in hotels, generally they do not eat large meals in Venice, and often they have their own guide.

TOURISM IN THE LLANTHONY VALLEY, BRECON BEACONS – A RURAL HOTSPOT

The Brecon Beacons National Park is located in the south of Wales and is one of the closest national parks to people living in cities such as London, Birmingham and Bristol. The Llanthony Valley is a microcosm of all that is bad about tourists. They bring little or no benefit to an area but cause disruption, irritation and problems. Farmers find it difficult at times to move animals and large machinery as they find their gates blocked. For the tourist, the trip is merely a pleasurable drive and they gain little or no understanding about the community, the landscape or the heritage that they have passed through.

Environmental impacts of tourism

Stressor activities	Stress	Primary environmental response
Permanent environmental restructuring Major construction activity • urban expansion • transport network • tourist facilities • marinas, ski-lifts	Restructuring of local environments Expansion of built environments Land taken out of primary production	Change in habitat Change in population of species Change in health and welfare of humans Changes in visual quality
Generation of waste • tourism, recreation and leisure activities • transport	Pollution loadings • emissions • effluent discharge • solid waste disposal • noise (traffic, aircraft)	Change in quality of environmental media • air • water • soil Health of organisms Health of humans
Tourist activities • skiing • hunting • walking • trail-bike riding	Trampling of vegetation and soils Destruction of species	Change in habitat Change in population of species
Effect on population dynamics • population growth	Population density (seasonal)	Congestion Demand for natural resources • land and water • energy

CHECK YOUR UNDERSTANDING

25. Distinguish between physical carrying capacity and perceptual carrying capacity.
26. Briefly explain why Venice does not benefit greatly from cruise ship tourism.

Sustainable tourism

SUSTAINABLE DEVELOPMENT

Sustainable development is defined as development that meets the needs of the present without compromising the ability of future generations to meet their own needs. Sustainable tourism therefore needs to:

- ensure that renewable sources are not consumed at a rate that is faster than the rate of natural replacement
- maintain biodiversity
- respect local cultures, livelihoods and customs
- involve local people in development processes.

ECOTOURISM

Ecotourism is a "green" or "alternative" form of sustainable tourism. It includes tourism that is related to ecosystems such as game parks, nature reserves, coral reefs and forest parks. It aims to give people a first-hand experience of natural environments and to show them the importance of conservation. Its characteristics include:

- increasing involvement of local communities
- being appropriate to the local area
- balancing conservation and development.

CASE STUDY

THE MONTEVERDE CLOUD FOREST, COSTA RICA

The Monteverde Cloud Forest Reserve was established in 1972. Initially, it covered 328 ha but now covers around 14,200 ha. There are over 100 species of mammals, 400 species of birds, 1,200 species of reptiles and amphibians, and several thousand species of insect. Monteverde attracts around 70,000 visitors a year. Monteverde now accounts for about 18% of Costa Rica's total tourist revenue. The growth and development of tourism came at a time when there was a long-term decline in agriculture in Costa Rica. At the same time, ecotourism absorbed some of the displaced agricultural workers in their own villages. Much of this development was on a small scale. For example, 70% of the hotels in Costa Rica have less than 20 rooms.

New businesses have been created in Monteverde, including hotels, bed and breakfasts, restaurants, craft stalls, riding stables, private reserves, hiking trails, hummingbird galleries, frog ponds and a butterfly and botanical garden. The Butterfly Garden consists of a biodiversity centre, a medicinal garden, four climate-controlled butterfly gardens and a leafcutter-ant colony. Many of these are locally owned. Over 400 full-time and 140 part-time jobs have been created. In addition, there are indirect employment opportunities and multiplier effects. There are also canopy walks and suspension bridges. Local farmers provide much of the food consumed by tourists in the area. Farmers' markets every Saturday also attract tourists. The Monteverde Coffee Tour provides a guided tour of the production of coffee from the field to

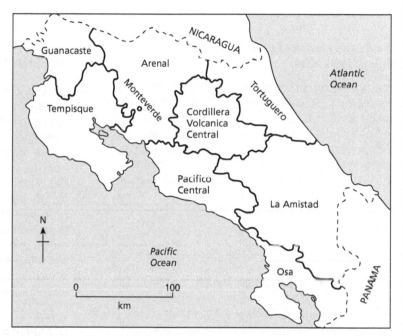

the cup. The proceeds go towards supporting local farmers and families involved in the production of coffee.

Controlled access to the cloud forest, and the use of locals employed as guides means that not only are jobs created, but there is a reduction in visitor impact on parts of the forest. Local arts and crafts have been rejuvenated, and jobs in accommodation, transport, food and communications have been created. The increase in small businesses means that income should be more evenly distributed. Formal and informal education programmes have been strengthened, and the local community are even more aware of the value of their natural resources than they were when they were farmers.

EXAM TIP

Sustainable tourism and ecotourism are often considered to be interchangeable terms. They are similar but ecotourism generally occurs in remote areas with low population densities.

CHECK YOUR UNDERSTANDING

27. Compare sustainable tourism with ecotourism.
28. Outline the advantages of ecotourism at the Monteverde Cloud Forest.

Future international tourism

TRENDS AND PROJECTIONS IN TOURISM

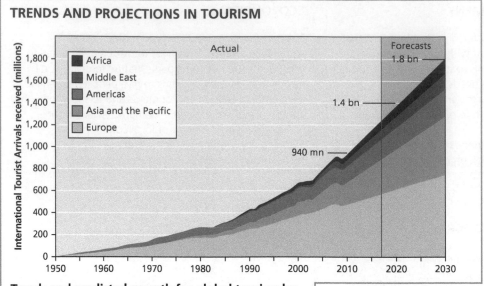

Trends and predicted growth for global tourism by region, 1950–2030

International tourist arrivals worldwide are expected to increase by 3.3% a year over the period 2010 to 2030 to reach 1.8 billion by 2030. By 2030, arrivals in emerging destinations are expected to increase at twice the rate of those in advanced economies. The market share of emerging economies increased from 30% in 1980 to 45% in 2014, and is expected to reach 57% by 2030, equivalent to over 1 billion international tourist arrivals.

FORECASTS

International tourist arrivals in the emerging economy destinations of Asia, Latin America, Central and Eastern Europe, Eastern Mediterranean Europe, the Middle East and Africa will grow at double the rate of advanced economy destinations. Consequently, arrivals in emerging economies are expected to exceed those in advanced economies before 2020.

The strongest growth by region will be seen in Asia and the Pacific.

SOCIAL MEDIA

Social media has made a huge impact on tourism and travel. It has changed how people research trips, make decisions and share their experiences. Tourism relies on favourable opinions and recommendations. Social media sites such as TripAdvisor (50 million users) and Facebook (800 million users) allow customers to easily share tips and suggestions. In one survey 92% of consumers said that they trusted social media more than any other form of advertising.

INTERNATIONAL SECURITY

The impact of terrorism on the travel and tourism industry could lead to long-term decline. Any terrorist threats to airlines, hotels and restaurants are a cause of major concern for many governments and TNCs.

The success of travel and tourism does not shield it from the impact of terrorism. The 11 September 2001 (9/11) terrorist attacks in the USA highlighted the need for safe travel.

At the national level, governments can also do a lot to implement tighter security. Many experts believe that certain parts of the Middle East, Pakistan, Afghanistan and sub-Saharan Africa are turning out to be the main power centres for terrorism, and so limit the potential for tourism there.

DIASPORA TOURISM

Diaspora tourism is an increasingly important form of niche tourism, and it is one that has distinctive features and potential value. Diaspora tourism is an important sub-set of VFR tourism (visiting friends and relatives). Diaspora tourists are more likely than most international tourists to have or make connections with the local economy. They are more likely to stay with relatives, or in locally owned businesses (e.g. bed and breakfast accommodation), to eat in local restaurants, go to local pubs and so on. Although they might not spend as much money as international tourists, it is more likely to go directly to local businesses. Thus, diaspora tourism can have positive development potential. Moreover, diaspora tourism is not as seasonal as international tourism, and may be spread more evenly throughout the year.

COMMON MISTAKE

✗ *All tourism predictions are correct.*

✓ Do not expect all predictions to be correct. Tourism is very volatile and depends on economic and political stability which might not occur in all regions.

CHECK YOUR UNDERSTANDING
29. Describe the trends in international tourism up to 2030.
30. Explain briefly how diaspora tourism differs from other forms of tourism.

Political and cultural influences on international sport participation

The success of competitive sport has social, economic and political benefits within making it very important in many countries. However, many countries have limited budgets for sport, and there has been a widespread decline in participation in sport among young people.

Increasingly, sport has been used by nations for a variety of reasons. Some governments may wish to increase participation in sport for the health benefits that it could bring. Others may wish to host a major event to help develop part of their economy, or to regenerate urban areas.

The evidence is far from conclusive. The economic benefits from mega-events are hard to quantify, and the link between government involvement in sport, and success or participation in sport, is unclear. The increase in sports participation for many sporting events is short term.

in sport. For Muslim women, for example, there are additional challenges.

Islam promotes good health and fitness. However, some Muslim women cannot participate in mixed-gender sports. In the UK, less Asian women than white women participate in sport each week. In contrast, in traditional Muslim countries, for example, in the Middle East, attitudes to physical exercise among Muslim women are much more positive than in the UK. In 2012, the UN Women and the International Olympic Committee agreed to promote women's empowerment through sports. Sport has the potential to increase psychological well-being, leadership skills and empowerment of girls and women. This coincides with the aims of UN Women, which are to advance gender equality and the empowerment of women worldwide.

INCLUSION VIA CHANGING GENDER ROLES

The world of sport is male dominated and women often face considerable challenges, sometimes just to participate

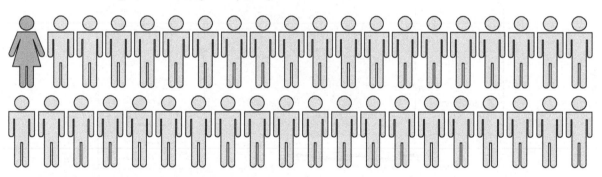

In 1900 women started competing in the Olympics, there were 22 women, out of 997 athletes

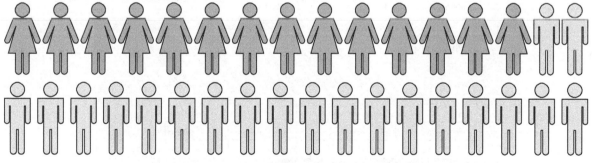

In 2012, there were 4,676 women out 11,000 athletes

Women in the olympics

PARALYMPIC GAMES

The Paralympic Games is an international competition for athletes with a range of physical disabilities. It occurs

immediately after the Olympic Games and takes place in the Olympic Stadium and other stadia.

CHECK YOUR UNDERSTANDING

31. Outline the UN's role in the development of participation in sport.

32. Outline the changing gender ratios among Olympic athletes.

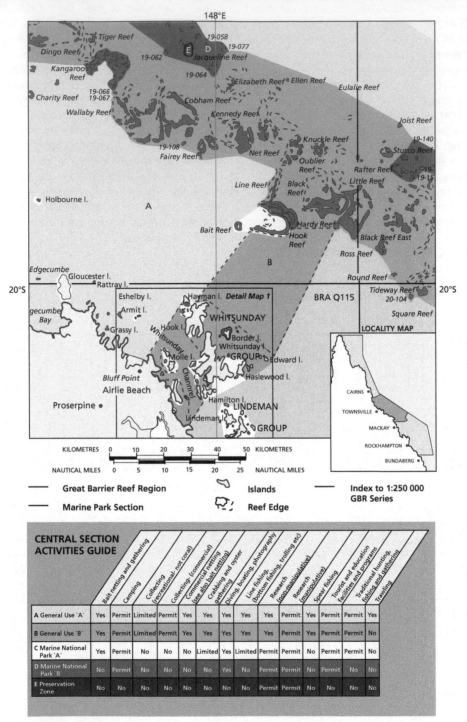

(a) (i) Outline the main tourist activities associated with the Great Barrier Reef. (2 marks)

(ii) Suggest reasons why this area is attractive to tourists. (2 marks)

(b) Briefly explain one advantage and two disadvantages of tourism in this area. (2 + 2 + 2 marks)

(c) Either

Evaluate the role of TNCs in the provision of touristic infrastructure and facilities. (10 marks)

Or

Examine the view that sustainable development is desirable, but, ultimately impossible to achieve. (10 marks)

1 MEASURING FOOD AND HEALTH

Global patterns in food and nutrition

FOOD SECURITY

Food security for a population exists when all its people, always have access to sufficient, safe and nutritious food to meet their dietary needs and food preferences. Food security for a household means access by all its members at all times to enough food for an active, healthy life (UN Food and Agriculture Organization). Food security includes, at a minimum: (i) the ready availability of nutritionally adequate and safe foods; and (ii) an assured ability to acquire acceptable foods in socially acceptable ways (that is, without resorting to emergency food supplies, scavenging, stealing, or other coping strategies).

GLOBAL HUNGER INDEX

The Global Hunger Index (GHI) ranks countries on a 100-point scale, with 0 being the best score (no hunger) and 100 being the worst. The GHI is based on four component indicators:

1. Undernourishment – the proportion of undernourished people as a percentage of the population (the share of the population with insufficient caloric intake).
2. Child wasting – the proportion of children under the age of 5 who suffer from wasting (that is, low weight for their height, reflecting acute undernutrition).
3. Child stunting – the proportion of children under the age of 5 who suffer from stunting (that is, low height for their age, reflecting chronic undernutrition).
4. Child mortality – the mortality rate of children under the age of five (partially reflecting the fatal synergy of inadequate nutrition and unhealthy environments).

According to the 2015 GHI, among regions, hunger is highest in Africa south of the Sahara and South Asia. Africa south of the Sahara has a GHI of 32.2, while South Asia's is 29.4. Both regions' GHI scores reflect "serious hunger". The food and hunger situation in several countries (Central African Republic, Chad, Zambia, Timor-Leste, Sierra Leone, Haiti, Madagascar and Afghanistan) is still "alarming".

CHECK YOUR UNDERSTANDING
1. Distinguish between child wasting and child stunting.
2. Identify the areas where hunger is most serious.

CALORIE INTAKE

Calorie intake is the amount of food (measured in calories) a person consumes. Average daily calorie intake per person varies by country. The world average is about 2,780 kcal/person per day, and the minimum recommended amount is around 1,800 kcal/person/day. However, this varies with age, gender, type of work, amount of physical activity and climate. Countries with the highest daily intake include Austria and the USA, with 3,800 and 3,750 kcal/person/day, respectively. In general, high-income countries have an intake of around 3,400 kcal/person/day. In contrast, people in low-income countries have an intake of around 2,600 kcal/person/day. However, in sub-Saharan Africa the intake is down to 2,240 kcal/person/day and in Central Africa it is just 1,820 kcal/person/day. In Burundi and Eritrea, daily calorie intakes are as low as 1,680 and 1,590 kcal, respectively.

INDICATORS OF MALNUTRITION

Malnutrition means poor nourishment, and refers to a diet that lacking or with too many nutrients. There are many types of malnutrition:
- Deficiency diseases result from a lack of specific vitamins or minerals.
- Kwashiorkor is a lack of protein in the diet.
- Marasmus is a lack of calories.
- Obesity results from eating too much food.
- Starvation refers to a limited or non-existent intake of food.
- Temporary hunger can occur when there is a short-term decline in the availability of food to a population.
- Famine occurs when there is a long-term decline in the availability of food.

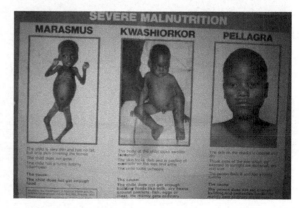

Types of malnutrition

The nutrition transition

As income increases in countries there is an increase and a change in food consumption. People in LICs generally derive their food energy from carbohydrates, while the contribution of fats, meat and dairy is small. In Bangladesh, an LIC, people derive 80% of their nutritional energy from carbohydrates and 11% from fats. In contrast, people in HICs generally derive most of their food energy from carbohydrates and fat, with a substantial contribution from meat and dairy.

Worldwide a **nutrition transition** is taking place, in which people are shifting towards more affluent food consumption patterns. For LICs, a small increase in income may lead to a large increase in calorie intake, while for HICs increases in income may not lead to an increase in calorie intake.

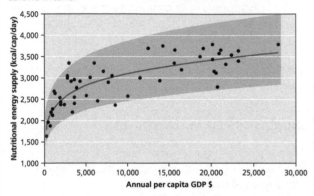

The relationship between GDP/head and energy consumption (kcal/person/day)

In HICs, the main dietary changes since the 1970s have been the reduction in cereals, while mainly vegetable oil and, to a smaller extent, meat intake increased. In LICs the diet has diversified since the 1970s. Cereals, including rice, as well as vegetable oil, sugar, meat and dairy intake are higher compared to 1961–73, although in more recent periods cereal intake is stagnating and even declining.

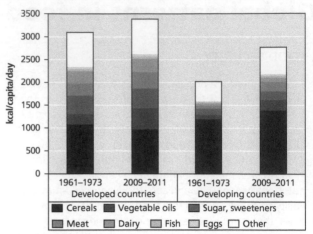

Changes in per capita calorie availability in high-income and low-income countries

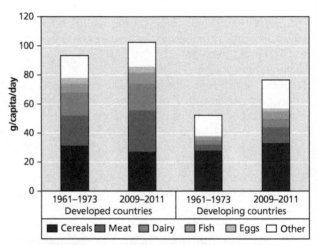

Changes in per capita protein availability in high-income and low-income countries

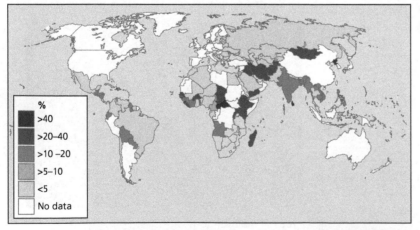

Global variation in undernourished population, 2016

COMMON MISTAKE

✗ *All people in the same country have the same diet.*

✓ There are major differences in the diet of people within a country. For example, there are major differences between rich and poor, urban versus rural, and between generations.

❗ CHECK YOUR UNDERSTANDING

3. Describe the main differences in the diets of HICs (developed countries) and LICs (developing countries) in 2009–11.
4. Describe the main changes in the diets between HICs (developed countries) and LICs (developing countries) between 1961–73 and 2009–11.

Global patterns in health indicators

HEALTH-ADJUSTED LIFE EXPECTANCY (HALE)

HALE is an indicator of the overall health of a population. It combines measures of both age- and sex-specific health data, and age- and sex-specific mortality data into a single statistic. HALE indicates the number of expected years of life according to years lived in full health, based on the average experience in a population. Thus, HALE is a measure of quantity of life and also of quality of life.

A number of observations have been made about HALE:

- The burden of ill health is higher for women than for men.
- The burden of ill health has a bigger impact on those in early old age, for example, between 60 and 70 years old.
- Sensory problems and pain are the largest components of the burden of ill health.
- Higher socio-economic status confers a dual advantage – longer life expectancy and a lower burden of ill health.

In Canada, the bottom one-third income group was associated with a loss of health-adjusted life expectancy at birth of 3.2 years for women and 4.7 years for men.

INFANT MORTALITY RATE

The infant mortality rate (IMR) is the number of deaths in children under the age of 1 per 1,000 live births per year.

Infant mortality rate (IMR) =

$$\frac{\text{Total no. of deaths of children} <1 \text{ year old}}{\text{Total no. of live births}} \times 1,000 \text{ per year (‰)}$$

It is an age-specific mortality rate, that is, it is comparing the death rates among the same ages. Infant mortality rates vary from a low of under 2 per thousand in Monaco and 2‰ in Iceland, Japan and Singapore to over 100‰ in Afghanistan and Mali. After Afghanistan (115‰), the 24 next highest IMRs are all from sub-Saharan Africa.

MATERNAL MORTALITY RATES

The maternal mortality rate (MMR) is the annual number of female deaths per 100,000 live births. The Millennium Development Goals (MDGs) 2000–15 included the goal of reducing maternal mortality by three-quarters by 2015. Although there has been much progress, there is still high mortality in sub-Saharan Africa and moderate mortality in Southeastern and Eastern Asia and in Oceania. In 2016, South Sudan had the highest MMR with 2,054 deaths per 100,000 live births, followed by Chad (1,100) and Somalia (1,000). The highest MMRs are found in sub-Saharan Africa and parts of South Asia. Haiti and Guyana are anomalies as they are not located in sub-Saharan Africa or South Asia. In contrast, the lowest MMRs are found in high-income countries of the EU, Australia and Singapore.

ACCESS TO SANITATION

According to the World Health Organization, 68% of the world's population now uses an improved sanitation facility, 9 percentage points below the MDG target. In 2015 it was estimated that 2.4 billion people globally had no access to improved sanitation facilities. The number of people in sub-Saharan Africa without access to sanitation has increased since 1990. In addition, there are rural and urban contrasts: whereas 80% of the urban population has access to improved sanitation facilities, only 50% of the rural population do.

ACCESS TO HEALTH SERVICES

Access to health services is usually measured in the number of people per doctor or per hospital. Access to health services varies from one doctor per 100,000 in Burundi and one doctor per 50,000 people in Mozambique to one doctor per 280 people in Hungary and Iceland. In China, there are 610 people per doctor and in India it is 1,960 per doctor. However, inequalities in health services are not merely a question of the number of doctors or beds per person but also concern the facilities available in hospitals and clinics.

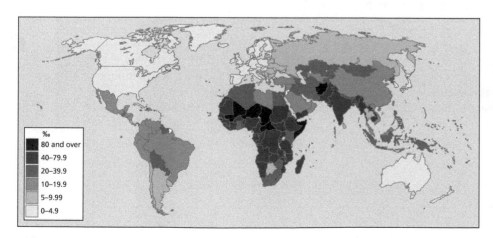

Global variations in the infant mortality rate, 2016

‰
80 and over
40–79.9
20–39.9
10–19.9
5–9.99
0–4.9

CHECK YOUR UNDERSTANDING

5. Explain why the infant mortality rate is a good indicator of the level of development in a country.
6. Explain briefly why the number of doctors/person is **not** a reliable measure of the quality of health care systems.

The epidemiological transition

As a country develops, there is a change in the health profile away from **infectious or contagious communicable diseases** (epidemics) to **non-communicable diseases** that cause a gradual worsening in the health of an individual (**degenerative diseases**). These are also referred to as "diseases of poverty" and "diseases of affluence". This is known as the epidemiological transition. For example, a poor country would be expected to have most deaths and illnesses from infectious diseases such as respiratory diseases, measles and diarrhoea. By contrast, an HIC would have more deaths and illnesses from heart attack, stroke and cancers.

Heart disease becomes more common with the epidemiological transition. As mortality decreases, nutrition improves and infections are controlled, high blood pressure, heart disease and strokes all become more common.

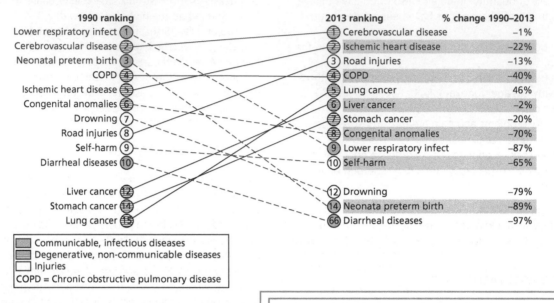

1990 ranking		2013 ranking	% change 1990–2013
Lower respiratory infect ①		① Cerebrovascular disease	–1%
Cerebrovascular disease ②		② Ischemic heart disease	–22%
Neonatal preterm birth ③		③ Road injuries	–13%
COPD ④		④ COPD	–40%
Ischemic heart disease ⑤		⑤ Lung cancer	46%
Congenital anomalies ⑥		⑥ Liver cancer	–2%
Drowning ⑦		⑦ Stomach cancer	–20%
Road injuries ⑧		⑧ Congenital anomalies	–70%
Self-harm ⑨		⑨ Lower respiratory infect	–87%
Diarrheal diseases ⑩		⑩ Self-harm	–65%
Liver cancer ⑫		⑫ Drowning	–79%
Stomach cancer ⑭		⑭ Neonata preterm birth	–89%
Lung cancer ⑮		⑯ Diarrheal diseases	–97%

Legend:
- ▨ Communicable, infectious diseases
- ▤ Degenerative, non-communicable diseases
- ☐ Injuries
- COPD = Chronic obstructive pulmonary disease

The leading causes of premature death in China, 1990 and 2013
Source: http://www.healthdata.org/china
In China the main causes of premature death were cerebrovascular disease, ischaemic heart disease and road injuries. Poor diet, high systolic blood pressure and air pollution were the leading risk factors in 2013. Females aged 28–36 years (84%) experienced the greatest reduction in mortality rate from all causes.

AGEING AND THE DISEASE BURDEN

The worldwide epidemic of chronic diseases is strongly linked to population ageing. In high-income countries, population ageing persists as fertility continues to fall and life expectancy increases slowly. For many middle-income countries, mortality has decreased over much of the 20th century and decreasing fertility is leading to a rapidly ageing population. The doubling of the population aged over 65 years – from 7 to 14% – took 46 years in the UK and 68 years in the USA, but will take just 26 years in China and 21 years in Brazil.

Cardiovascular diseases become more prevalent with the epidemiological transition. As mortality decreases, nutrition improves and infections are controlled, hypertension (high blood pressure), heart disease and strokes all become more common, with heart disease contributing most to mortality. However, as high-income countries advance into the age of "delayed degenerative diseases", age-adjusted mortality due to heart disease decreases as a result of better prevention and treatment. In low-and middle-income countries, heart disease episodes are occurring at younger ages than in high-income countries.

EXAM TIP
Use data from countries that you are interested in. Use the website http://www.healthdata.org/ to find out the epidemiological transition in two contrasting countries.

CHECK YOUR UNDERSTANDING
7. Briefly explain the terms "diseases of poverty" and "diseases of affluence".
8. Outline the evidence to suggest that China is experiencing an epidemiological transition.

A systems approach to food production

Due to the many factors influencing farming and types of farming, farming systems are complex operations. In order to simplify farming types, a systems approach is used showing inputs, processes and outputs. This allows us to compare different aspects of the farming types. The diagram below shows one example of a systems approach.

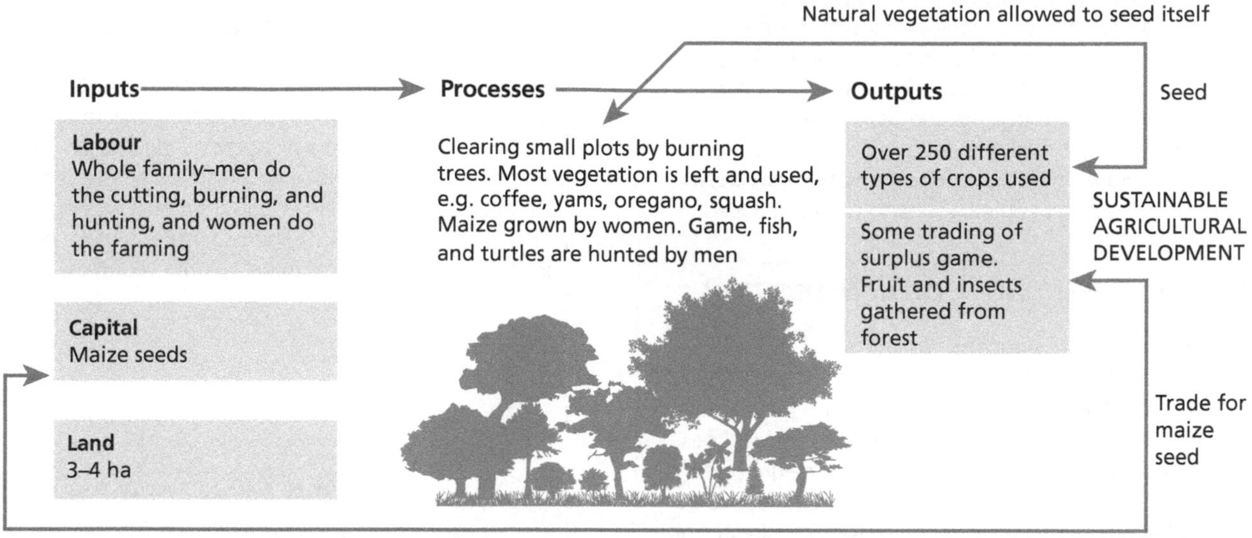

Natural vegetation allowed to seed itself

Inputs ⟶ **Processes** ⟶ **Outputs** | Seed

Labour
Whole family–men do the cutting, burning, and hunting, and women do the farming

Capital
Maize seeds

Land
3–4 ha

Clearing small plots by burning trees. Most vegetation is left and used, e.g. coffee, yams, oregano, squash. Maize grown by women. Game, fish, and turtles are hunted by men

Over 250 different types of crops used

Some trading of surplus game. Fruit and insects gathered from forest

SUSTAINABLE AGRICULTURAL DEVELOPMENT

Trade for maize seed

Shifting cultivation: Popoluca Indians, Mexico

PHOTOSYNTHETIC EFFICIENCY

Agriculture seeks to improve the productivity of ecosystems by adding fertilizers, water and/or pesticides. These may have a negative impact on the local environment.

A comparison of photosynthetic efficiency for types of vegetation and selected crops

Crop or ecosystem	Location	Growth period (days)	Photosynthetic efficiency (%)
Natural ecosystem			
Tropical rainforest	Ivory Coast	365	0.32
Deciduous forest	UK	180	1.07
Crops			
Sugar cane	Hawaii	365	1.95
Maize (two crops)	Uganda	135 + 135	2.35
Soya beans (two crops)	Uganda	135 + 135	0.95
Rice	Japan	180	1.93

ENERGY EFFICIENCY RATIOS

The energy efficiency ratio (EER) is a measure of the amount of energy inputs into a system compared with the outputs. In a traditional agroforestry system the inputs are very low, but the outputs may be quite high. In contrast, the inputs into intensive pastoral farming or greenhouse cultivation may be very great but the returns may be relatively low.

Energy efficiency ratios for selected farming systems (per unit of human/fossil fuel energy input)

Agroforestry	65
Hunter-gatherers	7.8
UK cereal farm	1.9
UK allotment	1.3
UK dairy farm	0.38
Broiler hens	0.1
Greenhouse lettuces	0.002

WATER FOOTPRINTS

Water footprints are a measure of how much water is used in human activities, such as for producing food.

SUSTAINABLE YIELD

The **sustainable yield** is the amount of food (yield) that can be taken from the land without reducing the ability of the land to produce the same amount of goods in the future, without any additional inputs.

CHECK YOUR UNDERSTANDING

9. Suggest why the agriculture practised by the Popolucas can be considered sustainable.
10. Suggest why the energy efficiency ratio of agroforestry and hunter gatherers is greater than that of greenhouse lettuces.

Variations in food consumption

FOOD AVAILABILITY AND FOOD ENTITLEMENT

Food availability deficit (FAD) suggests that food deficiencies are caused by local shortages due to physical factors. More recently, political and economic factors have been seen to be important. Sen (1981) observed that not all food shortages caused hunger, and that increased hunger occurred in areas where food production was, in fact, increasing. This has been seen in India, Ethiopia and Sudan. **Food entitlement deficit** (FED) occurs when people's access to food is reduced.

FACTORS AFFECTING FOOD CONSUMPTION

Income and level of education influence food choice through the resources available to purchase a higher-quality food. For a low-income family, price plays a larger role than taste and quality in whether the food will be purchased. The variety of foods carried in neighbourhood stores may also influence diet. Some people may live in a "food desert", where there is an absence of fresh fruit and meat, and on over-reliance on processed foods.

FACTORS AFFECTING FOOD CONSUMPTION IN THE MIDDLE EAST

Many factors have influenced food consumption in the Arab world. Food consumption patterns dramatically changed in some Arab countries due to the increase in income from oil revenues. Sociocultural factors such as religion, gender discrimination, education and women's employment have also had a noticeable influence on food consumption patterns. Mass media, especially televised food advertisements, play an important role in modifying dietary habits. Migration movements have also had a big impact on the food practices in many Arab countries.

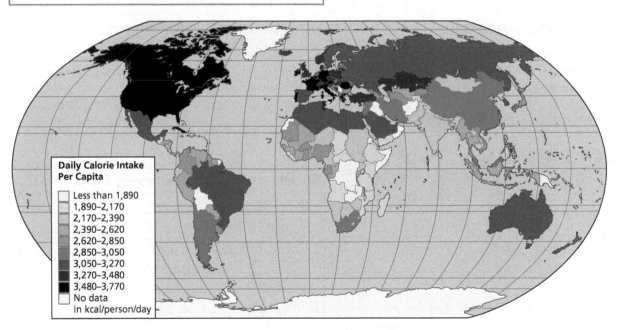

Daily calorie intake per person by country

✔

EXAM TIP
Think about the range of foods available in your home country. Identify foods that have been introduced by migrants or through advertising on media such as television.

CHECK YOUR UNDERSTANDING
11. Distinguish between food availability and food entitlement.
12. Outline how migration may affect food consumption.

❗

COMMON MISTAKE
✗ *Many students think that famines only occur when there are food shortages.*
✔ A famine may occur because much of the food is exported or the food has been taken to feed the military or political supporters. People may lose their entitlement to food, for example, through unemployment, and may not be able to buy food.

❓

Diffusion

DIFFUSION AND THE SPREAD OF AGRICULTURAL INNOVATIONS

The diffusion of innovations depends upon numerous factors, including information about innovations, financial security, the psychological make-up of the adopter, and physical proximity to other adopters.

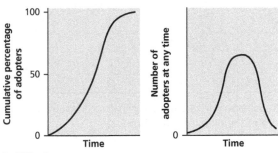

A diffusion curve

As information becomes more widespread, and as the cost is reduced, more people adopt the idea. However, some people take a long time to accept a new technique, so inequalities may develop.

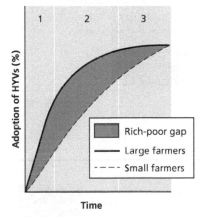

1 Rapid adoption by farmers with plentiful land and/or money. Land is used as security to buy seeds, irrigation pumps, fertilizers, pesticides, HYVs, etc. hence small farmers cannot benefit at first.
2 Adoption by smaller farmers caused by:
  a. government-backed agricultural development projects
  b. new seeds targeted for more environments
  c. continued population pressure creating extra demand for more food.
3 Diffusion of new techniques to most farmers. Widespread adoption.

The green revolution, diffusion and changing inequality over time

DISEASE DIFFUSION

Disease diffusion refers to the spread of a disease into new locations. It occurs when a disease spreads out from an initial source. The **frictional effect of distance**, or **distance decay**, suggests that areas closer to the source are more likely to be affected by it.

Five main patterns of disease diffusion have been identified:
- **Expansion diffusion** occurs when the disease spreads into new areas.
- **Relocation diffusion** occurs when, for example, a person carries a disease into a new area.
- **Contagious diffusion** is the spread of a disease through the direct contact of individuals.
- **Hierarchical diffusion** occurs when a disease spreads through an ordered sequence, for example from cities to large urban areas to small urban areas.
- **Network diffusion** occurs when a disease spreads via transport networks.

Some physical features act as a barrier to diffusion, including mountains and water bodies, while political boundaries and economic boundaries may also limit the spread of disease

THE ZIKA VIRUS

Zika, a mosquito-borne virus that can also be transmitted through sexual contact, arrived in Brazil in May 2014. It has since spread to 21 other countries in the Americas. The mosquito is a poor flier (it can fly only about 400 m), but people are transporting the virus by travelling to and from areas with the disease (relocation diffusion).

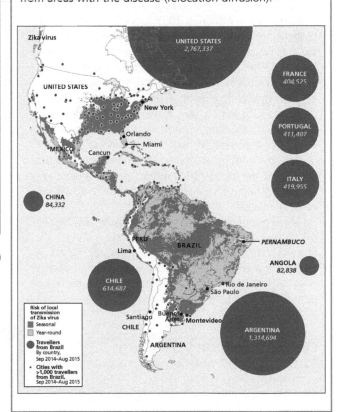

❓

CHECK YOUR UNDERSTANDING

13. Suggest why the diffusion of new agricultural techniques may follow an S-shaped curve.
14. Explain why the diffusion of the Zika virus was spread by relocation diffusion rather than by expansion diffusion.

Vector-borne and water-borne diseases

MALARIA – A VECTOR-BORNE DISEASE

Malaria is a life-threatening disease of humans caused by the plasmodium parasite and transmitted to people via the bite of the female *Anopheles* mosquito.

- In 2015, 95 countries were affected by malaria.
- Almost half the world's population is at risk of malaria.
- Between 2000 and 2015, the number of new cases fell by 37% globally and the malaria death rate fell by 60%.

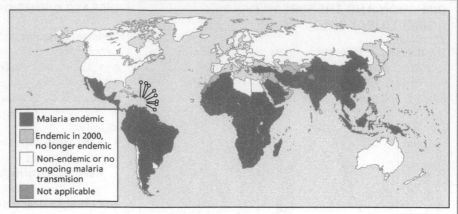

- Malaria endemic
- Endemic in 2000, no longer endemic
- Non-endemic or no ongoing malaria transmision
- Not applicable

- Sub-Saharan Africa carries a disproportionately high share of the global malaria burden.

ENVIRONMENTAL PREFERENCES

The malarial parasite thrives in humid tropics where a minimum temperature of 20°C allows it to complete its life cycle. The mosquito vector is a primary host and the human is secondary. The mosquito's ideal environment is stagnant water.

The disease is often triggered by natural events such as cyclones and flooding or by human conflicts such as war, which often results in refugees. The refugees may be forced to live in temporary camps with inadequate drainage, which are ideal breeding grounds for the mosquito.

IMPACTS

Global financing for malaria control increased from an estimated $960 million in 2005 to $2.5 billion in 2014. Malaria interventions led to health service savings of $900 million in sub-Saharan Africa between 2001 and 2014, owing to the reduced number of cases.

The direct cost of malaria to individual households includes medication, doctors' fees and preventative measures such as bed nets, which help to reduce transmission. Infected individuals are unable to work, which can reduce family incomes during the attacks.

Children with severe malaria frequently develop one or more of the following symptoms: severe anaemia, respiratory distress in relation to metabolic acidosis, or cerebral malaria.

Some population groups are at considerably higher risk of contracting malaria, and developing severe disease, than others. These include infants, children under five years of age, pregnant women and patients with HIV/AIDS, as well as non-immune migrants, mobile populations and travellers.

CHOLERA – A WATER-BORNE DISEASE

Cholera is an infection that is caused by the bacterium *Vibrio cholerae*. It is a water-borne disease, although it can also be transmitted by contaminated food. Cholera outbreaks require two conditions:
- a significant decline of water sanitation leading to contamination by the bacterium *Vibrio cholerae*
- the presence of cholera in the population.

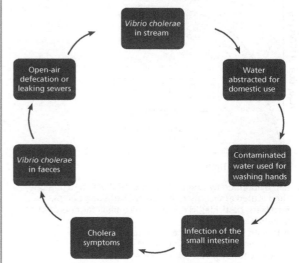

- Vibrio cholerae in stream
- Water abstracted for domestic use
- Contaminated water used for washing hands
- Infection of the small intestine
- Cholera symptoms
- Vibrio cholerae in faeces
- Open-air defecation or leaking sewers

The transmission of *Vibrio cholerae*

The infection mainly causes mild gastroenteritis (diarrhoea). However, about 5% of infected persons develop severe dehydration, which can kill within hours. There are 28,000–142,000 deaths from cholera every year.

RISK FACTORS

Transmission of *Vibrio cholerae* is relatively easy in areas of poor housing and sanitation. It particularly affects displaced populations living in overcrowded camps. Children under five are especially vulnerable.

CHECK YOUR UNDERSTANDING
15. Explain two physical factors for the distribution of malaria.
16. Outline the main impacts of cholera.

International organizations, governments and NGOs in combating food insecurity and disease

COMBATING FOOD INSECURITY

Global food security is made difficult because:
- a significant share of the world's population is malnourished
- the global population continues to grow
- climate change and other environmental changes threaten future food production
- the food system itself is a major contributor to climate change and other environmental harms.

UNITED NATIONS FOOD AND AGRICULTURE ORGANIZATION (FAO)

The FAO aims to eradicate hunger, food insecurity and malnutrition; to eliminate poverty; to use natural resources in a sustainable way; and to increase the resilience of people to threats and crises.

WORLD FOOD PROGRAMME (WFP)

The WFP aims to end global hunger by providing food assistance for the poorest and most vulnerable. Its plan has four main aims:
- To save lives and protect livelihoods in emergencies.
- To support food security and nutrition and (re)build livelihoods.
- To reduce risk and promote self-sufficiency.
- To reduce undernutrition and break the intergenerational cycle of hunger.

NATIONAL GOVERNMENTS AND FOOD SUBSIDIES

Subsidized agriculture in HICs is one of the biggest obstacles to economic growth in LICs. OECD countries spend over $300 billion a year on crop price supports, production payments and other farm programmes. These subsidies encourage overproduction. Markets are flooded with surplus crops that are sold below the cost of production, reducing world prices.

By 2013 many NICs had established their own agricultural subsidies. The BRIIC countries (Brazil, Russia, India, Indonesia and China) have increased theirs the fastest. China's agricultural subsidies are now greater than those of the USA and the EU combined. Global subsidies may also lead producers to overuse fertilizers or pesticides, which can have negative environmental impacts.

NON-GOVERNMENTAL ORGANIZATIONS (NGOs)

Many NGOs help to deliver food to those with insufficient access to food. Many of these are in low-income and middle-income countries, such as Operation Hunger in South Africa, but others operate in high-income countries such as the food banks in the UK.

WORLD HEALTH ORGANIZATION (WHO)

The World Health Organization (WHO) is the part of the UN that deals with health issues. It has 7,000 people from more than 150 countries working in 150 country offices. The WHO has a number of objectives:
- providing leadership on matters critical to health and engaging in partnerships where joint action is needed
- setting norms and standards and promoting and monitoring their implementation
- providing technical support, catalysing change, and building sustainable institutional capacity
- monitoring the health situation and assessing health trends.

There are other initiatives, for example, the Global Fund to Fight AIDS, Tuberculosis and Malaria was launched in 2002. The Global Alliance for Vaccines and Immunizations (GAVI) provides international financing for immunization coverage.

AN NGO: MÉDECINS SANS FRONTIÈRES (MSF) (DOCTORS WITHOUT BORDERS)

MSF was founded in 1971 to provide emergency medical aid. MSF is a worldwide movement owned and run by its staff. In 2015 it provided medical aid in over 70 countries through more than 30,000, mostly local, doctors, nurses and other medical professionals. Private donors provide about 90% of the organization's funding, giving MSF an annual budget of approximately $750 million.

MSF highlights (2014):
- Over two million cases of malaria treated.
- Over 200,000 severely malnourished children admitted to in-patient or out-patient feeding programmes.
- Nearly 200,000 women delivered babies, including Caesarean sections.
- Over 1.5 million people vaccinated against measles.

COMMON MISTAKE
✗ *Some students believe that only HICs offer subsidies for their farmers.*
✓ Many NICs offer subsidies and China's farm subsidies are the largest in the world.

CHECK YOUR UNDERSTANDING
17. Suggest what is meant by the intergenerational cycle of hunger.
18. Explain how farm subsidies lead to inequalities in farm production.

The influence of TNCs and the media in shaping food consumption habits

The policies and practices of transnational companies (TNCs) are steadily displacing traditional food systems around the world. The term "big food" refers to the food and drink TNCs that increasingly control the production and distribution of processed food and drink around the world. The products are generally high in sugars, fats and oils. They are generally packaged as "fast foods" or "convenience foods".

The main interest of a TNC and its shareholders is to make money. Their market penetration in HICs has largely peaked, and TNCs have now developed "healthy drinks" to satisfy the demand for healthier diets in HICs. However, TNCs have increasingly looked to LICs to sell convenience foods since incomes are rising and many people are leaving rural areas for urban ones. The potential for food TNCs to achieve greater market share is therefore immense.

FOOD INDUSTRY PLAYER

Company	2008 Sales (Billion)
Dole (Produce supplier)	7.62
Whole Foods	7.9
Dean Foods (Dairy)	9.8
Agrium (fertilizers)	10
Starbucks	10.4
Monsanto	11.8
Kellogg (food processing)	12.8
HE Butt (US grocery retailer)	13.5
General Mills (food processing)	13.7
CASE IH (equipment)	17.4
McDonald's	23.5
John Deere (equipment)	28.4
Coca Cola	31.9
Costco (food only)	34.1
Kraft (food processing)	42
Safeway	42.3
Pepsico	43.3
ADM (grain trade)	69.8
Kroger (intenational food retail)	70.4
Cargill (grain trade)	120.4
Carrefour (French food retail)	128
Walmart (food only)	149.6

PEPSICO AND GLOCALIZATION

Frito Lay – a glocalized product

PepsiCo has two core brands – Pepsi Cola in the soft-drinks brand and Frito-Lay in the packaged fast-food industry. PepsiCo adopted glocalization, that is, adapting its products to local tastes: Frito-Lay in Mexico sells chips (potato crisps) with chilli flavours, while Frito-Lay in China sells crab- or duck-flavoured chips. By using local suppliers, PepsiCo appeals to consumers' nationalist sentiments while also impacting local food-production systems.

CHANGING DIETARY PATTERNS IN BRAZIL, AN NIC

Traditionally, Brazil had a varied dietary pattern. Rice, beans and cassava were common, with added oils, spices and herbs. However, TV and internet advertising, as well as changes in lifestyle, including urbanization, have led to an increase in the consumption of convenience foods.

It would be simplistic to suggest that the traditional Brazilian diet was good whereas the modern one is poor. Food and drink TNCs are changing the diet in Brazil. TNCs are marketing their products as best they can and there has been a huge increase in the availability of snack foods in Brazil.

EXAM TIP

It is very easy to state McDonald's, for example, when talking about a fast-food chain and the impact on diet. Try to bring in specific details, as in the case of PepsiCo and Frito-Lays in Mexico and China, and specific details about marketing strategy.

CHECK YOUR UNDERSTANDING

19. Describe the main changes in diet related to TNC production of food.
20. Briefly explain the increasing consumption of fast food.

Gender roles related to food and health

GENDER, FOOD SECURITY AND NUTRITION

In many LICs, rural women and men play different roles in providing food security in their household and communities. While men grow mainly field crops, women are usually responsible for growing and preparing most of the food and rearing small livestock, which provides protein. Rural women also do most of the cooking. Women represent about half the food-producing workforce in South East Asia and sub-Saharan Africa, but often as unpaid workers.

GENDER AND FOOD PRODUCTION IN THE DEVELOPING WORLD

The positive impact of closing the gender gap in agricultural production cannot be overstated.

- Studies suggest that if women had the same access to productive resources as men, they would increase yields by 20 to 30%, raising overall agricultural production in developing countries by 2.5 to 4%.
- Women's access to education is a determining factor in levels of nutrition. Studies from Africa show that children of mothers who have spent five years in education are 40% more likely to live beyond the age of 5.
- Good nutrition and health depend on the safety of the food consumed. Contamination can lead to diarrhoea, a major cause of illness and death in children. Training women in hygiene and sanitation brings an immediate improvement in household health.
- Malnutrition in young children results from lack of dietary knowledge and leads to wasting and stunted growth.
- Access to land and land ownership for women is very limited in most parts of the developing world. This means that they have little control over their land and the size of their holding is only large enough for subsistence farming.
- A woman's time is spent fetching water, cooking, washing, cleaning and tending to the needs of children and livestock. For example, in Ghana and Tanzania, 77% of their energy is expended on load-carrying activities involving wood, grain and water. Diseases such as HIV/AIDS in the family mean that women have to assume a greater caretaking role, leaving them less time and energy to grow and prepare food.
- Access to most financial services such as credit and insurance is only available to men in rural areas. This greatly limits food security and farming innovation for women.
- Women tend to be engaged in the production of traditional subsistence crops of inferior quality and are often denied the opportunity to sell their produce at the market and engage in the cash economy. They are also limited by lack of time and poor access through limited transport.

INEQUALITY AND DISCRIMINATION AGAINST WOMEN

Women and girls make up around 60% of the world's chronically hungry. Women face discrimination in education and employment opportunities and within households. Women also face discrimination in access to land, training, finance and other services. Achieving gender equality is instrumental in ending malnutrition. This is because women tend to be responsible for food preparation and childcare within the family and are more likely to spend their income on food and their children's needs. A child's chances of survival increase by 20% when the mother controls the household budget. Women thus play a decisive role in future food security and children's health.

FEMALE CARERS

Many women have a "double burden" of paid and unpaid work. Domestic roles include bearing and rearing children, and caring for other dependents such as the elderly. Even in HICs, gender equality has proceeded at a faster pace outside the domestic sphere than within it.

Women in LICs continue to be the prime carers of children and the use of childcare facilities outside the family is limited.

AGEING POPULATIONS AND FEMALE CARERS

The impact of the growth of elderly dependents falls disproportionately on women. The trend towards delayed childbearing may mean that the care required by old or frail dependents increasingly coincides with that of children: "the sandwich generation" is stuck in the middle of caring for both their children and their parents. In China, children are legally bound to look after their parents. As a result of the one-child policy, two one-child children who marry could end up caring for five people, that is, their four parents and their one child.

FEMALE CARERS IN THE USA

Women are the major providers of long-term care in the USA, and they also have long-term care needs of their own. Women live longer than men, tend to outlive their spouses, and have less access to retirement savings such as pensions. For many women there is a conflict between work and care of the elderly. For example:
- 33% of working women reduced their work hours
- 29% passed up a job promotion, training or assignment
- 20% switched from full-time to part-time employment
- 13% retired early.

CHECK YOUR UNDERSTANDING
21. Outline the main discrimination in food production/ security against women.
22. Briefly explain the term 'sandwich generation' and oultine why it is particularly important in China.

Factors affecting the severity of famine

FACTORS THAT AFFECT FAMINE

Famine refers to a long-term decline in the availability of a food in a region. There are many interrelated factors affecting a famine's severity, including:
- the length and severity of drought
- governance – poor governance increases the risk of food scarcity
- the power of the media
- access to international aid
- population growth
- unemployment and entitlement to food
- civil unrest, including war
- access to land and production of food.

Emergency aid can help those suffering the effects of famine in the short term, but long-term plans need to be put in place to reduce the effects. Aid should be designed to empower poor people to build productive assets such as water-harvesting tanks, dams and irrigation projects. Foreign companies should not be allowed to grow crops for export only.

CASE STUDY

FAMINE IN ETHIOPIA, 2015–16

In 2015 and 2016, the seasonal rains that usually fall between June and September in north-eastern, central and southern Ethiopia did not arrive. According to the UN, this was Ethiopia's worst drought in 50 years. Around 90% of cereal production is harvested in autumn, after the summer long rainy season, and the rest at the end of spring after the end of the short rainy season. As a result, more than 18 million people were in need of aid.

These days, early-warning systems alert the government when famine threatens, and in 2015 the government was able to respond quickly to the crisis.

Famine early warning system: drought conditions in Ethiopia, March–September 2015

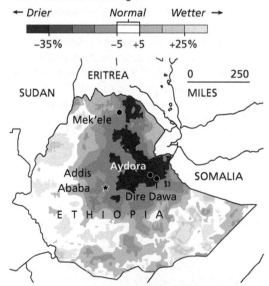

March to Sept. 2015 rainfall compared with 1981–2014 average

There is evidence that in 2015–16 the Ethiopian government made provision to mitigate the impact of poor harvests, such as establishing a social security net so that poorer farmers could access funds for public works such as digging water holes.

The drought was caused by the El Niño weather system, and resulted in a 90% reduction in crop yields and the deaths of over one million cattle. The famine, though, was brought about by factors including poor governance and state neglect. The key factors driving the famine, then and now, include the selling off of land to international corporations for industrial farming, that is, "land grabs".

In 2015–16 international donors were distracted by a string of humanitarian disasters around the world, such as in Syria and South Sudan. This meant that there was much less media coverage.

Events in Syria were more dramatic than the failure of the rains, although the failure of the rains probably caused more deaths.

Nevertheless, a number of organizations are working in Ethiopia, including the World Food Programme (WFP), Save the Children and the FAO. The WFP is helping to feed the refugees and also supporting the government's second five-year Growth and Transformation Plan (GTP), a school meals programme and a vulnerability and mapping unit (VAM). Save the Children Ethiopia reported substantial livestock losses in the Afar region. It had also mobilized $100 million, while the FAO announced a $50 million plan to assist agriculture- and livestock-dependent households.

CHECK YOUR UNDERSTANDING

23. Describe the pattern of Famine Early warning in Ethiopia, March–September 2015.

24. Outline two reasons why the amount of aid promised to Ethiopia in 2015–16 was lower than might have been expected.

Possible solutions to food insecurity

There are a number of possibilities for reducing food insecurity. These include short-term, medium-term and long-term measures.

SHORT-TERM MEASURES

- Increasing production and reducing set-aside: high market prices encourage more food production.
- Increasing food aid. The World Food Programme (WFP) reaches only about 80 million of the most desperate, mostly refugees from conflicts and natural disasters. There are 700 million more chronically hungry people scattered around the world.
- Greater use of seeds and fertilizer. As well as needing food to survive, the rural poor urgently need help planting next season's crops if there is to be an end to the food crisis. Millions have been forced to eat next season's seeds to survive.
- Export bans on food from countries experiencing food shortages.

MEDIUM-TERM MEASURES

- Free trade: reducing protectionism should help poor farmers in the future.
- Biofuels: the food crisis has triggered a backlash against plant-derived fuels.

LONG-TERM MEASURES

- Agricultural investment: yields in Africa could be increased up to fourfold with the right help.
- GM crops.
- Sustainability: campaigners argue that the world cannot feed its population if China, India and other emerging economies want to eat like people in the West. The only long-term solution, they argue, is rethinking western lifestyles and expectations.

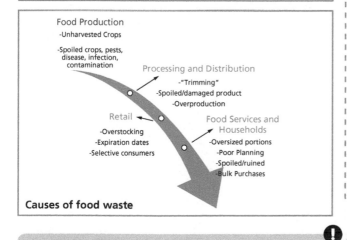

Causes of food waste

FOOD WASTE

Approximately one-third of food is thrown away in HICs every year. In LICs, up to 80% may be wasted before it reaches the market/shops. More efficient farming practices and storage would ensure that a larger proportion of the food produced reaches markets and consumers. However, produce is often wasted through retail and customer behaviour. Major supermarkets often reject perfectly edible food because it does not meet exacting marketing standards. Globally, retailers generate 1.6 million tonnes of food waste annually in this way.

In LICs wastage tends to occur primarily at the farmer–producer end of the supply chain. Mould and pests destroy or at least degrade large quantities of food material. Substantial amounts of foodstuffs simply spill from badly maintained vehicles or are bruised as vehicles travel over poorly maintained roads.

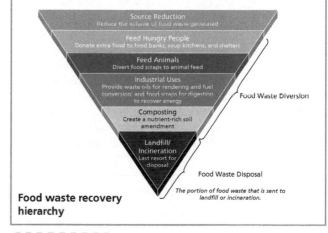

Food waste recovery hierarchy

COMMON MISTAKE

✗ *Some students believe all food waste comes from HICs.*
✓ A lot of food waste is produced by HICs, but there is also a great deal of food waste in LICs and NICs, especially in the production, storage and transport of food to market in LICs.

CHECK YOUR UNDERSTANDING

25. Describe the main differences in food waste between high income countries and low income countries.
26. Outline the ways in which Bangladesh has increased food security in recent years.

Contemporary approaches to food production

GENETICALLY MODIFIED FOOD (GM FOOD)

GM technology has helped farmers to increase yields by protecting crops against pests and weeds. Genetic engineering involves adding traits to a plant to make it more nutritious or more resistant to disease or pesticides.

More than 93% of corn, soy and cotton in the USA is now genetically modified. In 2014, GMO crops made up 94% of US soya bean acreage, 93% of all corn planted, and 96% of all cotton. The use of more herbicides raises the risk of creating herbicide-resistant weeds and may lead to the decline of some species, such as the monarch butterfly.

Globally, GMOs were planted on 175 million hectares in 2013 – or roughly 12% of global farmland. Most GM crops are grown in just five countries: the USA, Argentina, Brazil, Canada and India. Growth seems to be slowing, and one reason for that is saturation.

VERTICAL FARMING

By 2050 the world's population will be over 9 billion and food production will need to increase by 70%. Vertical farms aim to grow year-round in high-rise urban buildings.

It could make food supplies more secure as well, because production can continue even in extreme weather. If farmers are careful to protect their indoor "fields" from pests, vertical farming needs no herbicides or insecticides.

Japan is the leading producer in vertical farming because much of the region's farmland was affected by radiation from the Fukushima-Daiichi nuclear explosion. The plant trays in a vertical farm can be fed nutrients by water-conserving, soil-free hydroponic systems and are lit by LEDs. However, the electricity bills can add up quickly.

IN VITRO MEAT

In vitro meat, also known as cultured meat or synthetic meat, is a meat product that has never been part of a living animal. In 2013, the world's first lab-grown burger was cooked and eaten at a cost of €250,000 to produce.

It has been claimed that two months of *in vitro* meat production could deliver up to 50,000 tonnes of meat from 10 pork muscle cells. However, difficulties of scale and cost need to be overcome before *in vitro* meat becomes commercially available.

Cultured meat has significantly fewer environmental impacts than normally reared beef. In theory, *in vitro* meat is potentially much more efficient and environmentally friendly, generating only 4% greenhouse gas emissions, reducing the energy needs of meat generation by up to 45%, and requiring only 2% of the land that the global meat/livestock industry does.

CHECK YOUR UNDERSTANDING
27. Outline the advantages of GM food.
28. Outline the advantages and disadvantages of *in vitro* food.

Vertical farming

Prevention and treatment of disease

PREVENTATIVE TREATMENT

Preventative treatment means adopting policies and lifestyles that will reduce the risk of disease. This may range from people having a healthy diet to not smoking or drinking to excess to reduce the risk of cancer, heart attacks and strokes.

CURATIVE TREATMENT

Curative treatment is required to treat diseases such as cancer, heart disease and stroke. This is much more expensive that preventative health care and may involve lengthy hospitalization.

HEALTH CARE IN HICS AND IN LICS

In many cases, poor communities receive primary health care whereas wealthier individuals receive curative hospitalized care. Critics argue that investment in clean water and sanitation would have a greater impact on health among poor communities than health care, as most of the diseases are those of poverty. Scientific, curative medicine is more suited to degenerative diseases. Scientific interventions, that is, treating the disease, are far more expensive than preventative treatments. It may be most cost-effective, not just in terms of medical treatment but also in lost labour productivity, for more people to follow a healthy lifestyle, rather than develop degenerative diseases. Some progress has been made in some countries – smoking is less prevalent in HICs although it remains high in NICs. The AIDS awareness campaign, and new treatments, have reduced deaths from AIDS, although there are still over 2 million new infections every year. The availability of antiretroviral drugs may encourage some people to take risks in their sexual activities.

Health care reflects the nature of society, in terms of political developments and market forces. In HICs,

health care is based on curative medicine and the use of high-technology techniques. By contrast, health care in many LICs is based on low-technology, preventative measures. The aims of primary health care (PHC) is prevention rather than cure and include:

- growth monitoring
- oral rehydration
- breast feeding
- immunizations
- food supplementation
- female education
- food fortification.

In theory, demand for health care should increase with more pregnant women, elderly and very young in a population, namely the vulnerable.

Hart's inverse case law (1971) states that those who can afford it (the well off) are those who least need it, whereas those who need it most (the vulnerable) cannot afford health care or have health insurance, and therefore do not get it. Consequently, health care is often inversely related to those in most need.

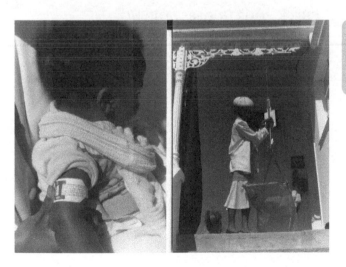

Growth monitoring – part of primary health care

CHECK YOUR UNDERSTANDING

29. Outline the main strategies of primary health care.
30. Suggest two disadvantages of the scientific, curative model of health care provision.

Pandemics

MANAGING PANDEMICS

Pandemics are global epidemics. Their large scale makes them difficult to manage and they may also involve new diseases, or relatively unknown diseases, such as Ebola and the Zika virus. If the disease is new, there may not be any recognized vaccinations.

To manage pandemics there must be a coordinated effort among global communities. For example, following the outbreak of Ebola in West Africa, the Nigerian government established a massive public health campaign. Containment was the key to ending Ebola. Nigeria set up a centralized emergency operations centre, staffed with public health experts who had earlier worked on a polio eradication programme. TV broadcasts and social media were used to reassure people. In Sierra Leone, gatherings were banned. Markets and schools were closed, and school lessons were given over the radio.

THE GLOBAL DIABETES PANDEMIC

The number of people worldwide with diabetes is over 400 million. In 2012 1.5 million people died from the disease.

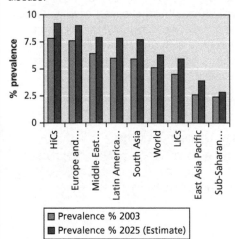

The global prevalence of diabetes

Diabetes is a chronic, lifelong condition and a major cause of blindness, kidney failure, heart attacks, stroke and lower limb amputation. The disease reduces both a person's quality of life and life expectancy.

Normally, the hormone insulin controls blood sugar levels, so lack of insulin in the blood leads to hyperglycemia (high blood sugar). Hypoglycemia (a "hypo") is low blood sugar – the opposite of hyperglycemia – and symptoms include sweating, tremors and confusion.

TREATMENT AND CARE

Some problems with diabetic care can be solved without great expenditure and these solutions are suitable for LICs. They include blood sugar control, blood pressure control, foot care and preconception care for women. Insulin can be supplied to those requiring it.

Environmental factors that can lead to type 2 diabetes include obesity, physical inactivity, diet and increased affluence; the adoption of western lifestyles has been associated with an increase in the prevalence of diabetes in many indigenous populations.

THE ECONOMIC BURDEN OF DIABETES

Diabetes causes a large economic burden on national health care systems and affects national economies, families and individuals. Direct medical costs include resources used to treat the disease while indirect costs include lost productivity. Other costs are the reduced quality of life for people with diabetes and their families brought about by stress, pain and anxiety.

EXAM TIP

You could use the information here as evidence of the demographic transition, as many LICs and NICs are experiencing a rise in diabetes and becoming more similar to HICs.

CHECK YOUR UNDERSTANDING

31. Outline the factors that led to the successful containment of the Ebola virus.
32. Compare the projected prevalence of diabetes in 2025 with that of the observed prevalence in 2003.

Exam practice

The maps shows the diffusion of the Ebola virus between March 2014 and March 2015.

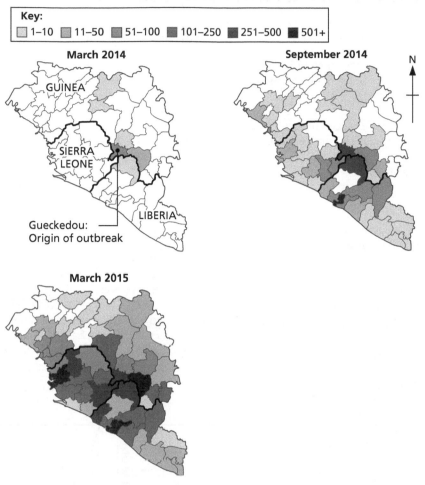

Key:
☐ 1–10 ☐ 11–50 ☐ 51–100 ☐ 101–250 ☐ 251–500 ☐ 501+

March 2014

GUINEA

SIERRA LEONE

Gueckedou:
Origin of outbreak

LIBERIA

September 2014

N

March 2015

Source: http://www.bbc.co.uk/news/world-africa-28755033

(a) **(i)** Describe the diffusion of the Ebola virus, as shown by the deaths from the virus. (3 marks)

   (ii) Outline the geographic factors that can facilitate the spread of the Ebola virus. (4 marks)

(b) Compare the relative advantages of preventative health care with that of curative health care. (2 + 2 marks)

(c) Either:

Discuss the view that physical factors are the most important ones in the development of famine. (10 marks)

Or

"Reducing waste is the most effective way of achieving food security." Discuss this statement. (10 marks)

1 THE VARIETY OF URBAN ENVIRONMENTS

Characteristics of urban places

URBAN PLACES

Many criteria are used to specify what an urban place is and it is not possible to give a single definition. However, an urban place is characterized by:
- population size
- specific features, such as a CBD and residential zones
- predominant economic activities, such as manufacturing and services
- an administrative function.

SITE

The **site** of a settlement is the actual land on which a settlement is built, whereas its **situation** or position refers to its relationship with its surrounding area. Favourable factors include a reliable supply of water; freedom from flooding; level sites to build on; timber for construction and fuel; sunny, south-facing slopes (in the northern hemisphere); fertile soils for cultivation; the potential for trade and commerce.

FUNCTION AND LAND USE

Functions change over time. Many settlements that were formerly fishing villages have become important tourist resorts. The Spanish Costas and many Caribbean settlements are good examples of this.

Most urban places have an industrial role (manufacturing and/or high-tech industry) and/or a service role (health care, education, retail, and so on). Urban places also have a very important residential role. There may be some agriculture in urban areas, such as allotments, but this uses only a small proportion of the land in the area.

MEGACITIES

Larger urban areas, such as cities, offer a wider range of services and goods, and more of them. A **millionaire city** is a city with over one million inhabitants whereas a **megacity** is a city with over 10 million inhabitants.

Megacities grow due to economic growth, rural–urban migration and high rates of natural increase. Due to people migrating to the city in search of jobs, megacities develop an age structure that is dominated by young adults. Thus, the city grows not only through migration but also because of the high birth rates associated with a younger population. As the cities grow, they swallow up rural areas and nearby towns and cities. They become multi-nuclei centres.

Urban growth can be planned, for example, by the government, or it can be spontaneous, for example, by migrants and the creation of squatter settlements. Many settlements grow due to trade potential, favourable accessibility (the two are linked), whereas others are new or planned, for example, Brasilia in Brazil and Canberra in Australia.

THE HIERARCHY OF SETTLEMENTS

The term "hierarchy" means order or importance. Settlement size is normally used as a measure of settlement hierarchy. Settlement hierarchy uses a number of concepts, including:
- **range** – the maximum distance that people are prepared to travel for a good or service
- **threshold** – the minimum number of people required for a good or service to stay in business
- **low-order goods** – goods or convenience goods bought frequently, such as bread, newspapers
- **high-order goods** – luxury goods bought or used infrequently, for example, watches, cars
- **sphere of influence** – the area served by a settlement.

Only **low-order** functions are found in a hamlet, while larger settlements (for example, market towns) support the same functions and services, as well as more specialized **high-order** functions. Market towns draw custom from surrounding villages and hamlets as well as serving their own population. The distinction between hamlet, village and town is not always very clear-cut and they all share features on a sliding scale (**continuum**) rather than being separate categories.

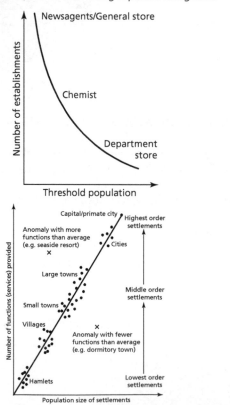

CHECK YOUR UNDERSTANDING

1. Define and illustrate the concept of settlement hierarchy.
2. Outline how the functions of settlements may change over time.

COMMON MISTAKE

✗ Site and situation mean the same thing.

✓ Site is the immediate location, what a settlement is built on, whereas situation is what it is close to.

Urban economic activities

FACTORS AFFECTING THE PATTERN OF URBAN ECONOMIC ACTIVITY IN CITIES

The value of land varies for different purposes – this is known as the **bid rent**. Land at the centre of a city is most expensive for two main reasons: it is (or was) the most accessible land to public transport, and there is only a small amount of it available. Land prices generally decrease away from the most central area, although there are minor peaks at the intersections of major transport routes.

RETAIL LAND USE

The hierarchy of retail outlets in cities is as follows:
1. Low-order goods concentrated in neighbourhood stores and shopping parades.
2. High-order goods in high street shops, department stores.
3. Out-of-town superstores and retail parks.

Central shopping areas or high streets are characterized by department stores, chain stores, specialist shops and, increasingly, by pedestrianized malls. Modern retailing has changed rapidly with the growth in the number of superstores and retail parks, built on edge-of-town sites with good accessibility and plenty of space for future expansion.

Many factors explain this change in the retail hierarchy, including:
• population change, such as smaller households, and more elderly people
• suburbanization of wealthier households
• technological change, such as more families owning deep freezes
• economic change, with increased standards of living, especially car ownership
• congestion and inflated land prices in city centres
• social changes, such as more women in paid work.

COMMERCIAL LAND USE – THE CBD

The central business district (CBD) is the commercial and economic core of a city. It is the most accessible area to public transport, and the location with the highest land values.

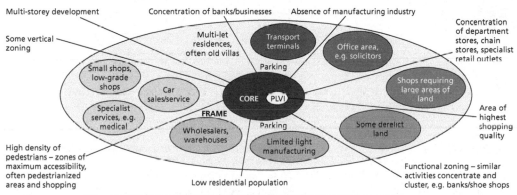

PLVI = peak land value intersection: the highest rated, busiest, most accessible part of CBD

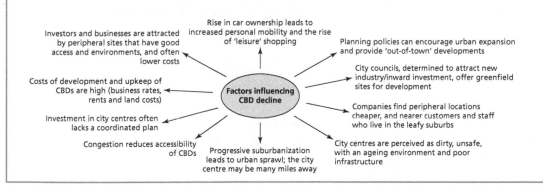

INDUSTRIAL ACTIVITY

There are many locations for industrial zones in most cities. These include:
• inner-city areas close to railways and/or canals
• brownfield suburban sites close to airports
• sites away from residential locations.

CHECK YOUR UNDERSTANDING
3. Outline the factors that influence the location of industrial areas in cities.
4. Explain why CBDs are characterized by higher-order retail outlets.

Factors affecting the location of urban residential areas

PHYSICAL FACTORS

In some urban areas, wealthier people live near attractive physical landscapes. In contrast, in some LICs the poor live close to rivers and areas at risk of flooding. Similarly, high ground in some urban areas in HICs may be attractive to wealthy households, whereas in LICs the poor may be forced to live in areas of steep relief, where mass movements are a risk.

ETHNICITY

Some ethnic groups may choose to live close together, and so end up forming neighbourhoods. For example, the majority of the South Korean population in London lives in New Malden, and it has a number of Korean restaurants and supermarkets. This is a form of **positive segregation**, whereby the ethnic group gains advantages by locating in one place: there are enough of them to support services. On the other hand, **negative segregation** occurs when population groups are excluded from certain areas.

Population in London

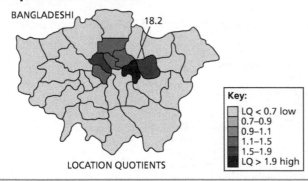

BANGLADESHI

18.2

LOCATION QUOTIENTS

Key:
- LQ < 0.7 low
- 0.7–0.9
- 0.9–1.1
- 1.1–1.5
- 1.5–1.9
- LQ > 1.9 high

URBAN RESIDENTIAL PLANNING

Planning is increasingly important in many cities. Authorities may plan for a balanced social mix by having a mixture of housing types located around the city. However, in many cities this does not occur. "Edge cities" – such as Barra da Tijuca, on the edge of Rio de Janeiro – are often gated communities of wealthy people who use security for self-segregation.

LAND VALUES

In most HIC cities the highest residential densities are in inner-city areas. Usually, residential density in the city centre is low because of high land values. With increasing distance from the city centre, residential density decreases. This reflects the greater availability of land in the suburbs.

The paradox of the poorest people being located on expensive inner-city land reflects their need to be close to sources of employment. In contrast, wealthier people live in the outer areas, in lower-density housing, where they are able to commute to work. However, land value is just one factor affecting the location of residential areas. Certain land uses may repel each other, and high-class and low-class residential areas may be separated by physical or built barriers.

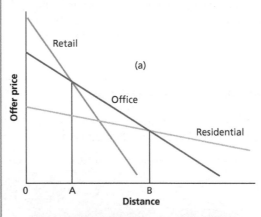

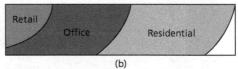

Offer prices of retail, office and residential uses with distance from the city centre:

(a) Section across the urban value surface.
(b) Plan of the urban value surface.

CHECK YOUR UNDERSTANDING

5. Briefly examine two reasons for the pattern of residential land-use.
6. Suggest how ethnicity may influence patterns of residential land-use.

FAMILY LIFE CYCLE

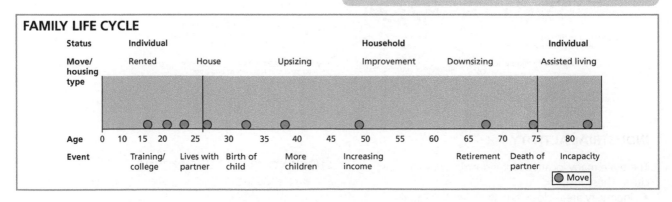

Poverty, deprivation and the informal sector

In most cities there is considerable variation in the quality of life. "Poor areas" are zones of deprivation, poverty and exclusion. In HICs these are often inner-city areas or ghettos, whereas in NICs/LICs it is frequently shanty towns that exhibit the worst conditions.

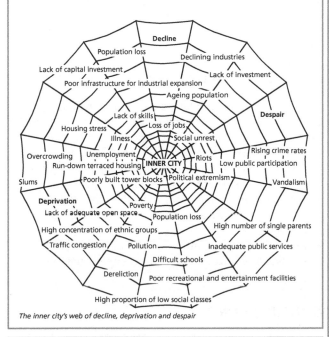

The inner city's web of decline, deprivation and despair

INFORMAL ECONOMY

The **dual economy** consists of a formal economy, complete with offices, factories and commercial buildings, and an informal economy, consisting of servants, gardeners, maids, cleaners, taxi drivers and a variety of other occupations. Much of the formal economy produces goods and services for an elite population. By contrast, the informal economy is small scale, locally owned and labour intensive.

Characteristics of the formal and informal economy

Informal sector	Formal sector
Ease of entry	Difficult entry
Indigenous inputs predominate	Overseas inputs
Family property predominates	Corporate property
Small scale of activity	Large scale of activity
Labour intensive	Capital intensive
Adapted technology	Imported technology
Skills from outside the school system	Formally acquired skills
Unregulated/competitive mark	Protected markets, for example, tariffs, quotas, licensing

SLUMS AND SQUATTER SETTLEMENTS

Around 1 billion peole live in slums. This represents about one-third of the world's urban population, but over three-quarters of the urban population is in LICs. Slums are typically located on land that planners do not want: steep slopes, floodplains, edge-of-town locations and/or close to major industrial complexes.

The UN defines a slum household as one that lacks one or more of the following five characteristics:
- a permanent housing structure that provides protection from extreme climatic conditions
- no more than three people sharing a room
- access to water that is sufficient, affordable and can be obtained without extreme effort
- access to a private toilet or a public one shared with a reasonable number of people
- protection against forced eviction.

The positives and negatives of living in a slum

Positive aspects	Negative aspects
They are points of assimilation for immigrants	Security of tenure is often lacking
Informal entrepreneurs can work here and have clienteles extending to the rest of the city	Basic services are absent, especially water and sanitation
Informal employment, based at home, avoids commuting	Overcrowding is common
	Sites are often hazardous
There is a strong sense of kinship and family support Crime rates are relatively low	Levels of hygiene and sanitation are poor, and disease is common

EXAM TIP ✓
Show in your answer that many people who work in the informal sector also work in the formal sector. For example, teachers who carry out private tutoring could be working in the informal sector, in addition to a job in the formal sector (teaching at a school or college).

CHECK YOUR UNDERSTANDING ?
7. Outline two advantages and two disadvantages of the informal economy.
8. Suggest how the distribution of poverty and deprivation differs in cities in HICs, NICs and LICs.

Urbanization, natural increase and population movements (1)

URBANIZATION

Urbanization is "an increase in the proportion of people living in urban areas". It can be caused by rural-to-urban migration, higher rates of natural increase in urban areas, or the reclassification of rural areas as urban areas. In many HICs, the process of urbanization is almost at an end and the proportion of urban dwellers is beginning to fall. For many HICs, they appear to have passed through a cycle of urbanization, suburbanization, counter-urbanization and re-urbanization.

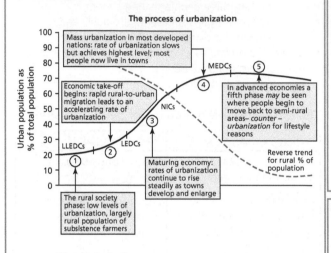

The process of urbanization

- ① The rural society phase: low levels of urbanization, largely rural population of subsistence farmers
- ② Economic take-off begins: rapid rural-to-urban migration leads to an accelerating rate of urbanization
- ③ Maturing economy: rates of urbanization continue to rise steadily as towns develop and enlarge
- ④
- ⑤ In advanced economies a fifth phase *may* be seen where people begin to move back to semi-rural areas– *counter-urbanization* for lifestyle reasons

Mass urbanization in most developed nations: rate of urbanization slows but achieves highest level; most people now live in towns

Reverse trend for rural % of population

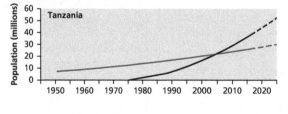

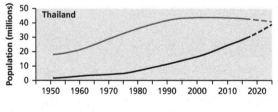

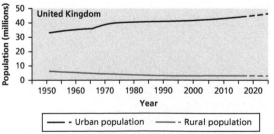

— Urban population   —— Rural population

NATURAL INCREASE

When the birth rate is higher than the death rate, natural increase occurs. Natural increase often occurs in cities due to their youthful age structure. Urban areas attract many young migrants and this makes the age structure younger. In contrast, the rural areas develop an ageing population, which means that rural birth rates are likely to fall.

RURAL–URBAN MIGRATION

Rural–urban migration refers to the long-term movement of people away from the countryside to towns and cities. This is a very important process, especially in LICs and NICs. It occurs because people believe they will be better off in urban areas than in the rural areas. There are many push and pull factors that explain rural-urban migration:

- **Push factors** are the negative features that cause a person to move away from rural areas, for example unemployment and low wages.
- **Pull factors** are the attractions (whether real or imagined) of urban areas such as better wages, more jobs and good schools.

GENTRIFICATION

Gentrification is the regeneration of inner-city areas by their residents, especially those who are young and upwardly mobile. It refers mainly to an improvement of residential areas. It is common in areas where there may be **brownfield sites** (abandoned or derelict industrial buildings). Gentrification may lead to the social displacement of poor people: as an area becomes gentrified, house prices rise and the poor are unable to afford the increased prices. As they move out, young, upwardly mobile populations move in.

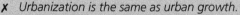

COMMON MISTAKE

✗ *Urbanization is the same as urban growth.*
✓ Urban growth is the increase in population size of an urban area whereas urbanization is an increase in the proportion of people living in urban areas.

CHECK YOUR UNDERSTANDING

9. Identify three reasons why urbanization occurs.
10. Briefly explain the process of gentrification.

Urbanization, natural increase and population movements (2)

CENTRIFUGAL POPULATION MOVEMENTS

Suburbanization

Suburbanization is the outward expansion of towns and cities, largely due to improvements in transport systems. By the early 20th century, railways, electric tramways and buses were critical to the growth of middle-class, residential suburbs. In addition, the price of farmland had declined dramatically and there was scope for urban expansion on a massive scale.

Moreover, rising wages and living standards were matched by rising expectations. Housing was now available, affordable and of a quality unimaginable only a few decades earlier.

The boom in private housebuilding was helped by:
- low costs of living
- very low interest rates
- expansion of building societies and increased availability of mortgages
- willingness of local authorities to provide utilities, such as sewers, electricity, gas and water
- improved public transport.

Counter-urbanization

Counter-urbanization is a process involving the movement of population away from larger urban areas to smaller urban areas and rural areas. There are many reasons why people may wish to leave large urban areas and move to smaller urban areas and rural areas. These include high land prices; congestion; pollution; high crime rates; a lack of community; and declining services in the urban area.

In contrast, there is a perception that smaller settlements have a better sense of community, a higher-quality environment and a safer location in which to raise a family.

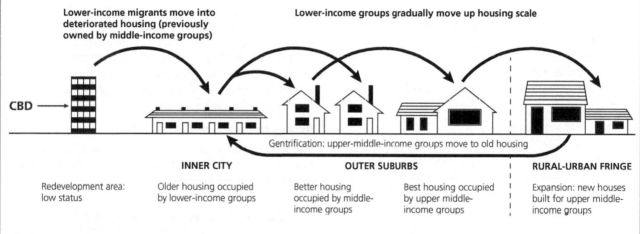

Lower-income migrants move into deteriorated housing (previously owned by middle-income groups)

Lower-income groups gradually move up housing scale

CBD →

Gentrification: upper-middle-income groups move to old housing

INNER CITY		OUTER SUBURBS		RURAL-URBAN FRINGE
Redevelopment area: low status	Older housing occupied by lower-income groups	Better housing occupied by middle-income groups	Best housing occupied by upper middle-income groups	Expansion: new houses built for upper middle-income groups

Filtering occurs as housing deteriorates and it moves downwards through the social groups.
Gentrification reverses this process as middle-income groups upgrade older city properties by renovating them.

Suburbanization and counter-urbanization

Urban sprawl

Urban sprawl is the uncontrolled growth of urban areas at their edges. Many of the world's largest cities, such as Tokyo, Seoul and Mexico City, have been characterized by urban sprawl. However, the existence of green belts prevents urban sprawl because there are limits on how far the urban area can grow.

EXAM TIP

Use examples of suburbanization, counter-urbanization and urban sprawl from your own country. You will find these easier to remember, and the examiner will find it more interesting.

CHECK YOUR UNDERSTANDING

11. Briefly explain the reasons why suburbanization occurred.
12. Outline the reasons why counter-urbanization may occur.

Urbanization, natural increase and population movements (3)

URBAN SYSTEM GROWTH

As urban areas grow, there is a greater demand for access to clean water, sanitation and waste disposal. If urban areas are to be successful, they need to expand their transport facilities and their telecommunications networks. Without these, the ability to attract new economic activity is limited.

CASE STUDY

SHANGHAI

Shanghai's population increased from 11 million in 1978 to over 24 million in 2015, and its population density rose from 1785/km^2 to 3809/km^2. Household size fell from an average of 3.77 in 1978 to 2.69 in 2015. In 2015, there were over 13 million workers in the city, with nearly 9 million working in services and over 3.5 million in manufacturing industries.

Transport

Rail transport is the key feature of Shanghai's public transport. The urban rail network developed in less than 20 years and carries over 5 million passengers daily. Approximately 25% of the city is covered by railways, serving 40% of the city's population.

In 2000 Shanghai had 6,600 km of roads – now it is over 18,000 km.

Water and sanitation

The total quantity of water supply was 2,400 million cubic m in 2000 – this increased to 3,100 million cubic m in 2015. However, per capita daily water intake remained approximately the same at around 112 litres per person per day. Over 99% of the population now have access to tap water. Around 80% of Shanghai's waste water is treated, and this is set to rise to 90% by 2020. Over 70% of households have access to sewerage services. Increasing demand, pollution and saltwater intrusion all threaten the city's water security.

Waste treatment

The amount of garbage produced in Shanghai has risen from over two million tons in 1978 to over 107 million tons in 2015. Shanghai had under 4,000 km of sewage pipes in 2000, but by 2015 it was over 20,000 km. Similarly, in 2000, less than half of urban sewage was treated, but by 2015 over 90% was treated. In the past, most of Shanghai's rubbish ended up in landfill sites or in unregulated heaps on the edge of the city. Increasingly, Shanghai is turning to incineration and generating electricity at "waste to energy" plants.

Access to telecommunications

Although the percentage of people with landlines has fallen, the number of people with access to mobile phones and/or the internet has risen.

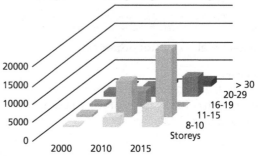

Telecommunications in Shanghai (per 10,000 households)

CHECK YOUR UNDERSTANDING

13. Briefly explain why there is a need for urban systems (transport, water, sanitation, waste collection and so on) to increase as urban areas grow.
14. Describe the changes in the number of fixed-line and mobile phone subscribers in Shanghai, 1985–2007.

The causes and consequences of urban deindustrialization

CASE STUDY

DEINDUSTRIALIZATION IN DETROIT, USA

Detroit was once the USA's fourth-largest city. Indeed, in 1960 it had the highest per capita income in the USA. Now up to a quarter of the city has been reclaimed by nature. Up to 40,000 buildings and parcels of land are vacant.

Between 1900 and 1950 Detroit prospered because General Motors (GM), Ford and Chrysler, which made most of the cars sold in the USA, were based there.

The causes of Detroit's troubles include:
- falling car sales and therefore less tax revenue from the city's large firms
- a shrinking population – many of the richer people have moved away
- high pension and social welfare costs – the city has an ageing population.

Population change in Detroit (millions)

Urban microclimates

CHANGES IN THE URBAN MICROCLIMATE
PROCESSES

Resultant processes

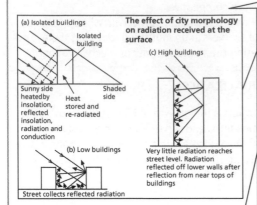

(a) Isolated buildings

The effect of city morphology on radiation received at the surface

Isolated building

(c) High buildings

Sunny side heated by insolation, reflected insolation, radiation and conduction

Heat stored and re-radiated

Shaded side

(b) Low buildings

Very little radiation reaches street level. Radiation reflected off lower walls after reflection from near tops of buildings

Street collects reflected radiation

1. Radiation and sunshine
Greater scattering of shorter-wave radiation by dust, but much higher absorption of longer waves owing to surfaces and CO_2. Hence more diffuse sky radiation with considerable local contrasts owing to variable screening by tall buildings in shaded, narrow streets. Reduced visibility arising from industrial haze.

2. Clouds and fogs
Higher incidence of thicker cloud covers in summer and radiation fogs or smogs in winter because of increased convection and air pollution respectively. Concentrations of hygroscopic particles accelerate the onset of condensation (see 5 below). Day temperatures are, on average, 0.6 °C warmer.

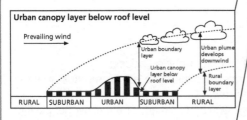

Urban canopy layer below roof level

Prevailing wind

Urban boundary layer

Urban plume develops downwind

Urban canopy layer below roof level

Rural boundary layer

RURAL SUBURBAN URBAN SUBURBAN RURAL

3. Temperatures
Stronger heat energy retention and release, including fuel combustion, gives significant temperature increases from suburbs into the centre of built-up areas, creating heat "islands". These can be up to 8 °C warmer during winter nights. Snow in rural areas increases albedo, thereby increasing the differences between urban and rural. Heating from below increases air mass instability overhead, notably during summer afternoons and evenings. Big local contrasts between sunny and shaded surfaces, especially in the spring.

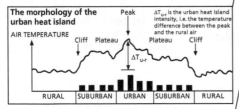

The morphology of the urban heat island

Peak

ΔT_{u-r} is the urban heat island intensity, i.e. the temperature difference between the peak and the rural air

AIR TEMPERATURE

Cliff Plateau Plateau Cliff

ΔT_{u-r}

RURAL SUBURBAN URBAN SUBURBAN RURAL

4. Pressure and winds
Severe gusting and turbulence around tall buildings, causing strong local pressure gradients from windward to leeward walls. Deep, narrow streets much calmer unless aligned with prevailing winds to funnel flows along them – the "canyon effect".

5. Humidity
Decreases in relative humidity occur in inner cities owing to lack of available moisture and higher temperatures there. Partly countered in very cold, stable conditions by early onset of condensation in low-lying districts and industrial zones (see 2 above).

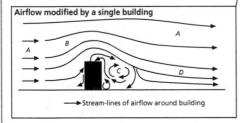

Airflow modified by a single building

A

B

A

C

D

Stream-lines of airflow around building

6. Precipitation
Perceptibly more intense storms, particularly during hot summer evenings and nights owing to greater instability and stronger convection above built-up areas. Probably higher incidence of thunder in appropriate locations. Less snow cover in urban areas even when left uncleared (e.g. road clearing).

REDUCING EMISSIONS

There are a number of ways to reduce air pollution from transport emissions, including:
- using more energy-efficient technologies such as hybrid/ electric cars
- using public transport

- using a car pooling scheme
- cycling or walking more
- using catalytic convertors to reduce emissions of NOx
- increasing enforcement of emissions standards.

CHECK YOUR UNDERSTANDING
17. Briefly explain the formation of the urban heat island.
18. Suggest reasons why relative humidity can be quite low in inner urban areas.

Traffic congestion patterns, trends and impacts

URBAN TRAFFIC

Urban traffic congestion varies with days of the week, time of day, weather and the seasons. The transport performance index (TPI) in a city may vary from 0 (excellent traffic flow) to 10 (serious congestion). There is more congestion on weekdays, especially during the peak flow times in the morning and evening.

Congestion is worse when the new school year starts, and festivals and national holidays also tend to increase the amount of traffic on the roads.

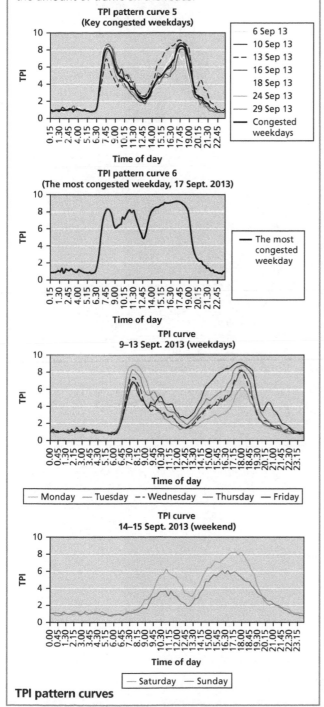

TPI pattern curves

IMPACTS OF TRAFFIC CONGESTION

Traffic congestion has many impacts. It can delay journeys and make people late for work, appointments and school. It increases fuel consumption and adds to vehicular emissions of greenhouse gases and other pollutants. It may lead to frustration and "road rage", and it can have a negative impact on people's health (stress and poor air quality). In 2013 the World Health Organization announced that air pollution could cause cancer. Diesel fumes are carcinogenic, and air pollution has a causal link with respiratory diseases.

In London, some 7,500 people die prematurely each year due to air pollution (this compares with around 8,400 for smoking). The long-term development of young children is affected by attending a school close to a busy road.

Noise pollution

Road traffic noise is related to traffic volume, traffic composition, speed, distance from the source of the noise, reflection of noise from barriers, weather conditions, terrain and road surface. Noise from road traffic has been associated with raised blood pressure, heart disease, psychological stress and annoyance, and sleep disturbance.

Exposure to street noise is believed to account for 4% of the average individual's annual noise dose. In the USA the Environmental Protection Agency (EPA) recommends 70 dB as a safe average for a 24-hour day. In New York City the mean level is 73.4 dB. The volume of traffic is the most important factor. In London, despite a reduction in air pollution, there was little change in noise levels close to main roads even after the introduction of congestion charging in 2003.

THE EFFECT OF AIR POLLUTION

New Delhi, India, is regarded as the most polluted city in the world due to the combination of diesel exhausts, construction dust, industrial emissions and the widespread burning of biofuels for cooking. Smaller Indian cities such as Gwalior, Patna and Raipur, each with populations of between 1 and 2 million people, also have dangerously high levels of PM2.5, which increases the risk of cancer.

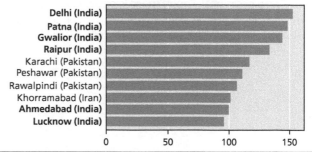

World most polluted cities
Particle matter (PM2.5) annual mean, ug/m³

CHECK YOUR UNDERSTANDING
19. Describe the main differences between traffic flow on a weekday and on a weekend.
20. Outline the impacts of noise pollution in urban areas.

EXAM TIP
Show that you are aware that patterns of traffic vary from city to city, and due to traffic measurements schemes. There is no reason why a city should follow the exact patterns shown on this page, but many do.

Contested land

CONTESTED LAND IN DHARAVI

The Dharavi slum in Mumbai is a prime site for development. The Indian property developer Mukesh Mehta wants to develop Dharavi into an international business destination. However, to do this would displace up to 1 million people to the edge of the city, in places that offer few economic prospects.

Dharavi is home to thousands of small-scale, informal industries, which generate around $650 million annually. For many people, working in Dharavi is a means of escaping poverty. The informal economy accounts for the overwhelming majority of India's economic growth and as much as 90% of employment.

SLUM CLEARANCE IN RIO DE JANEIRO

The 2016 Rio de Janeiro Olympic Games caused a property boom in some of the more central favelas. The Vidigal favela, for example, was transformed by the Olympic Games. The value of land rose threefold in three years, and land speculators bought properties. The average house price increased by 165% between 2012 and 2016.

Developers displaced around 170,000 people for games-related purposes. For example, the Favela do Metro, home to 1,000 residents, was destroyed to make way for parking facilities.

DEPLETION OF URBAN GREEN SPACE

Open spaces are important for physical and mental well-being. However, the amount of open space in urban areas varies enormously. For example, London has 50 m² of open space for each resident, whereas Mumbai has less than 2 m² of open space per resident.

It is not just green spaces that are important. In the 1960s in Bangalore, there were over 280 lakes but now there are fewer than 70, and many of these are biologically dead. The government converted some of them into parking spaces, a bus station and a sports stadium. In addition, between 2010 and 2014, over 50,000 trees in Bangalore were cut down to make way for road widening.

New open spaces are limited in many urban areas. However, occasionally open space is created in cities. This has happened in London with the creation of Queen Elizabeth Park, following the 2012 Olympic Games.

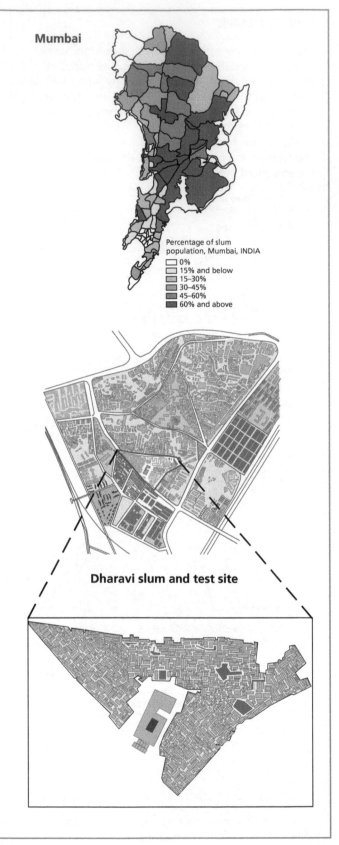

Mumbai

Percentage of slum population, Mumbai, INDIA

- 0%
- 15% and below
- 15–30%
- 30–45%
- 45–60%
- 60% and above

Dharavi slum and test site

CHECK YOUR UNDERSTANDING

21. Suggest why developers may wish to clear slums in Mumbai and Rio de Janeiro.
22. Outline the reasons for the depletion of green space.

Urban crime and social deprivation

MANAGING URBAN SOCIAL DEPRIVATION IN BARCELONA

Deprivation is associated with low incomes, a lack of employment opportunities, ill health, low educational achievement, reduced access to housing and services, high crime rates and a poor living environment.

In Barcelona there are two main areas of deprivation. One is the inner-city district and is associated with poor-quality housing built during the industrial era. The other is the edge-of-town locations, where social housing was built to accommodate migrants to the city during the 1960s.

During the 1970s and 1980s Barcelona experienced rapid deindustrialization, and unemployment in the city reached 20%, causing many locations to become derelict. However, Barcelona has since overcome much of this deprivation due to strong business and political leadership and diversification of the economy to include tourism, biomedicine, culture and ICT.

Employment in Barcelona increased by over 50% between 1995 and 2008, and 900,000 new jobs were created. More than 70% of the jobs were in the service sector, and employment in manufacturing rose by 25%.

Not only did Barcelona create many new jobs and upgrade its infrastructure, but it also provided new housing. For example, an Athletes' Village was built for the 1992 Olympic Games on an abandoned factory site. The combination of employment, housing and infrastructure was crucial for the regeneration of the city and for tackling deprivation.

Barcelona continues to develop, with the development of technical parks for high-tech industry contributing to the growth of Europe's so-called "sunrise belt"

URBAN CRIME

The majority of criminal activity is concentrated in the most urbanized and industrialized areas and where there is a concentration of pubs, clubs and retail outlets.

In the UK, the risk of being a victim of household crime is higher in more deprived areas. Rates of vandalism, burglary and vehicle-related theft are also higher in more deprived areas.

Common attributes of known offenders

Category	Indicator	Subgroup at risk
Demographic	Age	Young
	Sex	Male
	Marital status	Single
	Ethnic status	Minority group
	Family status	Broken home
Socio-economic	Family size	Large
	Income	Low
	Occupation	Unskilled
	Employment	Unemployed
Living conditions	Housing	Substandard
	Density	Overcrowded
	Tenure	Rented
	Permanence	Low

MANAGING URBAN CRIME

Crime can be tackled in a number of ways, including:
- having more police officers on patrol
- greater use of security cameras
- improved street lighting
- greater availability of taxi services around the closing time of clubs and bars.

❶

COMMON MISTAKE

✗ *All criminals are young, poor and live in overcrowded conditions.*

✓ Although most known offenders have the characteristics outlined in the table above, it does not mean that people with those characteristics will become offenders.

❓

CHECK YOUR UNDERSTANDING
23. Outline the conditions associated with deprivation.
24. Outline the main factors associated with known criminal offenders.

(a) By borough

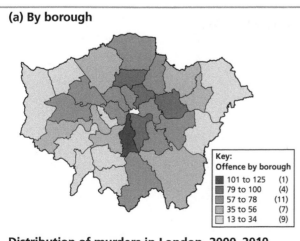

(b) by ward

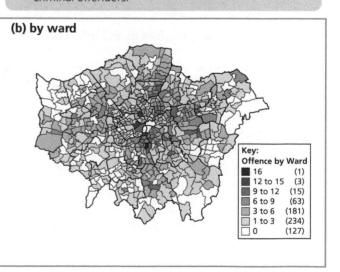

Distribution of murders in London, 2000–2010

Urban growth projections for 2050

GROWTH, 1950 TO 2050

In 1950, 30% of the world's population lived in urban areas. By 2014 this had risen to 54% and, by 2050, it could be 66%. Currently, North America is the most urbanized region, with 82% of the population living in urban areas.

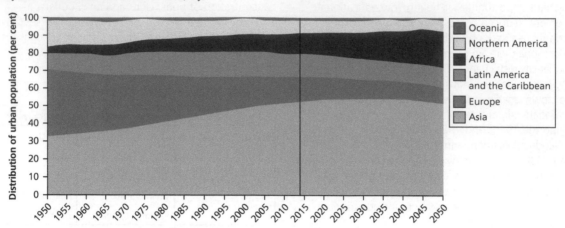

The origin/location of the world's urban population

The world's urban population has grown rapidly. By 2050, a further 2.5 billion people will live in urban areas. Asia contains the largest number of people living in urban areas, while India, China and Nigeria are expected to account for over 33% of the urban growth by 2050.

Currently, almost 50% of the world's urban population live in cities of fewer than 500,000, while about 12% live in megacities. Some cities in low-fertility countries in Europe and Asia have experienced decline. Cities of all sizes are growing, and megacities and large cities are growing slightly faster than medium-sized cities and smaller cities.

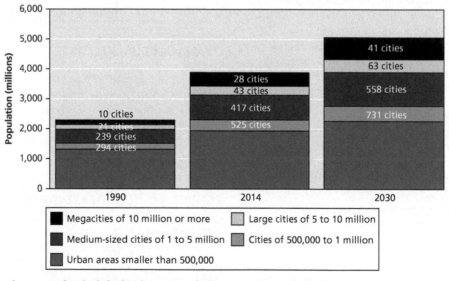

The growth of global urban populations in cities of all sizes

CHECK YOUR UNDERSTANDING

25. Identify the continent where the urban population is projected to grow most between 2015 and 2050.
26. Compare the growth of cities of different sizes between 2014 and 2030.

Resilient city design

Resilient cities are economically productive, socially inclusive and environmentally friendly. To function effectively, cities need a properly functioning transport network and an efficient energy, water and waste infrastructure. The high population densities associated with most cities increase their vulnerability to air and water pollution and contamination of land.

Large cities are often considered unsustainable because they consume huge amounts of resources and produce vast amounts of waste. Sustainable urban development meets the needs of the present generation without compromising the needs of future generations. The Rogers model (*Cities for a Small Planet*) compares a sustainable city with that of an unsustainable one. In the sustainable city, inputs are smaller and there is more recycling.

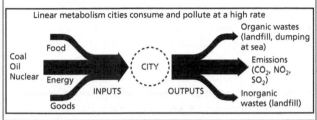

Linear metabolism cities consume and pollute at a high rate

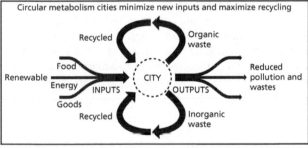

Circular metabolism cities minimize new inputs and maximize recycling

REDUCING ECOLOGICAL FOOTPRINTS

Cities can reduce their ecological footprint by burning less fossil fuel and increasing the number of resources that can be recycled, reused and reduced.

Densely populated cities have certain advantages over less dense cities: they tend to produce less CO_2 per capita and have greater potential for public transport. There are also savings in the heating and cooling of buildings. Emissions per person in apartments and terraced housing tend to be lower than emissions from a detached house. Some cities have been expanding their bus lanes, cycle lanes and footpaths, which should lead to a reduction in the amount of fossil fuel used. Bike-sharing schemes have been established in many cities.

In the USA, Portland and Seattle were the first cities to adopt sustainability plans. New York's CO_2 emissions are about one-third of the average in the USA. It expects the population to increase by 1 million by 2030, and is preparing for this by providing more affordable housing. It also plans to create more open space and clean up derelict sites.

MANAGING HAZARD RISK

In New York, following the flooding associated with Superstorm Sandy, electricity generators were removed from basements and transferred to the upper floors of buildings. In the Netherlands, tidal barriers protect polders (reclaimed land) but they are left open ordinarily to allow normal tidal flow and ecosystem functioning.

GEOPOLITICAL RISKS

Around much of the world, there has been rising nationalism, anti-globalisation sentiment, terrorist attacks, and anti-government protests. Many of these are expressed in urban areas, where there are significant concentrations of people. The World Economic Forum (2016) claims that geopolitical risks such as inter-state warfare, terrorist attacks and state collapse or crises are among the most likely risks to occur. Recent events in Sevastopol (Crimea), terrorist attacks in Paris, Brussels, Manchester and London and political change in the USA (linked to riots in Miami, Philadelphia and Atlanta) are examples of the geopolitical risks faced by cities. Interstate conflict is no longer physical but uses economic means and cyber warfare to attack people's privacy as well as national interests, many of which may be based in urban areas, for example, the alleged hacking of the USA by Russia during the 2016 presidential election.

CHECK YOUR UNDERSTANDING
27. Outline the ways in which New York City is attempting to manage climate change.
28. Outline the geopolitical risks that cities face.

Eco-city design

Compact cities minimize travel distances, use less space, require less infrastructure (pipes, cables, roads, and so on) and are easier to provide a public transport network for. This makes them more likely to be sustainable.

An eco-city or sustainable city is a city designed to have minimal environmental impact. To achieve sustainability, a number of options are available:

- reducing the use of fossil fuel
- keeping waste production to levels that can be treated locally
- providing sufficient green spaces
- reusing and reclaiming land
- conserving non-renewable resources
- using renewable resources.

THE BEDDINGTON ZERO ENERGY DEVELOPMENT (BEDZED)

BedZED, South London, is an environmentally friendly housing development built in 2000. The 99 homes and 1,405 square metres of workspace have a number of features. These include 777 m² of solar panels; south-facing houses that are triple-glazed and have high thermal insulation; building materials selected from renewable or recycled sources; and a location close to a tram line, a train line and bus routes.

The BedZED development has achieved a number of successes compared with UK averages:

- heating requirements are 88% less than the UK average
- hot-water consumption is reduced by 57%
- the electricity used is 25% less than the UK average, and 11% of it is produced by solar panels
- water consumption has been reduced by 50%.

However, the project cost around £15 million, this equates to approximately £150,000 per home, which is very expensive.

MASDAR CITY

Masdar City is a planned city project in the UAE. Work began in 2006 and will not be completed until around 2030, at a cost of about $22 billion. Up to 50,000 people will live there, at a cost of around $400,000 per resident. Eventually, it is hoped that 60,000 workers will commute daily to Masdar City, but by 2016 only 2,000 people were employed there.

The city is planned to be the world's most sustainable eco-city. It is powered by around 88,000 solar panels in a 20-hectare field. It is also connected to the public transport system: cars are not permitted in Masdar City. The International Renewable Energy Agency has its headquarters in Masdar City, and Siemens has a regional headquarters there too. Siemens' headquarters is believed to be the most energy-efficient building in Abu Dhabi.

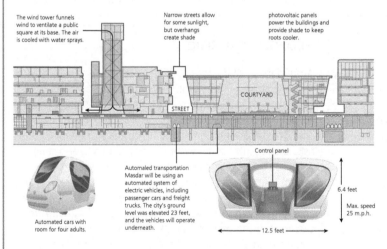

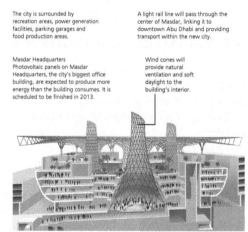

CHECK YOUR UNDERSTANDING

29. Outline the main advantages of eco-cities/sustainable cities, such as BedZed and Masdar City.
30. Outline the main disadvantages of eco-cities/sustainable cities, such as BedZed and Masdar City.

Smart cities

WHAT ARE SMART CITIES?

A city is defined as "smart" when investments in social and human capital, along with physical infrastructure and ICT, enable sustainable development and a high quality of life. Some smart cities are entirely new – as in the case of Songdo (South Korea) – whereas others have smart aspects added to them. The concept of the smart city is relatively new and is quite fashionable.

A smart city is a city that is performing well in six categories: economy, environment, people, living conditions, governance and mobility.

The success of urban development depends upon the physical infrastructure of cities (roads, railways, buildings, and so on) but also, increasingly, on the availability and quality of information communications technology and social infrastructure

Some see new smart cities as the solution to rising populations and dwindling resources. However, others believe that the majority of cities that will exist in the future already exist and so efforts should be directed at improving those cities rather than building new ones. The transformation of industrial waterfronts in New York and Oslo are examples of re-adaptation and upgrading of existing cities.

SONGDO INTERNATIONAL BUSINESS DISTRICT, SOUTH KOREA

Songdo in South Korea is a model for a brand-new smart city. It is a £23 billion project located on 600 hectares of reclaimed land, around 60 km from Seoul and about 10 km from Incheon International Airport. It was started in 2005, and it is home to 65,000 people and 300,000 workers. It has a range of sensors that control heating in houses and monitor traffic flow but also allow video conferencing and the delivery of health care, education and government services.

The city is the first LEED (Leadership in Energy and Environmental Design) certified district in South Korea. Over $10 billion has been invested in it, and it is expected that Songdo IBD will quickly become the central business hub in north-east Asia.

SONGDO CITY

The new Songdo City will be developed in three phases by 2020. The first phase, built by 2008, included the International Convention Centre, office buildings, deluxe hotels, shopping malls and a golf course. It also has a techno park and other research centres.

There is a wide range of smart technologies in use at Songdo, for example, video conferencing from every building. The city is compact and accessible and contains 25 km of bike lanes. Water conservation measures mean that commercial buildings use 30% less water than average. Smart meters measure energy consumption and there is micro-generation of wind power and photovoltaics. Plug-in hybrid electric vehicles can buy electricity when it is cheap and store it in batteries in the car. These schemes have enabled Songdo to reduce its ecological footprint.

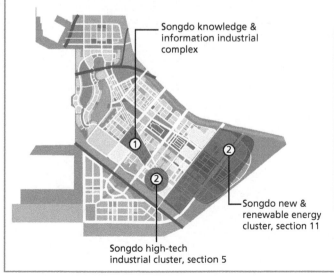

Songdo knowledge & information industrial complex

①

Songdo new & renewable energy cluster, section 11

Songdo high-tech industrial cluster, section 5

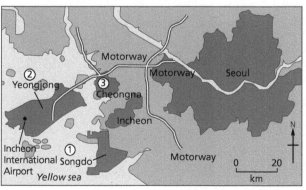

CHECK YOUR UNDERSTANDING
31. Comment on the accessibility of Songdo City.
32. Outline some of the characteristics of Songdo City that make it sustainable.

Exam practice

The diagram shows air quality in Mexico City.

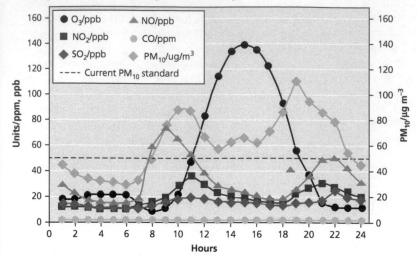

(a) (i)   Define air pollution.   (1 mark)
    (ii) Briefly describe daily variations in nitrogen oxide (NO) and nitrogen dioxide (NO$_2$). (3 marks)
    (iii) Outline the main sources of pollution in urban areas.   (2 marks)

(b) Briefly explain **two** ways in which urban pollution can be managed.   (2 + 2 marks)

(c) **Either**

Examine the view that there is no single solution to urban deprivation.   (10 marks)

Or

"The sustainable management of cities is an admirable goal, but ultimately unachievable."
Discuss the validity of this statement.   (10 marks)

1 POPULATION AND ECONOMIC DEVELOPMENT PATTERNS

Factors affecting population distribution at a global scale

Population distribution refers to where people live. On a global scale:

- most people live relatively close to the sea – some three-quarters of the world's population live within 1,000 km of the sea
- most people live on low ground – nearly 90% live in areas less than 500 m high
- over 80% live in the northern hemisphere.

The most favoured locations include:
- fertile river valleys
- places with a regular supply of water

- good communications and the potential for trade.

At a global scale, three major areas with high population density (over 200 people/km²) can be identified: South East Asia, north-east USA and Western Europe. Smaller concentrations include south-east Brazil, the Mexican plateau, the Nile Valley, California and Java.

Disadvantaged areas include those that are too dry, too steep, too cold and/or infertile. There is no such thing as a "best" climate; many people live in South East Asia and this has a monsoonal climate, with alternating seasons of flood and drought.

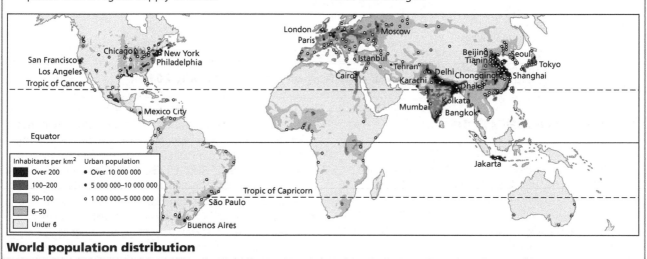

World population distribution

HUMAN FACTORS

Human factors are also important. For example:
- the distribution of raw materials, especially coal and other mineral resources, has led to large population clusters in parts of the USA, China, India, Germany and the UK
- government policy may lead to a redistribution of populations, such as in South Africa under the Apartheid policy or in Myanmar in relation to the Rohingya population
- new town policies, such as in Egypt, Hong Kong and the UK, may lead to a redistribution of population
- the influence of conflict and war may lead to some areas losing population, such as in Syria and northern Nigeria
- the migration of people (whether forced or voluntary) may lead to major redistributions of populations, for example, African migrants to South Africa.

LORENZ CURVES

Inequalities in population distribution can be shown by Lorenz Curves.

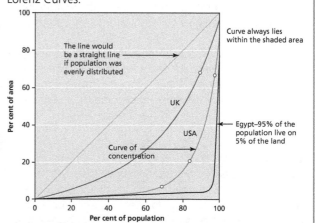

COMMON MISTAKE

✗ Distribution and density of population both refer to where people live.

✓ Distribution is where people live, whereas density is a measure of how many live there (per km²).

CHECK YOUR UNDERSTANDING

1. Define the term "population distribution".
2. Outline the most favourable conditions for human habitation.

Global patterns and classification of economic development

HICs, MICs AND LICs

According to the World Bank in 2014, people living in a high-income country (HIC) have an annual income of over $12,475. The term "HIC" is often used interchangeably with "developed" or "more developed". Low-income countries (LICs) are those where incomes are less than $1,025 per person per year, while that of middle-income countries (MICs) is between $1,025 and $12,475. A distinction can be made between lower middle-income countries ($1,025–$4,035) and upper middle-income countries ($4,036–$12,475). Some 5 billion people live in middle-income countries, and about one-third of global GDP is produced in MICs.

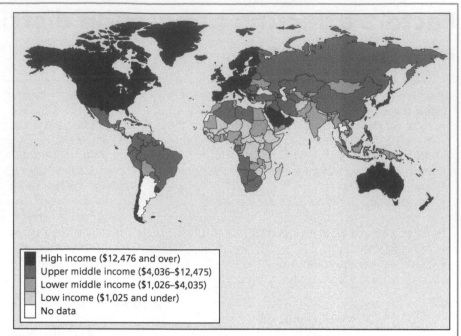

High income ($12,476 and over)
Upper middle income ($4,036–$12,475)
Lower middle income ($1,026–$4,035)
Low income ($1,025 and under)
No data

World Bank classification of countries, July 2016

ECONOMIC CLASSIFICATIONS

At its most basic, the world can be divided into rich and poor. There is evidence to suggest that the rich are getting richer and the poor relatively poorer, but such a dichotomous classification is simplistic. Nevertheless, it is still widely used in the media and by politicians and activists, and characterized by the use of the Brandt line or the North-South Divide.

A more detailed way of classifying countries is as follows:
- More economically developed countries (MEDC) such as Japan and Germany. These are the most developed countries and have a high standard of living. They are also called high-income countries (HICs).
- Newly industrializing countries (NICs) such as Malaysia and Taiwan. These are countries that have experienced rapid industrial, social and economic growth since 1960. Some NICs may develop and become HICs, such as

Singapore and South Korea. There are many categories of NICs, all showing rapid economic growth or potential for growth. These include
- BRICs – Brazil, Russia, India and China – who were joined by South Africa in 2010 to form the BRICS
- MINT – Mexico, Indonesia, Nigeria and Turkey
- Centrally planned economies (CPEs) such as North Korea. These are socialist countries under strict government control. Living standards are higher than in LEDCs, although personal freedom may be limited. Many former communist countries remain in this category.
- Oil-rich countries such as Saudi Arabia and the UAE are very rich in terms of income per head, although it may not be distributed very evenly.
- Less economically developed countries (LEDCs) such as Pakistan and Kenya are at a lower stage of development and have a lower quality of life. These are now called low-income countries (LICs).
- Least developed countries (LDCs) such as Afghanistan, Eritrea and Somalia have very low standards of living.

EXAM TIP
Most classifications of global economic groupings are related to levels of development, which is difficult to define. The categories should therefore be considered as a continuum, that is, there is a range of development from the very least developed to the most developed, rather than an either/or dichotomy. They refer to both economic and human development including economic growth, stable population growth, standards of living, and levels of technology, employment, health, nutrition, literacy, and GNI/capita.

CHECK YOUR UNDERSTANDING
3. Outline the main differences between LICs (LEDCs) and LDCs.
4. Outline the main difficulty in using classifications of development.

Population distribution and economic development at the national scale (1)

POPULATION DISTRIBUTION IN CHINA

China's population distribution is very uneven. It is mainly concentrated in the eastern part of the country, especially in coastal zones and the lower reaches of river valleys. Much of the rest of the country is characterized by desert and mountains. For example:

- More than 10% of the population live on less than 1% of the land.
- Half of the population live on less than one-tenth of the land.
- At the other end of the scale, less than one-twentieth of the population lives on half of the land in places such as Tibet and Inner Mongolia.

China's uneven population distribution is a result mainly of the country's physical geography. Only a small proportion of the country can provide for rain-fed agriculture. Moreover, the coastal and river locations are the more favoured sites for trade and commerce.

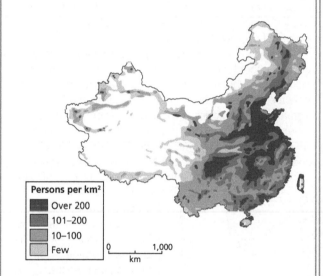

Persons per km²
- Over 200
- 101–200
- 10–100
- Few

Population density in China

One feature that characterizes population change in China is the increasing proportion of people living in megacities and millionaire cities. China now has more than 50 cities, each with over a million inhabitants, and it is predicted that there will be a further 350 million urban inhabitants by 2020. Another characteristic of this urbanization process is the creation of mega-regions, housing a number of megacities and millionaire cities.

INTERNAL MIGRATION IN CHINA

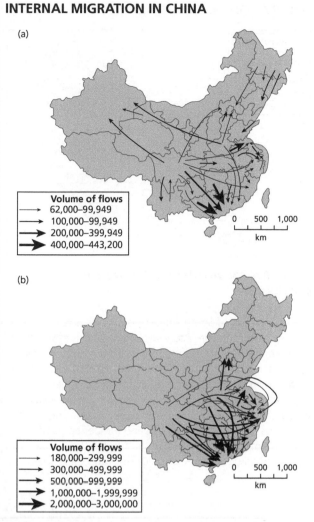

(a)

Volume of flows
- 62,000–99,949
- 100,000–99,949
- 200,000–399,949
- 400,000–443,200

(b)

Volume of flows
- 180,000–299,999
- 300,000–499,999
- 500,000–999,999
- 1,000,000–1,999,999
- 2,000,000–3,000,000

Migration flows in China, (a) 1990–95 and (b) 2000–2005
Source: Kam Wing Chan. 2013. *China, Internal Migration*, University of Washington

In recent decades, China has experienced the world's largest internal population movement. Some 160 million migrants have left rural areas to seek work in urban areas, mainly to work in better-paid jobs. Between 1990 and 2005, over 80 million people moved to the urban-industrial complexes on the coast. Internal migration has been good for the Chinese economy, and generally beneficial to the migrants, but at a personal and environmental cost. As migrants moved to the cities, land and labour costs have risen, and the Chinese government has attempted to direct recent industrial development, and the destination of internal migrants, to interior locations.

CHECK YOUR UNDERSTANDING
5. Suggest why only a small proportion of the Chinese population live in the western part of the country.
6. Comment on the main internal migration flows in China 1990–95 and 2000–05.

Population distribution and economic development at the national scale (2)

POPULATION DISTRIBUTION IN SOUTH AFRICA

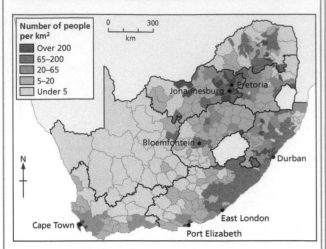

South Africa: population density and distribution

The distribution of South Africa's population is very uneven. Some parts of the core economic regions, such as Gauteng province, have population densities of over 1,000 people/km², whereas large areas of the periphery, such as Northern Cape Province have densities of less than 5 people/km². In general, high population densities are found in areas where there are good mineral resources, such as gold and diamonds around Johannesburg and Kimberley, good farming potential, for example, the Garden Route, and good trading potential, such as Durban and Cape Town.

INTERNAL AND INTERNATIONAL MIGRATION IN SOUTH AFRICA

The apartheid era (separate development for different racial groups) was characterized by the forced migration of many of the black population (over four million people) to the homelands, such as Transkei and KwaZulu.

However, since the end of the apartheid era, many blacks have migrated from the former homeland areas to large cities in search of work. South Africa's urban population grew from 55% in 1995 (the first post-apartheid census) to 65% in 2015. Most Africans fleeing violence or poverty stay on the continent. The African refugees heading to Europe received massive media coverage, but a far larger number sought security and prosperity elsewhere on their own continent. Of the world's 17 million displaced Africans, approximately 3% are in Europe; the majority remain in Africa.

New conflicts in Burundi, Libya, Niger and Nigeria, as well as unresolved crises in the Central African Republic, the Democratic Republic of the Congo and South Sudan have led to an overall increase of about 20% in the number of displaced people in Africa.

South Africa has attracted migrants from elsewhere in Africa for more than a century and the relative wealth and stability of the "Rainbow nation" still draws many migrants. There are believed to be between 1.5 million and 3.2 million migrants in South Africa, two-thirds of whom come from elsewhere on the continent, mainly neighbouring countries but also from as far away as Morocco or Eritrea. Most head towards Johannesburg. Neighbourhoods such as Mayfair in Johannesburg are almost entirely populated by east Africans.

There have been repeated outbreaks of violence towards migrants. Such attacks are fuelled by chronic shortage of jobs, housing and services.

South Africa has the highest number of pending asylum claims in the world, with over a million people waiting to be processed. Few believe that the flow of migrants to South Africa will diminish significantly soon, particularly as European resistance to accepting new migrants increases.

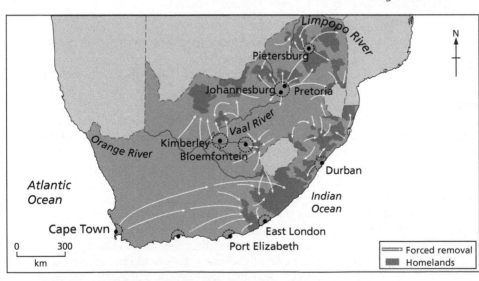

Forced migration within South Africa during the apartheid era

CHECK YOUR UNDERSTANDING
7. Describe the main features of the distribution of population in South Africa.
8. Describe the main pattern of internal migration in South Africa during the apartheid era.

Population change and demographic transition over time (1)

THE DEMOGRAPHIC TRANSITION MODEL (DTM)

Population change includes changes in birth rates, death rates and migration. The general demographic transition model shows changes in birth rates and death rates over time.

Stage 2
Early expanding:
- birth rates remain high but the death rate comes down rapidly
- population growth is rapid
- Afghanistan, Sudan and Libya are at this stage
- UK passed through this stage by 1850

Stage 3
Late expanding:
- birth rate drops and the death rate remains low
- population growth continues but at a smaller rate
- Brazil and Argentina are at this stage
- UK passed through this stage in 1950

Stage 4
Low and variable:
- birth rates and death rates are low and variable
- population growth fluctuates
- UK and most developed countries are at this stage

Stage 1
High and variable:
- birth rates and death rates are high and variable
- population growth fluctuates
- no countries, only some indigenous (primitive) tribes still at this stage
- UK at this stage until about 1750

High birth and death rates
Parents want children:
- for labour
- to look after them in old age
- to continue the family name
- prestige
- to replace other children who have died

People die from:
- lack of clean water
- lack of food
- poor hygiene and sanitation
- overcrowding
- contagious diseases
- poverty

Stage 5
Low declining:
- the birth rate is lower than the death rate
- the population declines
- Japan is at this stage

Low birth and death rates
Birth rates decline because:
- children are very costly
- the government looks after people through pensions and health services
- more women want their own career
- there is more widespread use of family planning
- as the infant mortality rate comes down there is less need for replacement children

Death rates decline because of:
- clean water
- reliable food supply
- good hygiene and sanitation
- lower population densities
- better vacations and healthcare
- rising standards of living

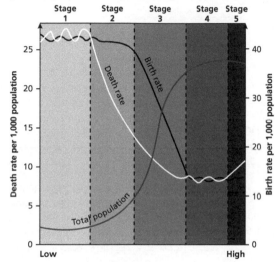

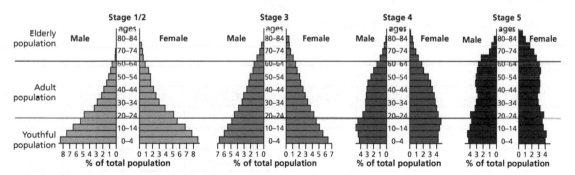

The demographic transition model

The DTM indicates that death rates fall before birth rates, and that the total population increases. Eventually birth and death rates fall to a low level, although the death rate increases as the population ages. However, the DTM is based on the data from just three countries – England, Wales and Sweden. Not only is the timescale for the DTM in these countries longer than in many LICs, but there are other types of DTM.

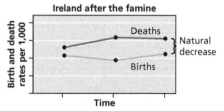

Alternative demographic transitions

Ireland's DTM shows falling birth rates and rising death rates due to emigration following the 1845–9 famine.

CHECK YOUR UNDERSTANDING

9. Describe the changes in the death rate that occur in the demographic transition model.
10. Explain the differences in Ireland's demographic transition model compared with the standard demographic transition.

Population change and demographic transition over time (2)

NATURAL CHANGE

Natural increase can be calculated by subtracting the crude death rate from the crude birth rate. It is expressed as a percentage. Natural decrease occurs when the death rate exceeds the birth rate. Natural change (increase or decrease) does not take into account migration gains or losses.

TOTAL FERTILITY RATE

The **total fertility rate (TFR)** is the average number of births per thousand women of childbearing age. It is the completed family size if fertility rates remain constant. In general, LICs have the highest fertility rates. Most HICs, by contrast, have much lower fertility rates.

Total fertility rate is influenced by many factors including the status of women, level of education and material ambition, religion, the health of the mother, economic prosperity, the need for children and social and cultural pressure.

LIFE EXPECTANCY

Life expectancy is the average number of years that a person can be expected to live, usually from birth, assuming that demographic factors remain unchanged.

Life expectancy varies – from over 80 years in some HICs to between 50 and 55 years for LDCs. Life expectancy in many sub-Saharan countries is declining due to a combination of poverty, conflict and the AIDS virus. In the 40 countries with the lowest life expectancy, only two are not in sub-Saharan Africa, namely Haiti and Afghanistan.

Life expectancy is often higher for women than for men due in part to a higher retirement age for men, heavy physical labour for men, greater likelihood of involvement in conflict, and more "self-destructive" lifestyles such as smoking and/or drinking to excess.

DEPENDENCY RATIO

The **dependency ratio** relates the working population to the dependent population. It is worked out by a formula:

$$\frac{\text{Number of Dependents}}{\text{Population (Ages 15–64)}} \times 100\%$$

where dependents are the population aged <15 + population aged >64

It is a very crude measure – for example, many people stay on at school after the age of 15 and many people work after the age of 64 – but it provides a simple measure for comparing countries or tracking changes over time.
- In HICs there is a high proportion of elderly.
- In LICs there is a high proportion of youth.

Population indicators

	China	India	Cuba	DR Congo	USA	UK	Germany
Population (millions)	1,367	1,251	11	79	321	64	81
Natural increase (%)	0.5	1.5	0.2	3.0	0.4	0.3	−0.2
Life expectancy (years)	75.4	68.1	78.7	56.9	79.7	80.5	80.6
Total fertility rate	1.6	2.48	1.47	4.66	1.87	1.89	1.44
Age structure							
0–14 years	17	28	16	43	19	17	13
15–64 years	73	66	70	54	66	65	65
≥65 years	10	6	14	3	15	18	22
Dependency ratio (per 100 workers)	37	52	43	85	52	54	54

COMMON MISTAKE

✗ *A longer life expectancy equates with a greater standard of living/quality of life.*

✓ A longer life expectancy does not necessarily mean a better standard of living. Many of the elderly have disabilities such as Alzheimer's disease, or physical incapacities such as arthritis; many also experience loneliness following the death of their partner.

EXAM TIP

Although it is important to support your points with data, do not worry about the exact figures – they will have changed by the time you sit your exams – but make sure you have a figure that is relatively close to the current data.

CHECK YOUR UNDERSTANDING

11. Identify the indicator shown in the table that is not a measure of level of development.

12. Briefly explain why the dependency ratio is limited as a demographic indicator.

Population change and demographic transition over time (3)

AGE/SEX PYRAMIDS

Population structure or composition refers to any measurable characteristic of a population such as age, sex and ethnicity of the population. Population pyramids are used to show these characteristics. Population pyramids tell us a great deal about the age and sex structure of a population:
- A wide base indicates a high birth rate.
- A narrowing base suggests a falling birth rate.

- Near-vertical sides indicate low death rates.
- Concave slopes suggest high death rates.
- Bulges in the slope suggest high rates of in-migration (for instance, excess males 20–35 years will be economic migrants looking for work).
- "Slices" in the slope indicate emigration or out-migration or age-specific or sex-specific deaths (epidemics, war).

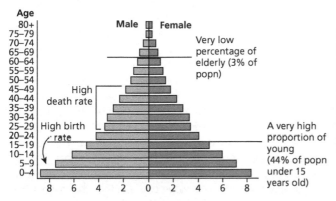

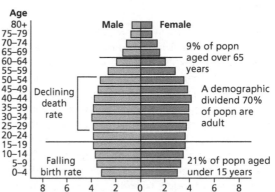

Population pyramids – Thailand from pyramid to pillar
Source: Adapted from http://www.prb.org/Publications/Articles/2013/population-pyramids.aspx

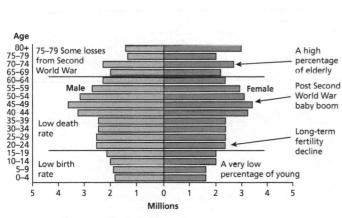

Germany, an ageing HIC
Source: Adapted from http://www.prb.org/Publications/Articles/2013/population-pyramids.aspx

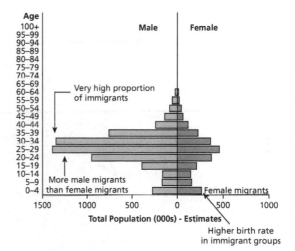

UAE, a country of immigration
Source: US Census Bureau

CHECK YOUR UNDERSTANDING
13. Describe the main changes in Thailand's population pyramid between 1970 and 2010.
14. Compare Germany's population pyramid with that of the UAE.

The consequences of megacity growth for individuals and societies

The rapid growth of the number of megacities around the world has been one of the most important geographical phenomena of the late 20th and early 21st centuries. For individuals and families, megacities offer the prospects of a job, a home and an opportunity to improve their standard of living. For others, migration to megacities may result in unemployment, poor-quality housing and deprivation. For governments, it is easier to provide housing and health care to large numbers of people living close together. However, if there are too many people, the provision of services may be inadequate. Megacity growth is associated with urban sprawl, increased traffic congestion and declining air and water quality. Nevertheless, the perception of megacities as a place of potential economic benefit for migrants fuels their growth.

CASE STUDY

MUMBAI

Mumbai is India's largest city, with a population of about 18 million. Until the 1970s, Mumbai's economy was largely based on textiles and imports/exports through the port. Since then, it has diversified and includes such industries as aerospace, engineering, computers and electronic equipment. It is also the financial, commercial and entertainment centre of India. It accounts for 25% of India's industrial output and 40% of its foreign trade. Many TNCs are based in Mumbai, such as the Tata Group, and it is home to the Bombay Stock Exchange. It contains many of India's scientific and nuclear industries. It is also the centre of the Bollywood film industry. Mumbai has more millionaires and billionaires than any other Indian city but is also home to millions of impoverished people.

According to the 2011 Census, 75% of Mumbai's population lived in slums. The gender ratio was 838 females per 1,000 males, partly reflecting the greater incidence of male migration to the city. Per capita income in Mumbai is about three times the national average.

Mumbai experiences many of the problems resulting from megacity growth – poverty, unemployment, limited access to health care and education, and poor sanitation and access to electricity. Residents living in slums also have limited security of tenure. Dharavi is an area of about 2 km², and home to up to 1 million people. Due to its proximity to Mumbai's financial and commercial district, there is great pressure to clear parts of Dharavi for modern developments. On the other hand, Dharavi has many informal activities that provide a livelihood for many of its residents. Up to 85% of Dharavi adults work locally, and there are major recycling industries. However, the working conditions for the recycling industry can be very dangerous.

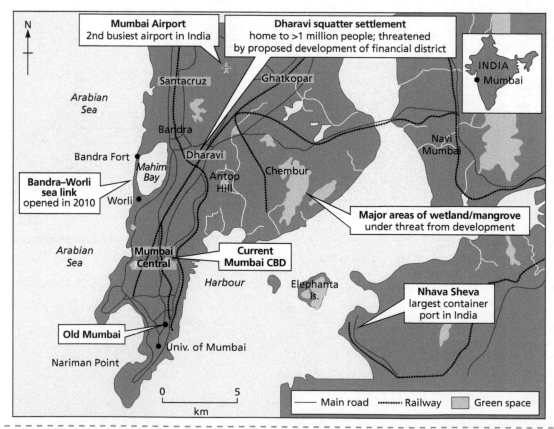

CHECK YOUR UNDERSTANDING

15. Outline the potential advantages of megacities for populations.

16. Outline the potential disadvantages of megacity growth for populations.

The causes and consequences of forced migration

DEFINITION

Forced migration relates to the movement of refugees and internally displaced people (those displaced by conflicts) as well as people displaced by natural or environmental disasters, chemical or nuclear disasters, famine or development projects.

TYPES OF FORCED MIGRATION

- Conflict-induced displacement includes people who are forced to move due to armed conflict such as civil war, violence or persecution.
- Development-induced displacement includes people forced to move due to large-scale infrastructure projects such as dams.
- Disaster-induced displacement includes natural disasters such as volcanoes and hurricanes, and human-induced disasters such as releases of radiation and chemicals.

TYPES OF FORCED MIGRANT

- A refugee is a "person residing outside his or her country of nationality, who is unable or unwilling to return because of a well-founded fear of persecution due to race, religion, nationality, membership in a political social group or political opinion" (International Association for the Study of Forced Migration).
- Asylum seekers are people who have left their country of origin in search of protection in another country, but whose claim for refugee status has not been decided.
- Internally displaced persons (IDPs) are groups of people who have been "forced to flee their home, due to armed conflict, internal strife, systematic violations of human rights or natural or man-made disasters, and who are still living within their own country" (International Association for the Study of Forced Migration).

CASE STUDY

FORCED MIGRATION IN AND FROM SYRIA

In Syria, the repression of the ruling government and the emergence of the fundamentalist Islamic group Isis have led to the displacement of over 10 million Syrians. Syrians near Damascus feared that the government were using the Syrian army to kill their own people, whereas in other parts of Syria, people feared they would be killed, captured or forced to live under the harsh Islamic rule of Isis. There are over 4 million IDPs in Syria and 4.5 million Syrian refugees in five other neighbouring countries: Turkey, Lebanon, Iraq, Egypt and Jordan. The Zaatari refugee camp in Jordan, just 8 km from the Syrian border, is the currently the world's second-largest refugee camp.

The impact of Syrian refugees on Lebanon

There are nearly 1.5 million Syrian refugees in Lebanon, about half of whom are children. Most live in poverty and are dependent on aid for survival. Lebanon has the highest number of refugees per capita in the world and has suffered severe economic and environmental consequences. These include strain on the health and education services, electricity, water and sanitation systems.

CHECK YOUR UNDERSTANDING

17. Distinguish between refugees, IDPs and asylum seekers.
18. Compare the impacts of forced migration in Syria with internal displacement in Nigeria.

CASE STUDY

FORCED MIGRATION IN NIGERIA

In 2014 the terrorist group Boko Haram kidnapped 276 schoolgirls from northern Nigeria. The violence associated with Boko Haram has caused many people to flee the region. Over 250,000 have been displaced by Boko Haram. This includes some 60% of the region's farmers, thus people in the region have been unable to feed themselves as less land is being farmed and less produce harvested. Over a third of health care facilities have been closed, and health workers have been abducted and killed. People in the region lack access to fresh water and sanitation. Homes, services and infrastructure have been extensively damaged.

In 2017, 82 of the Chibok schoolgirls held captive for more than three years were released in exchange for five militant leaders. However, there are still more than 100 schoolgirls held by Boko Haram. Some of the girls were forced to carry bombs to busy areas and explode them, killing themselves and hundreds of other civilians.

Thousands of women and other girls were seized by Boko Haram but the Chibok girls gained international recognition when the hashtag #BringBackOurGirls was promoted by Michelle Obama and other celebrities on social media. The unintended consequence of the girls' fame was that their value to Boko Haram immediately increased. The NGO Amnesty International claimed that the girls did not "deserve to put through a publicity stunt". Some of the abducted women were subjected to torture, rape, marriage, and indoctrination and were forced to fight for Boko Haram.

Ageing populations

OLDER DEPENDENCY RATIO

An ageing population is one with an increasing number of elderly people. The **older dependency ratio**, or ODR, relates the number of working-age people to the older population that they support. It varies widely, from just 6 in Kenya to 33 in Italy and Japan. Countries with a high ODR have to fund retirement and health care for their older population. In France, men live, on average, for 21 years after retirement, and women for 26 years.

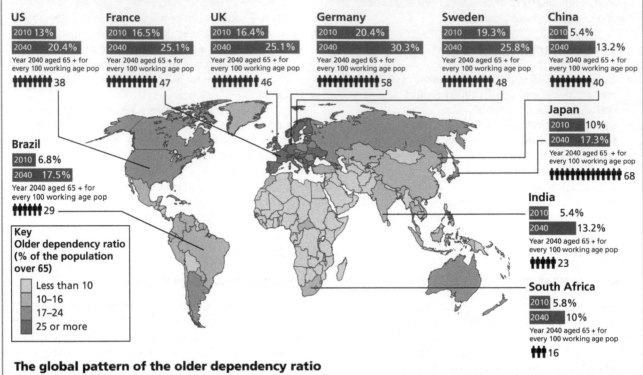

US
| 2010 | 13% |
| 2040 | 20.4% |
Year 2040 aged 65 + for every 100 working age pop
38

France
| 2010 | 16.5% |
| 2040 | 25.1% |
Year 2040 aged 65 + for every 100 working age pop
47

UK
| 2010 | 16.4% |
| 2040 | 25.1% |
Year 2040 aged 65 + for every 100 working age pop
46

Germany
| 2010 | 20.4% |
| 2040 | 30.3% |
Year 2040 aged 65 + for every 100 working age pop
58

Sweden
| 2010 | 19.3% |
| 2040 | 25.8% |
Year 2040 aged 65 + for every 100 working age pop
48

China
| 2010 | 5.4% |
| 2040 | 13.2% |
Year 2040 aged 65 + for every 100 working age pop
40

Japan
| 2010 | 10% |
| 2040 | 17.3% |
Year 2040 aged 65 + for every 100 working age pop
68

Brazil
| 2010 | 6.8% |
| 2040 | 17.5% |
Year 2040 aged 65 + for every 100 working age pop
29

India
| 2010 | 5.4% |
| 2040 | 13.2% |
Year 2040 aged 65 + for every 100 working age pop
23

South Africa
| 2010 | 5.8% |
| 2040 | 10% |
Year 2040 aged 65 + for every 100 working age pop
16

Key
Older dependency ratio (% of the population over 65)
- Less than 10
- 10–16
- 17–24
- 25 or more

The global pattern of the older dependency ratio

SEX RATIO IN INDIA

Selective abortion has accelerated in a globalizing India. Wealthier and better-educated Indians still want sons: a survey revealed that female foeticide was highest among women with a university degree. The urban middle classes can afford the ultrasound tests to determine the sex of the foetus. In 2003–05 there were just 880 girls born for every 1000 boys. Tamil Nadu has the highest number of girls per 1000 boys at c. 950 whereas Punjab has the least, about 800.

The shortage of women has had negative social effects: unmarried young men have turned to crime, and violence against women has increased.

Changes in average household size

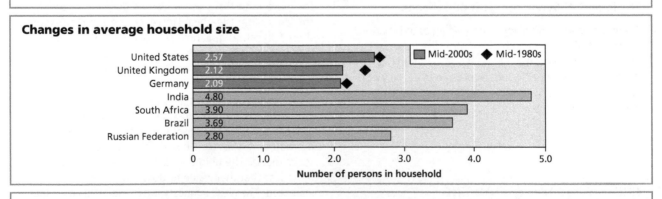

| | Mid-2000s | Mid-1980s |

- United States 2.57
- United Kingdom 2.12
- Germany 2.09
- India 4.80
- South Africa 3.90
- Brazil 3.69
- Russian Federation 2.80

Number of persons in household

In China, average family size has fallen rapidly as the country has modernized. In 1950 the average family size was 5.3, and this fell to 3.96 in 1990 and 3.10 in 2010 and 3.02 in 2015.

CHECK YOUR UNDERSTANDING

19. Identify the country that is predicted to see the biggest increase in the proportion of the elderly population between 2010 and 2040.

20. Suggest why average household size in some HICs is approaching 2.

Ageing populations

An ageing population has certain advantages. The elderly may have skills (including social skills) and training, and some employers, especially supermarkets and hardware/furniture stores, prefer them to younger workers. The elderly may look after their grandchildren and therefore allow both parents to work. This is important in Japan and in South Africa, where a "granny culture" occurs in many areas. In HICs, the elderly are often viewed as an important market – the "grey economy" – and many firms, ranging from holiday companies to health care providers, have developed to target this market.

CASE STUDY

JAPAN'S AGEING POPULATION

Since 1945 the age structure of Japan's population has greatly changed, largely due to a decrease in both birth and death rates. The population is ageing much more rapidly than that of other countries (although some European countries, such as Italy and Greece, are not far behind). The number of elderly people living alone in Japan increased from 0.8 million in 1975 to over 2.5 million in 2000. Since 1975 the percentage of young people has gradually declined, and by 2015 they accounted for only about 13% of the population.

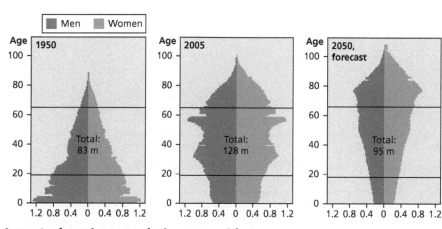

Japan's changing population pyramids

At present, 27% of the Japanese population are aged over 65 years. This is creating a huge burden on pension funds and social welfare programmes, especially health care. Problems include:
- inadequate nursing facilities
- depletion of the labour force
- deterioration of the economy
- a trade deficit
- migration of Japanese industry to other countries
- the high cost of funding pensions and health care
- falling demand for schools and teachers
- new jobs needed for the elderly
- new leisure facilities needed for the elderly
- an increase in the burden on the working population to serve the dependent population
- reduced demand for goods from the smaller working population
- a need for in-migration to fuel any increase in the workforce.

Options for the government include raising taxes, raising the retirement age, cutting back on social welfare programmes, and increasing care in people's homes, for example.

EXAM TIP ✓
Use data from https://www.populationpyramid.net/Japan/1950, https://www.populationpyramid.net/Japan/2005 and https://www.populationpyramid.net/Japan/2050 to calculate the approximate percentage of <15 year olds and >65 years old in Japan's population for these three dates. These data may be useful for you to quote in your exams.

COMMON MISTAKE ❗
✗ *An ageing population is an economic asset, as they have a large disposable income*
✓ *Not all of Japan's elderly are either healthy or wealthy. Depending on their levels of health and wealth, elderly people can be a major boost to a household or an economy, or they may be a burden.*

CHECK YOUR UNDERSTANDING
21. Describe the changes in Japan's population as shown in its population pyramids for 1950, 2005 and 2050.
22. Outline the challenges associated with Japan's ageing population.

Pro-natalist and anti-natalist policies (1)

Governments attempt to control population size in many ways. Their strategies will depend on whether the country wishes to increase its population size (**pro-natalist**) or limit it (**anti-natalist**). Family planning methods include contraceptives, forced sterilization and abortion.

CASE STUDY

CHINA'S ONE-CHILD POLICY

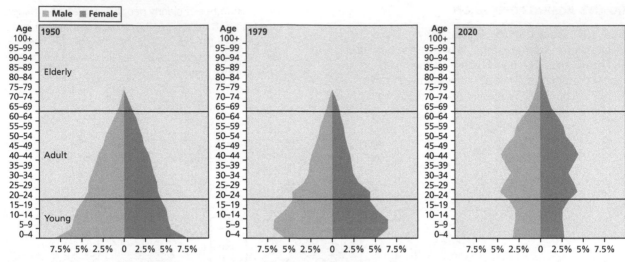

China's changing population structure (a) 1950 (b) 1979 (c) 2020

China's one-child policy was one of the world's most severe and controversial family planning programmes. The policy, imposed in 1979, is estimated to have reduced population growth in a country of 1.3 billion by as much as 300 million people over its first 20 years alone, and prevented as many as 400 million births. The Chinese government has predicted that the population will peak at 1.5 billion in 2033.

The policy has resulted in a ratio of 118 male to every 100 female births, above the global norm of 103–107 boys to every 100 girls. This reflects the fact that many people in China value baby girls less highly than baby boys, and millions of females were aborted or died due to neglect, abandonment or even infanticide.

The one-child policy was restricted to ethnic Han Chinese living in urban areas. While the growing middle classes in urban areas do not discriminate against daughters as much, people in the countryside remain traditionally focused on male heirs, and in most provincial rural areas couples could have two children if the first one was a girl.

The problems related to a one-child quota include a shrinking labour force and an ageing population. By 2050 China will have nearly 450 million people aged over 60; there will be just 1.6 working adults to support every person aged over 60 compared with more than seven in the 1970s.

Reform of the one-child policy

The prospect of an ageing society in which one worker is left to support two parents and four grandparents has led China to relax the one-child policy. In 2015, amid fears that the size of the working population was set to decline, the Chinese government changed the policy to allow couples to have two children.

Party conservatives still fear two things about loosening population controls: the population will grow beyond the country's capacity to feed itself, and relaxing rules too quickly will lead to a baby boom that will strain public services.

Factors other than the one-child policy have also encouraged couples to limit the number of children that they have. Education and housing costs as well as a lack of social security support make more than one child cost prohibitive for many.

CHECK YOUR UNDERSTANDING

23. Describe the changes in China's population structure between 1950 and 2020.
24. Outline the arguments against the one-child policy.

Pro-natalist and anti-natalist policies (2)

PRO-NATALIST POLICIES IN RUSSIA

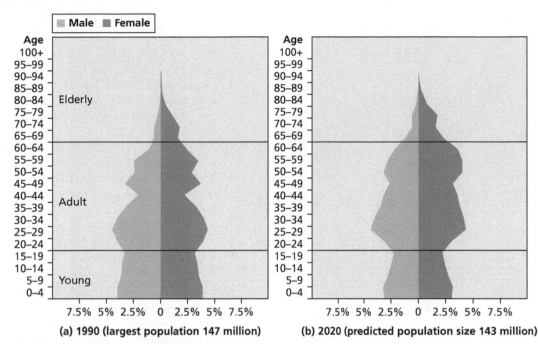

Russia's population structure

Low fertility has characterized Russia for many decades. There are many factors that help to explain this, such as poor reproductive health care services, a relative lack of modern contraceptives, widespread and unsafe abortions, high divorce rates, an ageing population structure, infertility and women choosing to have fewer children.

The Soviet Union and the Russian Federation have a long history of pro-natalist policies. These started in the 1930s, when families were rewarded for having a third or fourth child, and then later for having two children. The government even imposed a tax on childlessness between 1941 and 1990.

In the 1970s, fertility rates were below replacement level, which led to renewed pro-natalist policies in 1981. However, although the 1981 policy reduced the childbearing age it did not increase the number of children born.

By 2006, fertility had dropped to less than 1.3 births per woman, and further measures to increase the birth rate were introduced. These included:
- an increase in child benefits based on the number of children a family had
- increased parental leave
- increased payments to mothers of second and third children.

The government was also considering re-introducing the tax on childlessness.

Initially, these measures had a positive impact. Between 2006 and 2011, fertility increased by 21%. The increase in second births was 40% and in third children 60%. However, these increases were from a very low starting point. The effects of the policies wore off over the years. After five years of the policy, women's intentions to have another child had hardly changed.

In May 2012, the Russian government announced that it wanted Russia to have a TFR of 1.75 by 2018. Russia would have to double the state's financial support if this were to have any chance of success. Others believe that Russia will need to increase the immigrant population if it is to increase the fertility rate, but two-thirds of Russians want the number of immigrants reduced, not increased.

It is possible that Russia's population in 2100 will be similar to that of 1950: 100 million. As the population ages, the workforce is predicted to decline by about 15% as early as 2024. This could be reduced by increasing the retirement age.

CHECK YOUR UNDERSTANDING
25. Outline the reasons for low fertility in Russia.
26. Outline the measures that have been used in Russia to increase the birth rate.

Pro-natalist and anti-natalist policies (3)

GENDER EQUALITY ISSUES IN KERALA, INDIA

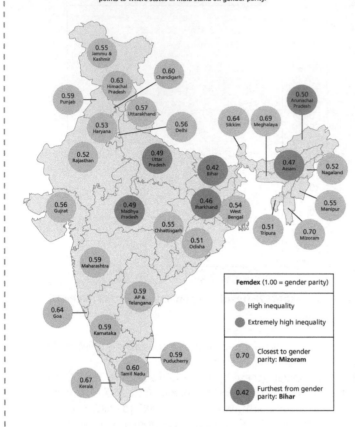

McKinsey Global Institute's Female Empowerment Index (Femdex) points to where states in India stand on gender parity.[1]

- Jobs have been open to women in health and education since the early 20th century.
- Women have autonomy in their personal life.

Statistics indicate that women in Kerala, are among the most "well-off" women in India. For example:

- Kerala has the highest female:male sex ratio in India, with 1,084 females to 1,000 males, compared with the national average of 940:1,000.
- Female literacy in Kerala increased from 86% to 92% between 1991 and 2011.
- Kerala has a low infant mortality rate of 13 per thousand (‰), compared with the national average of 80‰.
- Kerala has high life expectancy of 74 years for females and 70 years for males.
- The TFR in Kerala is low, just 1.9, below the replacement level.

Nevertheless, women are marginalized in the economic process and have a lack of control over resources. Women are concentrated in low-paid jobs such as farming, domestic services and informal economies.

Women's share in the labour market in Kerala (18%) is among the lowest in India. The reasons could be related to their role in looking after the household, collecting water and fuelwood, and looking after elderly relatives. While 18% of educated men are unemployed, 71% of educated women are unemployed. For those in work, a higher proportion of women have low-paid jobs and are more likely to be in the informal sector.

Many self-help groups (SHG) have been established to empower women from poor households. They also have micro-finance systems that all the participants pay into, and receive money from.

Nevertheless, life remains difficult for many women in Kerala. Although there is good progress in social development (health and education), women's role in decision-making is limited. The legal system, for example, is male dominated. Even in education, women account for 71% of primary school teachers but only 36% of university lecturers. Violence and sexual harassment against women is still common, much of it related to their husbands' alcohol consumption.

Overall, given the high educational level of women, their good health, and numerous government programmes, women cannot be considered completely equal to men. Although there have been some improvements in conditions for women, there is still a lot to do.

Kerala is the most densely populated area of India with an even spread of population and no large cities. It has had huge success in improving health, literacy and education, and in bringing down the birth rate. Its success is even more remarkable because it has a much lower per capita income than the rest of India. Many factors help to account for Kerala's success. These include:

- political stability
- social reform
- the status of women in society.

The role of women in Kerala

The status of women in Kerala has been extremely important in the region's development.

- There is a tradition of female employment, and girls are educated to the same standard as boys.
- There is open access to universities and colleges, and women often study to be doctors and nurses.

27. Outline the ways in which women in Kerala have equal or better opportunities than men there.

28. Outline the main areas in which women's rights in Kerala could be improved.

Trafficking and anti-trafficking policies

DEFINITION

The USA's Trafficking and Violence Protection Act (TVPA) defines "severe forms of trafficking in persons" as:
- sex trafficking in which a commercial sex act is induced by force, fraud, or coercion, or in which the person induced to perform such an act has not attained 18 years of age
- the recruitment, harbouring, transportation, provision, or obtaining of a person for labour or services, through the use of force, fraud, or coercion for the purpose of subjection to involuntary servitude, peonage, debt bondage, or slavery.

A victim need not be physically transported from one location to another for the crime to fall within these definitions.

GEOGRAPHIC PATTERNS

Human trafficking occurs in many countries and is often a transborder feature. In the past, trafficking may have been more likely between two nations, whereas now it is more likely to be a multinational feature. In 2015, anti-trafficking measures were incorporated into three of the 17 UN Sustainable Development Goals. The USA publishes a Trafficking in Persons Report annually and the United Nations has produced two reports on global trafficking. The USA's Trafficking and Violence Protection Act (TVPA 2000) encourages governments to join in the fight against human trafficking in three main components:
- Protection – increased efforts to protect all victims of trafficking.
- Prosecution – of traffickers.
- Prevention – to assist other governments to reduce trafficking.

POPULATIONS AT RISK OF TRAFFICKING

Certain populations are at an increased risk of trafficking. These include refugees and migrants; the stateless; lesbian, gay, bisexual, transgender and intersex (LGBTI) people; religious minorities; and people with disabilities.

CASE STUDY

TRAFFICKING OF NIGERIAN WOMEN TO EUROPE

Many women are trafficked between Nigeria and Western Europe for the purpose of sexual exploitation. It is one of Europe's more established trafficking flows, with Nigerian women frequently accounting for over 10% of the trafficked people. UN evidence suggests that corruption and blackmail are rife, and the prostitution gangs may also be engaged in the European drug trade.

ANTI-TRAFFICKING POLICIES

Attempts to reduce trafficking include increasing public awareness and designing policies to prevent trafficking. Governments can start by having up-to-date registration of births and migration into an area.

It is likely that no single measure, nor any single country, can end human trafficking. Approximately 90% of countries have now become parties to the UN 2003 Protocol to Prevent, Support and Punish Trafficking in Persons, Especially Women and Children. In 2003, around 65% of countries lacked specific offences that criminalized trafficking, but by 2006 this had fallen to just over 25%.

Overall, some two billion people live in areas where trafficking is still not criminalized.

Eight countries in Africa and the Middle East, some containing large populations, lack anti-trafficking legislation, so the number of people unprotected is large.

THE CRIMINAL JUSTICE RESPONSE

According to the UN, the number of convictions for trafficking remains very low – of the 128 countries covered in the latest UN report, 15% did not record a single conviction.

Many governments and NGOs believe that confiscating the proceeds of crime is appropriate and effective as a punishment and a deterrent. It also disrupts criminal activity by cutting off some funding, creates an image that crime does not pay, and helps win over public support. However, support for the victims is very limited.

CHECK YOUR UNDERSTANDING
29. Outline the main population groups at risk of trafficking.
30. Identify the geographic pattern of human trafficking.

The demographic dividend

The **demographic dividend** refers to the increase in the proportion of adults in a population. It occurs when fertility rates decline, but the adult population increases due to the large number of children reaching adulthood.

It has the potential to enable faster economic growth. As families recognize that fewer of their children will die during infancy or childhood, they begin to have fewer children.

THE BENEFITS OF THE DEMOGRAPHIC DIVIDEND

The demographic dividend may produce many benefits. First there is an increased labour supply. However, this depends on the ability of the economy to employ the extra workers. Then there is an increase in savings. As the number of dependents decreases, individuals can save more. Next, the decrease in fertility rates result in healthier women and fewer social and economic pressures at home. This allows parents to invest more resources per child, leading to better health and educational outcomes. Finally, the increasing domestic demand brought about by the increasing incomes per capita and the decreasing dependency ratio leads to greater consumer spending and economic growth.

AFTER THE DEMOGRAPHIC DIVIDEND

After the demographic dividend, the dependency ratio increases again. The population cohort that created the demographic dividend grows old and retires. With a disproportionate number of old people relying upon a smaller generation following behind them, the demographic dividend may become a liability. This is currently seen in Japan, with younger generations having a major economic burden to look after the older generation.

CASE STUDY

SOUTH KOREA'S DEMOGRAPHIC DIVIDEND

South Korea made a rapid transition from high to low fertility between 1960 and 1990, and this enabled economic growth. South Korea's success was due to addressing population issues while also investing in education and economic policies to create infrastructure and manufacturing. Korea invested in health centres to provide a range of services, including family planning. The government set a target of 45% of married couples using family planning. Between 1950 and 1975, fertility dropped from 5.4 children per woman to 2.9. By 2016, fertility had dropped to 1.25 children per woman.

Shifting the education strategy

The government also improved education. Between the 1950s and 1960s, South Korea's education strategy shifted to universal schooling and education, which provided people with the knowledge and skills they needed to achieve economic development.

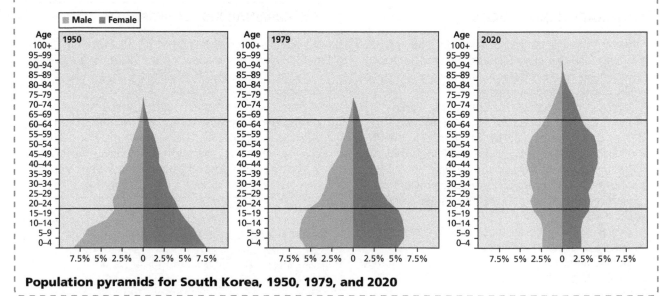

Population pyramids for South Korea, 1950, 1979, and 2020

CHECK YOUR UNDERSTANDING

31. Describe the changes in South Korea's population structure between 1950 and 2020.

32. Explain why the demographic dividend does not last.

Exam practice

The diagram shows changes in life expectancy, by quintiles (fifths of the population), in the USA between 1980 and 2010.

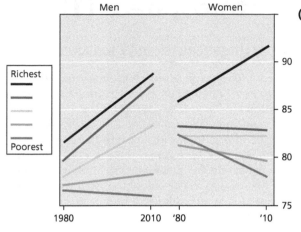

(a) (i) Define life expectancy. (2 marks)

(ii) Describe the changes in life expectancy for men and women between 1980 and 2010. (4 marks)

(iii) Suggest reasons for the differences in life expectancy between men and women, and for the differences in progress between the richest and poorest quintiles. (2 + 2 marks)

Sex trafficking: Sex trafficking victims are manipulated or forced to engage in sex acts for money. Sex traffickers may use violence, threats, manipulation, or the promise of love and affection to lure their victims. Truck stops, hotel rooms, rest areas, street corners, clubs and private residences are just some of the places where victims are forced to sell sex.

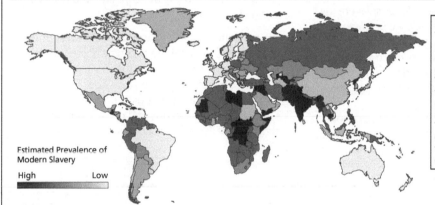

Forced labour:
Victims of forced labour may be found in factories, on farms, doing construction work, etc. Very often victims are forced to manufacture or grow products that we use and consume regularly. Through force, fraud or coercion, victims are forced to work for little or no pay.

Domestic servitude: Traffickers sometimes take a victim's identification papers and travel documents in order to limit their freedom. Victims of domestic servitude are hidden in plain sight, they are forced to work in homes across the USA. They are prisoners working as nannies, maids or domestic help.

(b) The maps and infographic show the prevalence of modern slavery, and some of the forms of modern slavery.

(i) Comment on the global spread of trafficking/slavery. (3 marks)

(ii) Comment on the forms of trafficking that occur. (3 marks)

(iii) Outline the ways in which trafficked people are abused. (4 marks)

(c) Either

Examine the view that megacities create more challenges than opportunities for individuals and societies. (10 marks)

Or

"Environmental causes of forced migrations are more important than political causes." Discuss this statement. (10 marks)

1 THE CAUSES OF GLOBAL CLIMATE CHANGE

The structure of the Earth's atmosphere (1)

The atmosphere consists of a mixture of solids, liquids and gases. Up to a height of around 80 km, the atmosphere consists of nitrogen, oxygen, argon and a variety of other trace gases such as carbon dioxide, helium and ozone. In addition, there is water vapour and solids such as dust, ash and soot.

Most "weather" occurs in the troposphere. Here, temperatures fall with height (on average 6.5°C per km). Different gases are concentrated at various heights. Most water vapour, for example, is contained in the lowest 15 km of the atmosphere. Above this, the atmosphere is too cold to hold water vapour.

At high altitude, there are significant concentrations of gases, such as ozone between 25 km and 35 km.

THE ATMOSPHERIC ENERGY BALANCE

The atmosphere is an open energy system receiving energy from both the Sun and Earth. **In**coming **sol**ar radi**ation** is referred to as **insolation**.

Solar energy drives all weather systems and climates. The Earth absorbs most of this energy in the tropical regions, whereas there is a loss of energy from temperate and polar regions. To compensate for this, there is also a redistribution of energy from lower latitudes to higher latitudes, driven by wind circulation and ocean currents.

THE ATMOSPHERIC ENERGY BUDGET

The Earth's atmosphere constantly receives solar energy, and yet until recently there was a balance between inputs (insolation) and outputs (re-radiation). This recent imbalance – known as global warming – has been linked with human activities such as land-use changes and the use of fossil fuels.

Under "natural" conditions, the balance is achieved in three main ways:
- **Radiation** – the emission of short waves and long waves; as the Sun is a very hot body, most of its radiation is in the form of very short wavelengths such as ultraviolet and visible light.
- **Convection** – the transfer of heat by the movement of a gas or liquid.
- **Conduction** – the transfer of heat by contact.

Of the short-wave radiation which reaches the ground (46 units),
- 14 are re-radiated as long-wave radiation to the atmosphere and to space
- 10 units pass to the atmosphere by conduction (contract heating) or the lower atmosphere only – since air is a poor conductor of heat
- 22 units are transferred by latent heat – the heat energy used by a substance change

form but not temperature; for example, water is evaporated into the atmosphere, and releases heat on condensation).

Since the atmosphere only absorbs 23 units of short-wave radiation the atmosphere is largely heated from below.
Of the solar energy,
- 46% is absorbed by the earth
- 22% drives the hydrological cycle
- 1% powers the winds and ocean currents
- 31% is reflected to space.

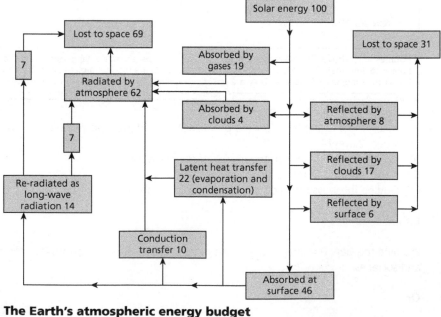

The Earth's atmospheric energy budget

CHECK YOUR UNDERSTANDING
1. Calculate the amount of solar energy that is reflected back to space.
2. State the proportion of solar energy that reaches the Earth's surface.

The structure of the Earth's atmosphere (2)

SHORT-WAVE AND LONG-WAVE RADIATION

Short-wave energy emitted by the Sun is re-radiated at long wavelength by the Earth. (While very hot bodies such as the Sun emit short-wave radiation, cold bodies such as the Earth emit long-wave radiation.) Clouds and the atmosphere absorb some of the energy and re-radiate it back to Earth. Evaporation and condensation also account for a loss of heat. In addition, a small amount of condensation occurs (carried up by turbulence).

The atmosphere is largely heated from below. Most of the incoming short-wave radiation is let through the atmosphere, but CO_2 traps the outgoing long-wave radiation, warming the atmosphere. This is known as the greenhouse effect.

Incoming (short-wave) solar radiation

Incoming solar radiation (insolation) is the main energy input and it varies according to latitude, season and cloud cover. The amount of insolation received varies with the angle of the Sun and with cloud type. The less cloud cover there is, and/or the higher the cloud, the more radiation reaches the Earth's surface. Incoming solar radiation is mostly in the visible wavelengths. These are not absorbed by the Earth's atmosphere; instead, they heat the Earth, which, in turn, heats the atmosphere by way of long-wave radiation.

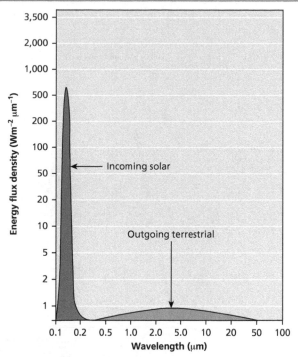

Solar radiation, terrestrial radiation and wavelength

Long-wave radiation

Long-wave radiation refers to the radiation of energy from the Earth into the atmosphere.

During a cloudless night there is a large loss of long-wave radiation from the Earth. Due to a lack of clouds, there is very little return of long-wave radiation from the atmosphere. Hence there is a net loss of energy from the surface. In contrast, on a cloudy night the clouds return some long-wave radiation to the surface, which reduces the overall loss of energy. For example, in hot desert areas, where there is a lack of cloud cover, the loss of energy at night is maximized. In contrast, in cloudy areas such as tropical rainforests, the loss of energy (and change in daytime and night-time temperatures) is less noticeable.

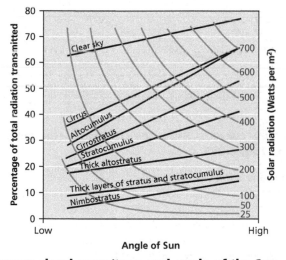

Energy, cloud cover/type and angle of the Sun

The greenhouse effect

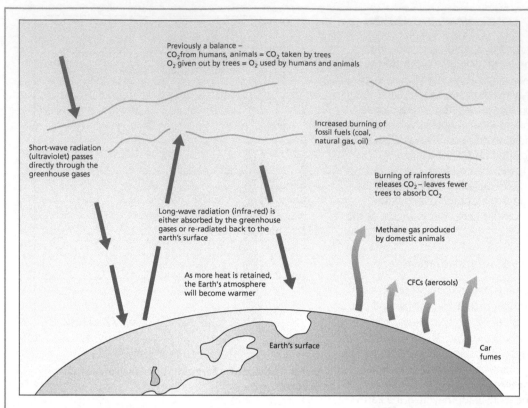

Previously a balance –
CO_2 from humans, animals = CO_2 taken by trees
O_2 given out by trees = O_2 used by humans and animals

Short-wave radiation (ultraviolet) passes directly through the greenhouse gases

Long-wave radiation (infra-red) is either absorbed by the greenhouse gases or re-radiated back to the earth's surface

As more heat is retained, the Earth's atmosphere will become warmer

Increased burning of fossil fuels (coal, natural gas, oil)

Burning of rainforests releases CO_2 – leaves fewer trees to absorb CO_2

Methane gas produced by domestic animals

CFCs (aerosols)

Earth's surface

Car fumes

is due to human activities, such as burning fossil fuel (coal, oil and natural gas) and land-use changes such as deforestation. Deforestation is a double blow, since it not only increases atmospheric CO_2 levels but it also removes the trees that convert CO_2 into oxygen and act as a major carbon store. Carbon dioxide accounts for about 20% of the greenhouse effect but an increased proportion of the enhanced greenhouse effect.

The greenhouse effect

The greenhouse effect is the process by which certain gases (greenhouse gases) allow short-wave, solar radiation to pass through the atmosphere but trap a proportion of outgoing long-wave radiation from the Earth. This radiation leads to a warming of the atmosphere. The greenhouse effect is a good thing, for without it there would be no life on Earth. For example, the moon is an airless planet that is almost the same distance from the Sun as the Earth. Average temperatures on the moon are about –18°C compared with about 15°C on Earth. The Earth's atmosphere therefore raises temperatures by about 33°C.

There are a number of greenhouse gases. **Water vapour** is the most common greenhouse gas, accounting for about 95% of greenhouse gases by volume and for about 50% of the natural greenhouse effect. However, the gases mainly implicated in global warming are carbon dioxide, methane and chlorofluorocarbons.

Carbon dioxide (CO_2) levels have risen from about 315 parts per million (ppm) in 1950 to over 400 ppm in 2015, and are predicted to reach 600 ppm by 2050. The increase

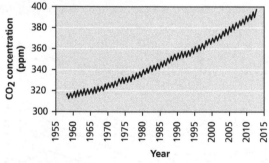

Keeling curve to show the change in atmospheric CO_2, 1960–2015

Methane is the second-largest contributor to global warming, and its presence in the atmosphere is increasing at a rate of 1% per annum. Cattle convert up to 10% of the food they eat into methane, and emit 100 million tonnes of methane into the atmosphere each year. Paddy fields emit up to 150 million tonnes of methane annually, while, as global warming proceeds, bogs trapped in permafrost will melt and release vast quantities of methane.

Chlorofluorocarbons (CFCs) are man-made chemicals that destroy ozone as well as absorbing long-wave radiation. CFCs, which are increasing at a rate of 6% per annum, are up to 10,000 times more efficient at trapping heat than CO_2.

CHECK YOUR UNDERSTANDING
5. Briefly explain the natural greenhouse effect.
6. Describe the changes in the Keeling curve (changes in atmospheric carbon dioxide).

Changes in the global energy balance (1)

VARIATIONS IN SOLAR RADIATION

The Earth's temperature changes for a number of reasons, such as changes in solar output. There is evidence of an 11-year cycle, and much longer periods of changes in the movement of the Earth, that is, Milankovitch cycles – variations in the Earth's orbit affect the seasonal and latitudinal distribution of solar radiation, and are responsible for initiating ice ages. On a shorter timescale, changes in atmospheric composition, for example, following a volcanic eruption, are linked to a decrease in global temperature. Changes in reflectivity (albedo) are affected by, and affect, global climate change, for example, as ice melts and is replaced by darker-coloured vegetation, the amount of insolation absorbed increases and temperatures rise.

ALBEDO VALUES

Surface	Albedo (%)
Fresh snow	75–90
Old snow	40–70
Black road surface	5–10
Grass	20–30
Coniferous forest	5–15
Tundra	15–20

Reflection from the Earth's surface (known as the **planetary albedo**) is generally about 7%. Albedo values vary for different types of ground cover, for example, over fresh snow albedo is 75–90% while over coniferous forest it is 5–15%.

GLOBAL DIMMING

Following the 9/11 terrorist attacks on the USA and the banning of air travel across the USA, average temperature rose by about 1.1°C in the absence of condensation trails (contrails). Global dimming is the cooling of air temperature due to pollution. It is possible that global dimming has been reducing what would be even faster global warming than is currently occurring. There are at least two timescales with global dimming, namely short-term cycles lasting less than a decade following volcanic eruptions and long-term changes related to anthropogenic (man-made) sources of pollution.

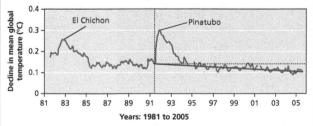

Decline in mean global temperature following volcanic eruptions (the straight line shows the global decline in sun-blocking aerosols following the eruption of Mt Pinatubo in 1991)

POLLUTION AND GLOBAL DIMMING

From the 1950s to the early 1990s, the level of solar energy reaching the Earth's surface had dropped over Antarctica, the USA, the UK and Russia. This was all due to high levels of pollution at that time. Natural particles in clean air provide condensation nuclei for water. Polluted air contains far more particles of ash, soot and sulphur dioxide than clean air and therefore provides many more sites for water to bind to. The droplets formed tend to be smaller than natural droplets, which means that polluted clouds contain many much smaller water droplets than naturally occurring clouds. Many small water droplets reflect more sunlight than fewer larger droplets, so polluted clouds reflect far more light back into space, thus preventing some insolation from getting through to the Earth's surface.

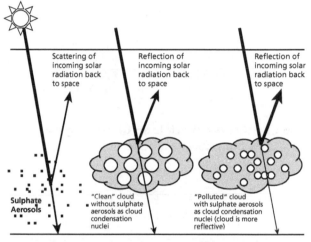

How sulphate aerosols affect solar radiation

Human activity and global dimming

Changes in the global energy balance (2)

FEEDBACK LOOPS

Feedback mechanisms play a key role in controlling the Earth's atmosphere. Both positive and negative feedback loops are associated with changes in global warming.

Positive feedback

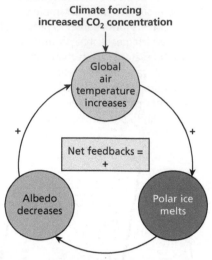

The positive feedback loop of melting ice reducing the planet's albedo

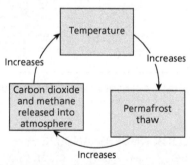

A positive feedback mechanism involving methane and enhancing climate change

The impacts of global warming may be greatest in tundra environments. These are regions of seasonal ice cover at the edges of glaciers and ice sheets. The effects will be most noticeable in terms of winter warming. Melting of the polar ice caps results in less ice and lowers planetary albedo. Since ice is more reflective than water, less ice leads to less reflection, increasing the amount of solar energy absorbed at the surface, and an increase in temperature.

Rotting vegetation trapped under permafrost in the tundra releases methane that is unable to escape due to the frozen subsurface. Increased thawing of permafrost will lead to an increase in the release of methane, adding to greenhouse gases in the atmosphere and thereby increasing global temperatures.

Other mechanisms of positive feedback include:
- increased carbon dioxide released from increased decomposition of biomass, especially in forests, caused by rising temperatures, leading to a further rise in temperature as greenhouse gases are added to the atmosphere
- increased forest cover in high latitudes, decreasing albedo and increasing warming.

Feedback mechanisms associated with global warming tend to involve very long time lags. By the time the effects appear, the factors responsible may be beyond the tipping point (the point of no return), and attempts to solve the problem may be futile.

Negative feedback

Increased evaporation in low latitudes due to higher temperatures, may lead to increased snowfall on the polar ice caps, reducing the mean global temperature. Similarly, an increase in carbon dioxide in the atmosphere leads to increased plant growth by allowing higher levels of photosynthesis. Increased plant biomass and productivity would increase the store of terrestrial carbon but reduce atmospheric concentrations of carbon dioxide.

Other mechanisms of negative feedback include:
- global dimming, that is, burning, leading to more aerosols and thus reduced solar radiation at the surface, thereby causing cooling
- increased evaporation in tropical and temperate latitudes leading to increased snowfalls in polar areas. This has happened in parts of Norway, with the growth of the Boyabreen Glacier. Growth of glaciers and ice caps, albeit localized, can reduce mean temperatures due to changes in albedo.

EXAM TIP

When writing about feedback loops, distinguish between positive and negative feedback. Positive feedback leads to increasing change (sometimes referred to as cumulative causation or a vicious circle) whereas negative feedback tends to produce long-term stability in a system.

CHECK YOUR UNDERSTANDING

9. Describe one terrestrial albedo change and its feedback loop.
10. Explain how global dimming could be interpreted as an example of negative feedback.

The enhanced greenhouse effect

The enhanced greenhouse effect is the impact of increasing levels of greenhouses gases in the atmosphere due to human activities. It is more frequently called global warming. Global climate change refers to the changes in the global patterns of rainfall and temperature, and the incidence of drought, floods and storms, resulting from changes in the Earth's atmosphere, caused mainly by the enhanced greenhouse effect.

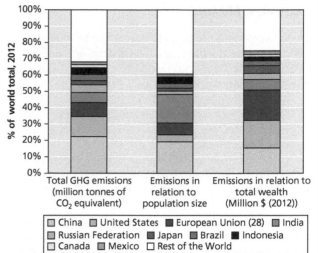

Total greenhouse gas emissions by the top ten emitters, 2012

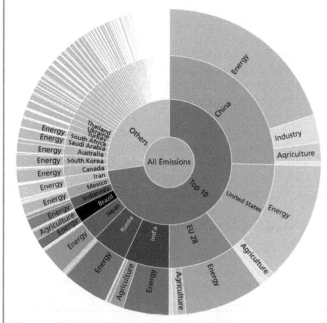

The main contributors to greenhouse gas emissions

The increase in greenhouse gases is linked to industrialization, trade and globalization. As industrialization has increased, so too has atmospheric CO_2. Many LICs and NICs are actively industrializing. Industrial activity among the NICs has great potential to add to atmospheric CO_2. Nevertheless, the per-capita emissions of greenhouse gases in HICs are responsible for much of the growth in atmospheric CO_2.

Greenhouse gas emissions by country

Country	Largest emitters, billion tonnes (rounded)	Emissions per person (tonnes, rounded)
China	10.2	6.6
USA	5.4	12.1
India	2.1	1.7
Russia	1.9	8.4
Japan	1.2	10.1
Germany	0.8	10.1
South Korea	0.7	10.1
Canada	0.65	11.2
Brazil	0.6	2.4
Indonesia	0.6	2.0
Saudi Arabia	0.5	12.4
United Kingdom	0.5	7.5
Mexico	0.4	3.9
Iran	0.3	5.2
Australia	0.2	12.5

COMMON MISTAKE

✗ The greenhouse effect and global warming are the same thing.

✓ The natural greenhouse effect is beneficial/essential for life on Earth, whereas the enhanced greenhouse effect (global warming) is causing the world to overheat and is having many negative impacts.

CHECK YOUR UNDERSTANDING

11. Outline the main sources of emissions that are contributing to the enhanced greenhouse effect.
12. Comment on the main contributors to emissions of greenhouse gases.

The implications of climate change (1)

GLOBAL WARMING

Global warming is predicted to have many far-reaching effects on the natural, social and economic environment:
- Sea levels will rise, causing flooding in low-lying areas.
- Storm activity will increase.
- Agricultural patterns will change, for example, the USA's grain belt will shrink and production is likely to decline, but Canada's growing season will get longer and probably more productive.
- There will be less rainfall over the USA, southern Europe and the Commonwealth of Independent States (CIS).
- Up to 40% of wildlife species will become extinct.

Nevertheless, there is a great amount of uncertainty, and nobody knows exactly what the impact of climate change will be. Different scenarios are based on different possible temperature changes. Certain areas might get colder, others will become warmer. The results may be very different from the predictions.

The potential impact of temperature increase on aspects of the environment and society

Feature	Effect or impact
Environmental features	
Ice and snow	Melting of polar ice caps and glaciers
Coastlines	Increase in sea level causing coastal flooding
Water cycle	Increased flooding; more rapid circulation
Ecosystems	Change in biome distribution and species composition, for example, poleward and altitudinal migration
Societal features	
Water resources	Severe water shortages and possibly wars over supply
Agriculture	May shift towards poles (away from drought areas)
Coastal residential locations	Relocation due to flooding and storms
Human health	Increased disease, for example, the risk of malaria

CHANGES TO THE HYDROSPHERE

The potential impacts of global climate change on the hydrosphere (freshwater, seawater and ice/glaciers) are great. Impacts could include:
- a rise in sea levels causing flooding in low-lying areas such as Bangladesh and Kiribati, displacing up to 200 million people
- floods from melting glaciers threatening 4 million km² of land and the reliability of water supply to millions of people.

As global temperatures increase, sea level is predicted to rise as the ice caps and glaciers melt. In addition, the steric effect – the expansion of water as it gets warmer – is leading to a slight rise in sea level. By 2100, it is estimated that sea levels will have risen by between 40 cm and 80 cm. Not only would sea levels rise, but the level of high tide and storm surges would increase. Cities such as New York and London would be at increased risk of coastal flooding and would need to build or raise barriers and sea walls.

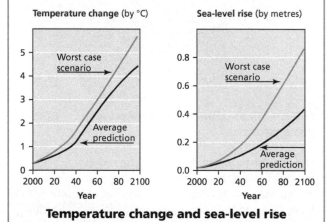

Temperature change and sea-level rise

The potential impact of climate change

❓

CHECK YOUR UNDERSTANDING

13. Outline the main ways in which global warming will affect climate change.

14. Outline two reasons for the rise in sea level due to global warming.

The implications of climate change (2)

CHANGES IN SEA ICE

Arctic sea ice has declined dramatically since the mid-1970s. The main reason for this is global warming. The Arctic is believed to be at its warmest for 40,000 years, and the length of the melting season has increased by nearly three weeks since 1979. During February 2016, the sea ice was the lowest recorded volume on record.

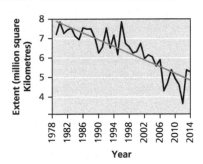

The extent of Arctic sea ice 1979–2014

The annual Arctic sea ice minimum usually occurs during September each year, and the maximum occurs during March. However, the overall volume, thickness and extent have been declining for decades. In addition, the age of the ice is changing. For example, in 1988, ice that was more than 4 years old accounted for over 25% of the Arctic sea ice, but by 2013 it was less than 8%.

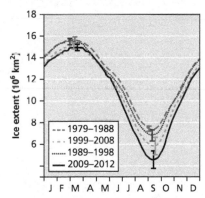

Seasonal variation in Arctic sea ice, 1979–2012

As the ice recedes, the potential for wave formation increases. In 2012, 5-metre waves were recorded in the Beaufort Sea. These waves helped break up the sea ice, thus establishing a positive feedback loop of disappearing sea ice and wave formation.

Scientists predict that the Arctic will become ice-free during summer, by 2040.

There are many impacts of sea ice decline. Another study has linked the decline in Arctic sea ice with wet summers in northern Europe, and with extreme weather in the northern mid-latitudes. Methane emissions from the tundra may increase due to the release of chlorine atoms from the sea.

Sea ice decline has been linked with increased primary productivity causing plankton blooms. However, the melting ice reduces the time that polar bears have to hunt seal pups, and they must spend more time on land. Their diets have become less nutritional, resulting in reduced body size and decreased reproductive success.

GLACIERS AND ICE CAPS

Many Himalayan glaciers are retreating. This may have a major impact on the region's water supply. Many of Asia's major river systems have their source in the Himalayas and provide water for drinking, irrigation, industry and other uses. Over 1.5 billion people depend on these rivers. At lower elevations, glacial retreat is unlikely to cause major water shortages in the near future (due to the monsoon rain system). However, other factors such as population growth and groundwater depletion could have a serious impact on water supplies in lowland areas.

For higher elevations, glacial retreat could alter stream-flow characteristics. For example, the Gangotri Glacier is one of the largest glaciers in the Himalayas. It has been in retreat since 1780, although the retreat has intensified since 1971. Since 1990, it has retreated more than 800 m. In Europe, the Gorner Glacier has retreated 2.5 km over the last 130 years.

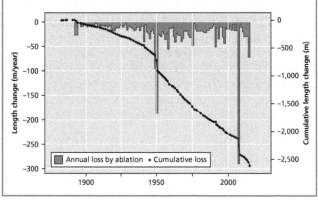

CHECK YOUR UNDERSTANDING
15. Describe the annual and long-term trends in the extent of sea ice cover in the Arctic Ocean.
16. Analyse the seasonal variations in the Arctic sea ice.

Changes in carbon stored in ice, oceans and the biosphere

ICE

Periglacial areas have seasonally low temperatures and permafrost (permanently frozen soil). In many areas, the permafrost, which contains large deposits of carbon, is beginning to thaw. The slow rate of decomposition has allowed the large accumulation of carbon to develop, in the form of dead organic matter (DOM). Periglacial environments contain some 400–500 gigatonnes (Gt) of carbon stored as DOM, compared with 700–800Gt of atmospheric carbon. Up to 30–40% of the global soil carbon storage is contained within periglacial soils.

The likely impact of global warming on periglacial environments is to increase net primary productivity, as long as a lack of nutrients does not limit productivity. Currently, low temperatures limit the release of nutrients. However, as temperatures rise, decomposition of DOM may increase, and release more nutrients.

Warming of periglacial environments will lead to increased methane emissions. Current emissions of methane are dominated by periglacial areas, in particular western Siberia and central Canada. About 25–40 megatonnes (Mt) of methane are released each year. Changes in snow cover and permafrost could have important feedback mechanisms. Decreased snow leads to decreased reflectivity of the surface, hence increased absorption of solar radiation. Melting of the permafrost could release methane which could lead to an increase in air temperature.

OCEANS

In pre-industrial times, the Earth's atmosphere contained 578 Pg (petragrams, 1 Pg = 1 billion tonnes) of carbon. This has increased by about 1%. In contrast, the oceans contain about 38,000–40,000 Pg of carbon, about 50 times more than the atmosphere.

However, with climate change, the atmospheric carbon content is causing changes in the ocean carbon content. Increased carbon in the atmosphere warms the Earth, and may make plants grow more, and store more carbon. In contrast, increased carbon in the oceans acidifies the water.

Since 1750, the oceanic pH has dropped by about 0.1, a 30% change in acidity. Some of the excess carbon dissolves in the ocean to become carbonic acid, which reacts with carbonate in the water to form bicarbonate. With less carbonate available, shell-building organisms such as lobsters and coral end up with thinner shells.

In addition, warmer oceans may decrease the abundance of phytoplankton, which grow more vigorously in cool, nutrient-rich waters.

BIOSPHERE

Terrestrial plants have absorbed approximately 25% of the carbon dioxide that humans have released into the atmosphere. For a doubling of carbon dioxide, plant growth increases significantly, providing there are no other limiting factors, such as water.

Agriculture has had a variable impact on the Earth's carbon cycle. When farmland is abandoned, the vegetation may revert to forest. However, by preventing wildfires, humans prevent carbon from entering the atmosphere, and allow carbon to build up in plants instead. However, in many areas, the use of fire to create new farmland is releasing considerable amounts of carbon into the atmosphere.

Until recently, many forests were sinks, converting carbon dioxide through photosynthesis into biomass. In addition, forest soils store large amounts of carbon. Deforestation leads to release of carbon from biomass and soils to the atmosphere.

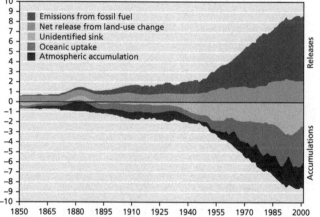

Flux of Carbon (Gt per Year)

COMMON MISTAKE

✗ Carbon dioxide and methane are two completely different substances.
✓ Both carbon dioxide (CO_2) and methane (CH_4) contain carbon, and both are greenhouse gases.

CHECK YOUR UNDERSTANDING
17. Briefly explain why global warming in periglacial areas may lead to positive feedback regarding climate change.
18. Suggest how fire influences the amount of atmospheric carbon in the biosphere.

Extreme weather events

Climate change can lead to changes in weather patterns and rainfall. Climates may become more extreme and unpredictable. Probable impacts include:

- up to 200 million people at risk of being driven from their homes by flood or drought by 2050
- an increase in hurricane frequency and intensity
- water shortages affecting up to 4 billion people.

The impacts of climate change will vary with the scale of the temperature change.

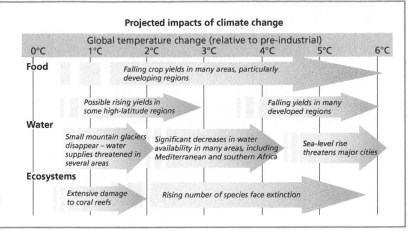

Projected impacts of climate change

UP TO 1°C INCREASE

A maximum rise in temperature of 1°C is vital for the survival of low-lying island states, but it is now thought to be virtually impossible to achieve. The Arctic sea ice is already disappearing and, with a 1°C global average temperature rise, it will disappear for good in the summer months. Heatwaves and forest fires will become more common in the subtropics, and the worst-hit areas will be the Mediterranean region, southern Africa, Australia and the south-west United States.

UP TO 2°C INCREASE

This is the temperature limit the scientists want. However, even a 2°C rise will have significant impacts. The heatwaves seen in Europe during 2003, which killed tens of thousands of people, will occur annually with a 2°C global average temperature rise. The Amazonian rainforest will turn into desert and grasslands, and increasing CO_2 levels in the atmosphere will make the oceans too acidic for any remaining coral reefs and thousands of other marine lifeforms. More than 60 million people, mainly in Africa, will be exposed to higher rates of malaria. Agricultural yields around the world will drop, exposing half a billion people to a greater risk of starvation. The west Antarctic ice sheet will collapse, the Greenland ice sheet will melt and the world's sea level will begin to rise by seven metres over the next few hundred years. Glaciers all over the world will recede, reducing the fresh water supply for major cities. Coastal flooding will affect more than 10 million extra people. A third of the world's species will become extinct if the 2°C rise changes their habitats too quickly for them to adapt.

UP TO 4°C INCREASE

This scenario is possible if countries agree to only an extremely weak climate deal. At this level of increase, the Arctic permafrost would enter the danger zone, releasing into the atmosphere much more of the methane and carbon dioxide currently locked in the permafrost. At the Arctic itself, the ice cover would disappear permanently, meaning extinction for polar bears and other native species that rely on the presence of ice. Further melting of Antarctic ice sheets would mean a further five-metre rise in sea level, submerging many island nations. Italy, Spain, Greece and Turkey would become deserts and central Europe would reach desert temperatures of almost 50°C in summer.

CASE STUDY

EXTREME WEATHER AND CLIMATE CHANGE IN THE UK

These are the likely impacts of climate change on the UK for the 2020s and beyond.

- Rising temperatures: Temperatures are expected to increase at a rate of about 0.2°C per decade becoming about 0.9°C warmer by the 2020s and about 1.6°C warmer by the 2050s. This is equivalent to warmer temperatures moving about 200 km north in the UK, for example, the temperature in Manchester would be similar to what is experienced currently on the south coast.
- Increased rainfall and wind speeds: Annual precipitation is predicted to rise by about 5% by the 2020s and by nearly 10% by the 2050s. There will be more intense rainfall events and extreme wind speeds, especially in the north – the frequency of gales will increase by about 30%.
- Droughts and floods: The currently dry south-east will tend to become drier and the moist north-west will get wetter. Drought in the south-east and flooding in the north-west will both become more common.

CHECK YOUR UNDERSTANDING

19. Outline the reasons why limiting temperature increase to 2°C might not be good for people and ecosystems.

20. Suggest why tropical storms are likely to become more frequent and/or intense as a result of global warming.

Changes in biomes

BIOME CHANGES

Climate change in the geological past can show how biomes might move with changes in future global temperatures. Models suggest a latitudinal shift in biomes relative to the Equator, and an altitudinal shift as biomes move up-slope. Low-lying biomes such as mangroves may be lost due to changes in sea level, and high altitude biomes may be lost as they have nowhere to move to.

Species composition in ecosystems is also likely to change. Climate change in the past has allowed species to adapt gradually to the new conditions. Current changes in temperature are happening very rapidly, so there is little time for organisms to adapt.

A reduction in biodiversity may also occur. Some species – especially those in high-altitude and high-latitude habitats – have fewer options for migration and so are more vulnerable to extinction. A rise of 2°C could lead to the extinction of up to 40% of wildlife species.

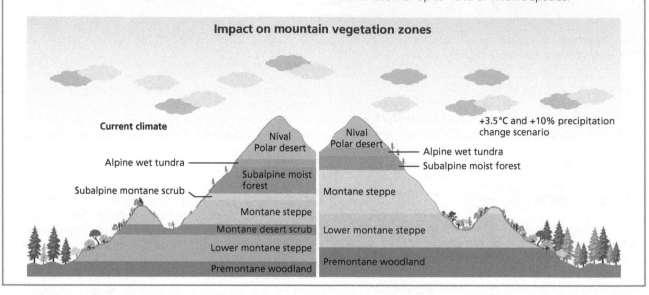

Impact on mountain vegetation zones

ANIMAL MIGRATIONS

Rising temperatures are forcing some plants and animals outside of their normal/preferred temperature range. To survive, they will need to move polewards or to higher elevation. In the USA, there are important wildlife corridors that allow plants and animals to migrate. One of these is the Appalachian Mountains, one of the least developed parts of eastern USA. Due to their higher elevation, they are also providing some shelter for climate change refugees. A network of national parks could provide a similar refuge.

SOIL EROSION

Global warming could lead to an increase in soil erosion, degradation, desertification and salinization. Warmer average annual temperatures and increased precipitation could double soil loss on moderate slopes. Soil is higher when rainfall is high and vegetation is low. Higher run-off occurs when fields have been harvested, as there is less interception of rainfall and less disruption of overland flow. In the north-west USA soil loss is predicted to increase to 0.76 tonnes/ha on ploughed soils and 0.13 tonnes/ha on a no-till system.

CHANGES TO AGRICULTURE

As global temperatures rise, changes in agricultural patterns are likely. A rise of 3°C could lead to a 35% drop in crop yields across Africa and the Middle East. A rise of 2°C could lead to 200 million more people experiencing hunger, while a rise of 3°C could lead to up to 550 million people being affected by hunger.

Changes in the location of crop-growing areas can be expected, with latitudinal movements away from the Equator. For example, viticulture – the growing of grapes to produce wine – will move polewards, as will corn and wheat. Many wheat-growing regions of the USA may become unviable by 2050, although there may be an increase in Canada's growing season. This would have a serious effect on the US economy.

The reduction in water resources will make it increasingly difficult for farmers in many areas to cultivate the crops that they currently grow. Crop types may need to change and the availability of water resources will determine what may be grown, and how successfully.

CHECK YOUR UNDERSTANDING
21. Outline the likely changes to global biomes as a result of global warming.
22. Outline the likely barriers to migration for some species.

The impact of climate change on people and places

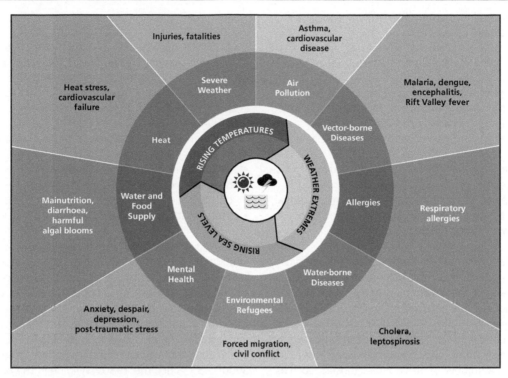

Impact of climate change on human health

The effects of global warming are varied. A rise in temperature of 2°C could expose up to 60 million more Africans to malaria. Mosquitoes would be able to breed in areas previously too cool for them. Other tropical diseases will spread as warmer conditions extend to higher latitudes.

Global warming may lead to an increase in human migration. Already, some communities are claiming to be environmental refugees due to rising sea levels. Residents of low-lying islands such as Kiribati in the South Pacific have abandoned their homes.

One benefit of climate change could be new sea routes as sea ice melts. Russia's Arctic coastline and the North-West Passage could open up new trade routes. However, there are geopolitical issues to be resolved before either of these sea routes is used for international shipping.

Tourism is also likely to change due to global warming. Summer seasons may be extended and coastal resorts may develop in more northerly locations. Winter sports holidays, however, may decline due to lack of snow and ice.

SOCIAL PROBLEMS

Globally climate change will lead to social problems such as hunger and conflict. National resource bases will change, which will drive economic, social and cultural change. For example, in Bangladesh 20% of gross domestic product (GDP) and 65% of the labour force is involved in agriculture that is threatened by floods. These issues are more likely to affect low-income countries (LICs) than high-income countries (HICs) because LICs are technologically and economically less able to cope. Moreover, a greater percentage of the population in LICs is already vulnerable to the effects of climate change.

Coastal flooding will particularly affect countries that have land below sea level, such as the Netherlands. LICs are more likely to have weak infrastructure, communications and emergency services, which will also make them less able to respond to the effects of climate change.

Climate change may also lead to a loss in ecosystem services. Ecosystems provide a range of services, for example, primary productivity, flood control and climate regulation, and these may be at risk if there are significant changes to climate.

EXAM TIP
Try to get across some of the complexity of the impact of climate change on human health. For example, people's health will not only be affected by the spread of infectious diseases, such as malaria, but many people will also be affected by severe dehydration due to prolonged drought, and fatigue, due to high temperatures, which may lead to an increase in illnesses (morbidity) and deaths (mortality).

CHECK YOUR UNDERSTANDING
23. Explain why the incidence of malaria may increase as a result of global warming.
24. Outline contrasting reasons why human migrations may increase due to global warming.

Disparities in exposure to climate change

There are uncertainties about the nature and scale of climate change. Levels of climate change risk and vulnerability will vary according to a person's location, wealth, age, gender, education and perception of risk.

Vulnerability to global climate change refers to the degree to which people are susceptible to, or unable to cope with, the negative impacts of climate change. There are three main factors associated with vulnerability:

- the degree to which people are exposed to climate change
- the degree to which they could be harmed by exposure to climate change
- the degree to which they could mitigate the potential harm by taking action to reduce their exposure or sensitivity to climate change.

Some population groups are more vulnerable to climate change than others. These include women, the very young, the elderly, those with disabilities, people with mobility problems, the poor, minority groups, refugees and indigenous people. Carers are also vulnerable because of their burden of caring for the young, the elderly and the sick. Single-parent households are often vulnerable as they may combine a number of at-risk characteristics, such as age, gender and poverty.

It is not just people who are affected by climate change. Institutions such as the emergency services, schools and transport services, as well as economic activities such as agriculture and tourism, may all be at risk.

Some locations are more at risk than others. These include low-lying islands, river mouths, coastal areas and regions that derive their water supplies from mountain glaciers. Many islands in the Indian Ocean and the Pacific Ocean are among the areas most vulnerable to climate change risks. Much of the infrastructure and socio-economic activities of these islands are located along the coastline.

The problems that low-lying areas face include:
- increased coastal erosion
- salt-water intrusion into groundwater
- damage to coral reefs
- out-migration of people
- a decline in economic activities and infrastructure.

Indigenous people often live in vulnerable environments. In addition, they often experience socio-economic problems such as low incomes and limited access to resources. Most indigenous populations have adapted their lifestyles to their environment, and they are therefore vulnerable to any changes in that environment. For example, the Inuit people in the Mackenzie Basin in Canada have experienced a rise in temperature of about 3.5°C since the 1980s. Wildlife in the Basin is a major source of food, clothing and income. However, the muskrat population has declined due to the lack of available water, and trapping, once a major economic activity, has disappeared.

Vulnerability to the Health Impacts of Climate Change at Different Life Stages

Mothers and babies

Adverse pregnancy outcomes such as low birth weight and preterm birth have been linked to extreme heat events, airborne particulate matter, and floods.

Infants and toddlers

Young children's biological sensitivity places them at greater risk from asthma, diarrhoeal illness, and heat-related illness.

School age and older children

The behaviors and activities of older children increase their risk of exposure to heat-related illness, vector-borne and waterborne disease, and respiratory effects from air pollution and allergens.

COMMON MISTAKE

✗ *Only the poor will be affected by climate change.*
✓ Although certain groups are more vulnerable to climate change e.g. the poor, indigenous peoples, refugees etc. – many middle income and high-income people will also be affected whether directly or indirectly. Food prices will rise, food scarcities will increase, insurance premiums will rise, and the likelihood of water shortages will increase too.

CHECK YOUR UNDERSTANDING

25. Identify the main population groups that are at risk of climate change.
26. Briefly explain why indigenous peoples are at risk of climate change.

Contrasting vulnerability: Bangladesh and Ghana

FLOODING IN BANGLADESH

Most of the country of Bangladesh forms a low-lying delta. Since 1970, the scale, intensity and duration of the floods have increased. Scientists predict that flooding will continue to increase considerably in the future. Due to sea level rise, the densely populated coastal zone of Bangladesh is also increasingly vulnerable to coastal floods.

The potential flood damage is likely to worsen as a result of increased intensity of extreme precipitation events. Monsoon rainfall is predicted to increase by 14–40% by the 2030s and 52–135% by the 2090s.

In 1988, the Government of Bangladesh developed its Flood Action Plan, with the aim of protecting the country from future flooding. Sluice gates were built on a number of rivers. These also provide protection from flooding by tidal waves and storm surges. The government has built about 5,700 km of embankments and dug nearly 5,000 km of drainage channels to divert floodwater away from buildings. It also constructed 200 flood shelters on stilts for the evacuation of people.

In Bangladesh's Padma River islands, frequent flooding makes the life and livelihoods of people extremely vulnerable. Since 1990 the frequency of extreme precipitation has increased.

Poor construction materials increase the potential impact of flooding. Many of the houses in Bangladesh are built from sand and mud. Income levels are generally very low and are not diversified enough to cope with the possible effects of global climate change.

VULNERABILITY IN GHANA

In Ghana, the population groups that are most vulnerable to climate change include the poor, the elderly, women, children, the infirm, indigenous groups, minority groups and refugees. Impacts include decreased water availability, increased illness and fatality, out-migration and increasing poverty, although these vary with type of environment.

Climate change impacts in different environmental regions of Ghana

Zone	Climate change impacts
Northern savanna	Increased morbidity and disease prevalence Increased vulnerability of the poor Increased out-migration and loss of human capital
Forest	Decreased food security Dry-ups of water bodies and underground water Population pressure on land
Coastal savanna	Decreased water availability and quality Higher burden on women Increased migration Increased cholera

Climate change is likely to lead to conditions outside the normal range of experience for many communities. In Ghana, types of adaptation include infrastructure development, ecosystem-based measures and capacity development.

- Infrastructure development includes dams, levees and sea walls.
- Ecosystem-based adaptation includes preserving and restoring natural habitats, such as mangrove swamps, to provide ecosystem services.
- Capacity development includes education and extension services to teach people new techniques and to empower people.

CHECK YOUR UNDERSTANDING

27. Outline three potential impacts of climate change in Bangladesh.
28. Suggest why the impacts of global warming in Ghana may lead to an increased burden on women.

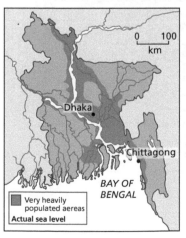

Very heavily populated aereas
Actual sea level

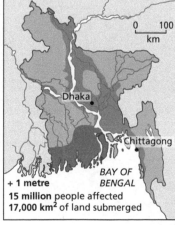

+ 1 metre
15 million people affected
17,000 km² of land submerged

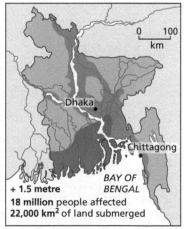

+ 1.5 metre
18 million people affected
22,000 km² of land submerged

Sources: Dacca University, Intergovernemntal pannel on Climate Change (PCC).

Government-led action on climate change

There are many obstacles to achieving a low-carbon world. Political obstacles are found nationally and internationally. The fossil fuel industry is one of the most powerful lobby groups in the USA, where coal, oil and gas interests have managed to limit climate control regulations. The main obstacle to a global agreement on climate change remains the bargaining power of the major fossil fuel countries such as the USA, Canada, China, Russia and countries in the Middle East.

THE KYOTO PROTOCOL

In 1997, 183 countries signed up to an agreement that called for the stabilization of greenhouse gas emissions at safe levels that would avoid serious climate change. The agreement (the Kyoto Protocol) aimed to cut greenhouse gas emissions by 5% of their 1990 levels by 2012. The Kyoto Protocol came into force in 2005 and was extended to 2015.

THE PARIS AGREEMENT, 2015

The 2015 UN Climate Change Conference was held in Paris. France was taken as an example of an HIC that had decarbonized its energy production – it generates over 90% of its energy from nuclear power, hydroelectric power and wind energy. One hundred and seventy-four countries signed the Paris Agreement on the Reduction of Climate Change. The key objective is to limit global warming to 2°C compared with pre-industrial levels. It also seeks for zero net anthropogenic greenhouse gas emissions between 2050 and 2100. Unlike the Kyoto Protocol, there are no country-specific goals or a detailed timetable for achieving the goals. Countries are expected to reduce their carbon usage "as soon as possible", although there is no mechanism to force a country to set a specific target, nor is there any measure to penalize countries if their targets are not met.

The 2015 conference wanted to achieve a universal agreement on climate change from all of the world's major countries.

ADAPTATION STRATEGIES

It is possible to reduce human emissions of greenhouse gases. The technologies exist, and measures such as energy efficiency, low-carbon electricity, and fuel switching are all possible. Nevertheless, even with these, CO_2 will continue to increase for a number of decades. Thus, as well as trying to mitigate climate change, humanity will need to adapt to climate change.

Examples of adaptations include flood defences, vaccination programmes, desalination plants and planting of crops in previously unsuitable climates. Adaptive capacity varies from place to place and can depend on financial and technological resources.

In agriculture, crop varieties can be made more resilient to higher temperatures and more frequent floods and droughts. Cities can be protected against rising ocean levels and the greater likelihood of storm surges and flooding. The geographic range of some diseases, such as malaria, will spread as temperatures rise, so more widespread vaccination programmes will be needed; to meet the demand for water, more desalination plants will be required.

Mitigation

- Sustainable transportation
- Energy conservation
- Building Code changes to improve energy efficiency
- Renewable energy
- Expand deep lake water cooling
- Improve vehicle fuel efficiency
- Capture and use landfill & digester gas

- Geothermal
- Solar thermal
- District heating
- Building design for natural ventilation
- Tree planting & care
- Local food production
- Water conservation
- Green roofs

Adaptation

- Infrastructure upgrades: sewers & culverts
- Residential programs: sewer backflow & downspout disconnection
- Health programs: West Nile, Lyme disease, Shade Policy, cooling centres, smog alerts, Air Quality Health Index
- Emergency & business continuity planning
- Help for vulnerable people

Mitigation: the globally responsible thing to do

Actions that reduce the emissions that contribute to climate change.

Adaptation: the locally responsible thing to do

Actions that minimize or prevent the negative impacts of climate change.

CHECK YOUR UNDERSTANDING
29. Distinguish between adaptation and mitigation.
30. Outline the main differences between the Paris Agreement and the Kyoto Protocol.

Mitigation strategies

WHAT MITIGATION INVOLVES

Mitigation involves the reduction and/or stabilization of greenhouse gas (GHG) emissions and their removal from the atmosphere. Mitigation strategies to reduce GHGs may include:
- reducing energy consumption
- using alternative sources of energy to fossil fuels
- geo-engineering.

CARBON CAPTURE AND SEQUESTRATION (CCS)

Currently, when fossil fuels are burned, the CO_2 enters the atmosphere, where it may reside for decades or centuries. One potential solution is to capture the CO_2 before it is released into the atmosphere.

There are two main ways to do this:
- Capture the CO_2 at the site where it is produced (the power plant) and then store it underground in a geologic deposit.
- Allow the CO_2 to enter the atmosphere but then remove it using specially designed removal processes. This approach is called "direct air capture" of CO_2.

However, relatively little research and development has been undertaken to test the technological, economic and geologic feasibility for large-scale CCS.

CARBON TAXES

Some countries are introducing carbon taxes to encourage producers to reduce emissions of carbon dioxide. Taxes could be imposed relative to the proportion of carbon burnt.

CO_2 imposes high costs on society (including future generations) but those who emit the CO_2 do not pay for the social costs that they impose. The result is the lack of an incentive to shift from fossil fuels to renewables. Users of fossil fuel could pay an extra "carbon tax" equal to the social cost of the CO_2 emitted by the fuel. This would raise the costs of coal, oil and gas relative to wind and solar, for example, thereby shifting energy use towards the low-carbon options.

CARBON TRADING

Carbon trading is an attempt to create a market in which permits issued by governments to emit carbon dioxide can be traded. In Europe, carbon permits are traded through the Emissions Trading System (ETS). Governments set targets for the amount of carbon dioxide that industries can emit. Plants that exceed that limit are forced to buy permits from others that do not.

CARBON OFFSET SCHEMES

Carbon offset schemes are designed to neutralize the effects of the carbon dioxide human activities produce by investing in projects that cut emissions elsewhere. Some climate experts say offsets are dangerous because they dissuade people from changing their behaviour.

GEO-ENGINEERING

Geo-engineering schemes are large-scale engineering schemes that alter natural processes. For example, sulphate aerosol particles in the air could be used to dim the incoming sunlight and thereby cool the planet. Another idea is to place giant mirrors in space to deflect some of the incoming solar radiation. These are fairly radical, expensive and perhaps unworkable ideas.

OCEAN FERTILIZATION

Carbon dioxide absorption can be increased by fertilizing the ocean with compounds of iron, nitrogen and phosphorus. This introduces nutrients to the upper oceans, increases marine food production and takes carbon dioxide from the atmosphere. It may trigger an algal bloom, which can trap carbon dioxide and sink to the ocean floor.

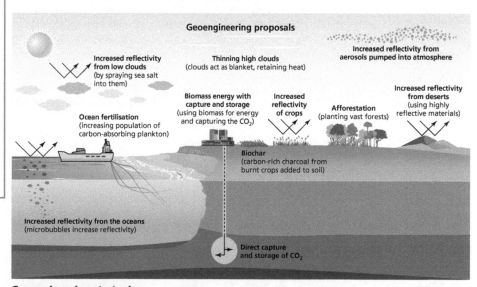

Geoengineering strategies

CHECK YOUR UNDERSTANDING

31. Outline the main constraints of climate mitigation strategies.

32. Explain how carbon taxes work.

Responding to climate change **151**

Civil society and corporate strategies to address global climate change

CIVIL SOCIETIES

There are many examples of civil societies engaged in global climate change, including the World Wide Fund for Nature (WWF) and Greenpeace. For example, the WWF is attempting to tackle climate change in a number of ways, by:
- pressurizing major economies and emerging economies to reduce greenhouse gas emissions
- calling on governments to sign up to international agreements to reduce the use of fossil fuels and work towards 100% renewable energy by 2050
- trying to encourage people to use new technologies and have a greener lifestyle.

One of WWF-UK's schemes is the One in Five Challenge, which aims to encourage companies and government agencies to reduce the environmental impact of their business travel and other activities. WWF claims it benefits the companies through:
- financial savings
- time savings
- improvements in productivity
- improvements in family life
- higher staff retention.

One company that took the One in Five Challenge was Vodafone. In 2010 it invested £600,000 in video-conferencing facilities. In the first five months after the investment, it spent 3,600 hours on video-conferencing and travelled 320,000 km less on business. The company saved about one-third of its previous costs of air travel.

CASE STUDY

CORPORATE CHANGE MITIGATION EFFORTS IN THE USA

In the USA, many corporations are becoming involved in climate mitigation schemes. In 2007, 28 companies formed the US Climate Action Partnership (USCAP), including Chrysler, General Electric, General Motors, Rio Tinto, Shell and Siemens. USCAP has lobbied the government to set legally binding CO2 emissions targets of 80% by 2050.

In 2008, Levi Strauss, Nike, Starbucks and others joined with the Coalition for Environmentally Responsible Economies to form the Business for Innovative Climate and Energy Policy (BICEP). BICEP calls for a reduction in greenhouse gas emissions to 25% below 1990 levels by 2020, and 80% by 2050.

Nevertheless, many US companies continue to lobby the government to block such measures, and many US citizens do not want their government to pursue policies that they feel would lead to a decline in US competitiveness and job losses.

WWF and IKEA

WWF and IKEA are currently working together on six climate projects:
- The Climate Positive project is an umbrella for all the climate projects that IKEA and WWF are working on. It is mapping various opportunities for IKEA to create a general positive climate impact in society, not only by reducing tis own emissions, but also by thinking and acting on how IKEA's impact effects others.

- Promoting a Sustainable Life at Home project will try different methods of helping customers reduce their CO_2 emissions. The project started in spring 2010 and through smart products and targeted communication it will help customers change their behaviour through more sustainable living at home.
- Improving the IKEA Food range from a climate perspective project aims to find solutions to reduce CO_2 emissions from the food products that IKEA offers in its restaurants and food markets and at the same time offer healthy and appetizing meals.
- Closing the loops project will investigate the impact of full recyclability and how cyclic systems can save scarce resources, minimize CO_2 emissions and optimally have a positive impact on the environment. It will explore how to close material loops, with specific focus on recyclable products and use of recycled materials in new products. It will also investigate how to secure that products can be reused, recycled or – from an environmental acceptable way – be returned to nature.
- Sustainable transportation of People (SToP) project aims to develop models and tools to help decrease CO_2 emissions from IKEA-customer transportation. The tools will offer guidance to local IKEA stores on strategies to achieve a more sustainable transportation of people.
- Developing Climate Positive Opportunities for Suppliers project aims at achieving significant reductions in energy efficiency at IKEAs suppliers. The goal is to remove barriers and to promote a low carbon IKEA supply chain.

https://www.yumpu.com/en/document/view/36410665/wwf-and-ikea-co-operation-climate-projects-global-hand

EXAM TIP
Try to keep up to date – changes in government can have significant impacts on climate change policy. At the start of 2017, President Trump in the USA signed deals to increase production of fossil fuels, whereas the Chinese President Xi Jinping indicated that China would lead to world in its attempts to increase production and use of clean energy.

CHECK YOUR UNDERSTANDING
33. Suggest how WWF is attempting to reduce the impacts of climate change.
34. To what extent have corporate strategies helped address global climate change?

Exam practice

Refer to the diagrams below and on page 27, which show the potential impact of climate at selected temperatures.

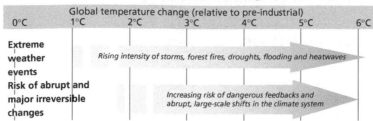

Projected impacts of climate change

(a) (i) Identify the impact that will occur with the least global warming. (1 mark)
(ii) Identify **one positive** impact of global warming. (1 mark)
(iii) State the temperature at which dangerous feedback and abrupt large-scale shifts in climate appear. (1 mark)
(iv) Briefly explain **two** problems associated with water and global warming. (2 + 2 marks)
(v) Suggest why farm yields may decrease in many developed regions. (3 marks)

The infographic shows the impacts of global warming that are already affecting King County, Washington, USA.

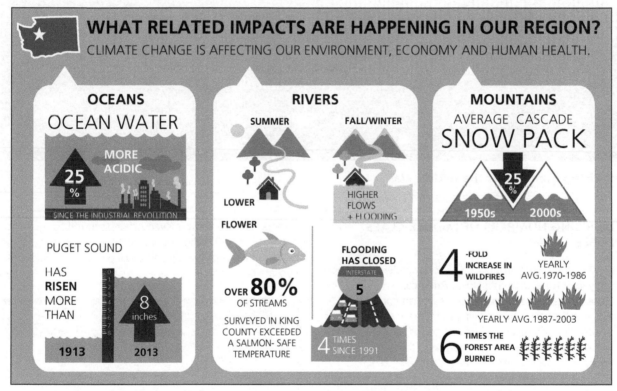

Source: http://kingcounty.gov/~/media/services/environment/climate/images/infographic/CC-inforgraphic-04.ashx?la=en

(b) (i) Outline the evidence for the increase in temperature in King County over the last century. (3 marks)
(ii) Suggest how global warming may have affected the economy of King County. (3 marks)
(iii) Outline the potential health effects of global warming on the residents of King County. (4 marks)

(c) **Either**

Examine the view that global climate change affects all people equally, whether rich or poor. (10 marks)

Or

Discuss the role of feedback loops in changes in the global energy balance. (10 marks)

1 GLOBAL TRENDS IN CONSUMPTION

Poverty reduction and the global middle class

MILLENNIUM DEVELOPMENT GOALS AND POVERTY REDUCTION

Between 1990 and 2015, the number of people living in extreme poverty fell from 1.9 billion to around 840 million.

The amount of people in LICs who were living in extreme poverty fell from 50% in 1990 to 14% in 2015. In contrast, the middle-class population in LICs – those living on at least $4/day – increased from 18% in 1900 to nearly 50% in 2015.

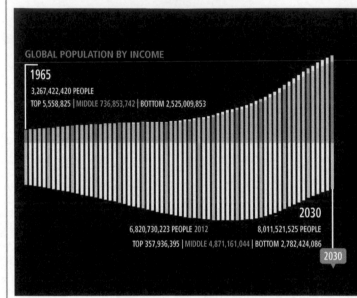

GLOBAL POPULATION BY INCOME

1965
3,267,422,420 PEOPLE
TOP 5,558,825 | MIDDLE 736,853,742 | BOTTOM 2,525,009,853

2030
6,820,730,223 PEOPLE 2012   8,011,521,525 PEOPLE
TOP 357,936,395 | MIDDLE 4,871,161,044 | BOTTOM 2,782,424,086

2030

TOP
The ranks of the world's wealthy will continue to get bigger as the emerging market nations create more millionaires.
Annual per capita expenditure over US$36,500 in 2005 price

MIDDLE
The biggest surge in new members of the middle class over the next 20 years will come from hundreds of millions of Chinese and Indians; the percentage of people in India and China below the middle will drop by 70% by 2030.
Annual per capita expenditure below US$36,500

BOTTOM
The percentage of poor people in the world has been on the rise for decades, but it will start to shrink as millions of Chinese and Indian citizens rise out of poverty.
Annual per capita expenditure below US$3,650

The rising numbers of the middle class

THE RISING NUMBERS OF MIDDLE-CLASS PEOPLE

In 2009, there were around 1.8 billion middle-class people, mainly in Europe, Asia and North America.

The increasing middle-class sector is an important economic feature, since it helps to increase sales of consumer goods such as electrical goods, mobile phones and cars. Sales of cars and motorbikes, for example, increased by over 800% between 2009 and 2015. However, continued growth is not always guaranteed. For example, during the 1960s Brazil and South Korea had similar incomes and economic growth rates. By the 1980s, Brazil's middle class accounted for less than 30% of the population, whereas Korea's was over 50%.

Not all the middle class has economic security. Many people on incomes of over $4/day remain vulnerable to unemployment and underemployment, especially those working in informal activities.

There has also been uneven progress in reducing poverty. Over 800 million people still live in extreme poverty and approximately half of all global workers work in unsafe conditions.

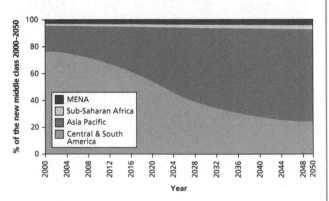

The growth of the middle class by world region

CHECK YOUR UNDERSTANDING
1. Define the term "middle class".
2. Identify the origin of the projected rise in middle class by 2050.

Global consumption of resources

ECOLOGICAL FOOTPRINTS

An ecological footprint is the hypothetical area of land required by a society, a group or an individual to fulfil all their resource needs and assimilate all their wastes. It is measured in global hectares (gha). A country with an ecological footprint of 3.2 gha is consuming resources and assimilating its wastes on a scale that would require a land area 3.2 times larger than the actual size of the country.

Ecological footprints can act as a model for monitoring environmental impact. They can also allow for direct comparisons between countries/areas, such as comparing LICs and HICs. They can highlight sustainable and unsustainable lifestyles, for example, populations with a larger footprint than their land area are living beyond sustainable limits. The available biological capacity for the population of the Earth is about 1.3 hectares of land per person (or 1.8 global hectares if marine areas are included as a source of productivity).

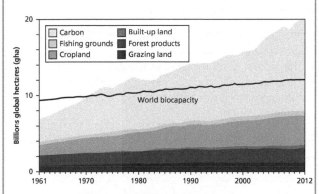

Changes in the components of the ecological footprints, 1961 to 2012

ECOLOGICAL FOOTPRINTS IN HICs AND LICs

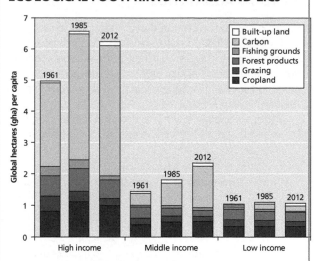

Changes in ecological footprints for HICs, MICs and LICs over time

LICs tend to have smaller ecological footprints than HICs because of their much smaller rates of resource consumption. In HICs, people have more disposable income, leading to greater demand for and consumption of energy resources. HICs' resource use is often wasteful and HICs produce far more waste and pollution. People in LICs, by contrast, have less to spend on consumption and the informal economy in LICs is responsible for recycling many resources. However, as LICs develop, their ecological footprint size increases.

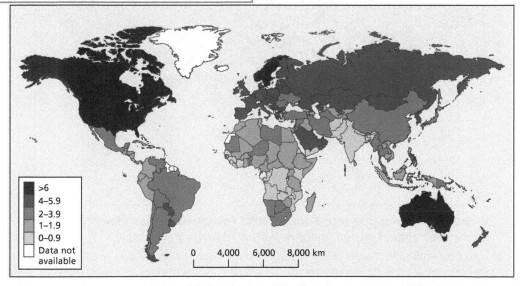

The ecological footprints (in global hectares) of countries around the world, 2014

EXAM TIP
Remember to use the units for ecological footprints – they are measured in global hectares (gha)

CHECK YOUR UNDERSTANDING
3. Describe how the component parts of the ecological footprint changed between 1961 and 2012.
4. Compare the size and composition of the ecological footprint of HICs, MICs and LICs between 1961 and 2012.

Water availability and consumption

PATTERNS AND TRENDS

Water is unevenly distributed over the world; over 780 million people do not have access to safe water. Demand for clean water will increase due to population growth and rising standards of living. The increased demand for water for renewable energy will further stretch the Earth's water resources.

Water availability is likely to decrease in many regions. For example, up to 300 million people in sub-Saharan Africa live in a water-scarce environment, and climate change will increase water stress in many other areas too, such as Central and Southern Europe.

HICs are maintaining or increasing their consumption of resources such as water. The average North American and Western European adult consumes 3 m³/day, compared with around 1.4 m³/day in Asia and 1.1 m³/day in Africa. An increasing proportion of this water is embedded in agricultural and manufactured products.

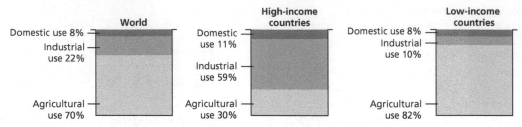

Water use in the two main income groups of countries and the world

More water will be required to produce food for the world's growing population, partly because of changes in diet. Much of the growth will be in LICs, many of which are already experiencing water stress.

A number of trends are increasing the pressure to manage water more efficiently. These include:
- population growth, which is predicted to reach 9 billion by 2050 and may eventually peak at 11 billion.
- the growing middle class – increasing affluence leads to greater water consumption
- the growth of tourism and recreation
- urbanization – urban areas require significant investment in water and sanitation facilities
- climate change.

VIRTUAL WATER

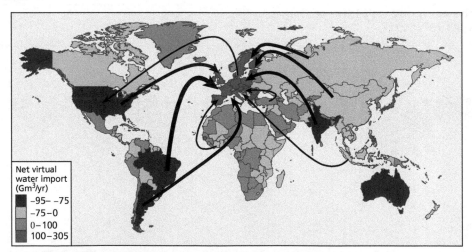

Virtual water imports into Europe, 2011 (negative values identify exports of virtual water, positive values identify imports)

The concept of virtual (or embedded) water refers to the way in which water is transferred from one country to another through its exports, such as in foods, flowers or manufactured goods. It allows countries with limited water resources to "import or outsource" their water from countries that have more water resources. It also allows a country to reduce the use of its own water resources by importing goods.

❓

Availability of land and food consumption

PATTERNS AND TRENDS

Food intake (measured in calorie intake) has steadily increased as the world's population has increased. Calorie intake has remained steady in sub-Saharan Africa, but it has increased dramatically in East Asia, the Middle East and North Africa. However, land availability/person has declined in many areas due to a combination of rapid population growth, land-use changes and land grabs by foreign companies. Increasing urbanization is another factor affecting the demand for meat and the availability of land.

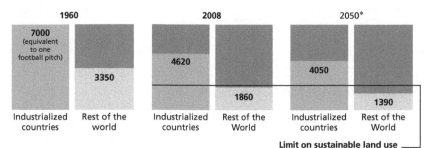

Changes in agricultural land use per capita (square metres)

As countries develop, there is a change in diet away from cereals towards a more varied diet including meat, vegetables and dairy products. For example, between 1964 and 1999, meat consumption per capita increased by 150% in LICs, and consumption of milk and dairy products rose by 60%. Global meat production was 218 million tonnes in 1998–9 and is predicted to reach 376 million tonnes by 2030.

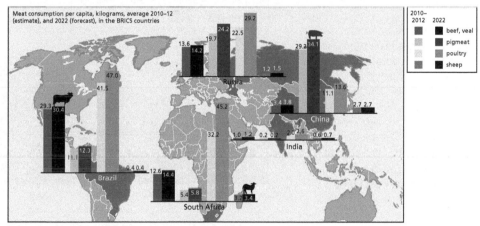

Demand for meat is satiated in HICs, but rising steeply in LICs

Since 1961, total fish supply and consumption have increased by about 3.6% per year while the world's population has grown by 1.8%. However, production from the world's ocean fisheries has levelled off since the 1970s, largely because of overfishing. However, the considerable growth in aquaculture has filled the gap left by the drop in wild fish stocks.

In recent years, the growth rates of food production and crop yields have been falling. Some of this reduction can be put down to natural hazards (fires, floods, drought), global climate change, and the use of land to produce biofuels. Food shortages have led to riots in Indonesia, Egypt and across North Africa. There are three main ways of increasing crop production:
- extensification – expanding the area farmed
- multicropping – harvesting two or more crops a year
- intensification, for example, using high-yielding varieties or genetically modified organisms.

Some scientists believe that all three of these may be reaching their limits. Others suggest that there are many other ways of increasing food production, including reducing food losses, improving food storage, in vitro farming and vertical farming.

As the world's population growth slows, and high rates of food consumption per person are reached, the growth in the demand for food will decrease. According to the FAO, the growth in the demand for food is likely to fall from 2.2% per year to 1.5% per year by 2050. However, this is still an increase.

CHECK YOUR UNDERSTANDING
7. Suggest why land availability/head is falling in many parts of the world.
8. Suggest why crop yields and crop production have fallen in some locations.

Availability and consumption of energy

PATTERNS AND TRENDS

The period from 1985 to 2003 was an era of energy security, but since 2004 there has been an era of energy insecurity. Following the energy crisis of 1973 and the Iraq War (1990–91), there was a period of low oil prices and energy security. However, insecurity has risen for many reasons, including:
- increased demand, especially from newly industrializing countries (NICs)
- decreased reserves as supplies are used up

- geopolitical developments, for example, the conflict between Russia and Ukraine
- global warming and natural disasters have increased awareness about the misuse of energy resources
- terrorist activity such as in Syria.

For most countries, a diversified energy mix offers the best energy security. Depending on a single source, especially from a single supplier, is more likely to lead to energy insecurity.

RENEWABLE AND NON-RENEWABLE RESOURCES

Energy can be generated from both renewable and non-renewable resources.

Non-renewable energy supplies include fossil fuels (such as coal, gas and oil). These provide most of mankind's energy supply and their use is expected to increase to meet global energy demand. Fossil fuels cannot be renewed at the same rate as they are used; this results in depletion of the stock. Nuclear power can be considered non-renewable because uranium is a non-renewable resource.

Renewable energy sources include solar, hydroelectric, wind and tidal schemes. They can be large scale or small scale, within single houses or communities. Renewable energy resources are sustainable because there is no depletion of natural capital.

RENEWABLES AND ALTERNATIVE ENERGY SOURCES

Sources of energy with lower carbon dioxide emissions than fossil fuels include renewable energy (solar, biomass, hydropower, wind, wave, tidal and geothermal) and their use is expected to increase. Nuclear power is a low-carbon, low-emission, non-renewable resource but it is controversial due to the problem of storing the radioactive waste, the cost of decommissioning nuclear power stations, and the potential scale of any accident.

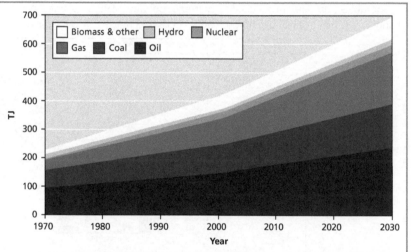

The world's fuel sources, 1970–2030 (1 TJ = 1,000 gigajoules)

CHANGES IN DEMAND

The major consumers of energy are the HICs and NICs, and consumption has been increasing rapidly. However, the pattern of demand is projected to shift from the OECD region to NICs. Energy resources are used in large quantities for manufacturing and transport. The world continues to use fossil fuels despite the growth in renewable energy sources. This is partly because resources are still available, the infrastructure is already in place, and, in some cases, the energy companies are important sources of revenue for governments.

The energy consumption of LICs will continue to grow faster than that of HICs, but the consumption of energy by LICs is still low in comparison with that of HICs.

COMMON MISTAKE
✗ *Nuclear power is a renewable source of energy.*
✓ Nuclear power is not a renewable source of energy – it is an alternative to fossil fuels. Carbon dioxide is released during the building of nuclear power stations but relatively little is released once the plant is built and energy is being produced.

CHECK YOUR UNDERSTANDING
9. Describe how the world's fuel sources are predicted to change between 2020 and 2030.
10. Outline the factors that have contributed to increased energy insecurity in many countries.

The water–food–energy nexus

DEFINITION

The water–food–energy nexus refers to the close links between the three sectors and the ways in which changes in one sector have an impact on one or both of the other sectors. The nexus approach stresses the need for stewardship of these resources.

- **Water security** is defined in the SDGs as "access to safe drinking water and sanitation".
- **Food security** is defined by the FAO as the "availability and access to sufficient, safe and nutritious food to meet the dietary needs and food preferences for an active and healthy life".
- **Energy security** refers to having access to clean, reliable and affordable energy sources for cooking, heating, lighting, communications and productive uses.

INTERACTIONS

There is a range of interactions between water, food and energy. For example, water is essential for the mining, refining and transport of energy sources, as well as being necessary to produce crops and rear livestock. Food production is the world's largest single use of water, accounting for about 70% of water usage. In turn, food production may affect water quality and quantity, through water extraction, water pollution (eutrophication, salinization) and land-use changes. Energy is needed for the extraction, transport and distribution of water, and for the construction of dams. Energy is needed for farming, for example, for fertilizers, for machinery and for transport. Agriculture accounts for about 30% of global energy use.

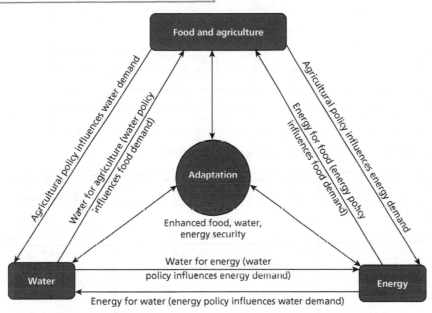

Interactions between water, food, energy and adaptation

Many countries face difficulties providing for the growing demand for water, food and energy. The difficulties are compounded by the uncertainties of climate change. Adapting to climate change will require effective use of scarce resources and a coordinated response. The rapid increase in resource use since the mid-1990s, particularly in the emerging economies, has accelerated the risk of resource scarcity.

The nexus approach can be applied at a number of scales, from small-scale local competition over access to water, to large-scale global interactions between energy, food and water providers. At times there are positive impacts, such as the increased use of fertilizers leading to increase total food production. More often, though, the impacts are negative – the increased use of fertilizers leads to higher energy use and greenhouse gas emissions, and water pollution.

EXAM TIP

When discussing or analysing the nexus provide specific examples of interactions between water, energy and food, for example, how does the construction of a dam affect the provision of water, energy and food?

CHECK YOUR UNDERSTANDING

11. Explain the meaning of the term "nexus".
12. Outline some of the interactions between water, food and energy.

Resource security (1)

WATER SECURITY IN SOUTH ASIA, A NEXUS APPROACH

South Asia faces the challenge of providing sufficient water and energy to grow enough food for its expanding population. The Hindu Kush Himalayan (HKH) region is vital for the promotion of food, water and energy security downstream. The issues and challenges in the food, water and energy sectors are interrelated in many ways. Moreover, there is a high degree of dependency of downstream communities on upstream ecosystem services such as water for irrigation, HEP and drinking water.

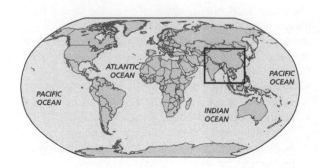

Key indicators in water security in South Asia

Indicator	2007	2050 projection
Population (millions)	1520	2242
Annual population growth rate (%)	1.5	0.53
Population below US$1.25/day (million)	596	14.1
Irrigated area (million ha)	104	135
Total water consumption in agriculture (km³)	1479	1922
Total water withdrawal for irrigation (km³)	1095	1817

Key features of, and challenges for, water security in South Asia

Key characteristics	Adaptation challenges	Interface among nexus resources and adaptation to climate change
Growing water stress		
Growing water demand for agriculture, energy, industry and human and livestock use: annual water demand is predicted to increase by 55% compared with 2005.	Providing access to safe drinking water in the face of increasing increasing variability in the water supply.	Water-intensive adaptation prectices leading to increased water pollution and water-borne diseases, high child mortality, poor human health.
Upstream-downstream dependence on water		
High dependence of downstream communities on the upstream for water to grow food and generate hydropower.	Need for enhanced upstream-downstream coordination and cooperation for sustainable development of Hindu Kush Himalayan (HKH) water resources.	HKH rivers are the lifeline for dry season water for irrigation, hydropower and major economic activities.
Increasing dependence on groundwater for food production		
About 70–80% of agricultural production depends on groundwater irrigation.	Adapting to declining water tables.	Groundwater pumping for irrigation requires excessive energy, which further increases electricity demand.

CHECK YOUR UNDERSTANDING

13. Outline why demand for water in the Hindu Kush Himalayan region is increasing.

14. Explain briefly how increasing use of groundwater has an impact on the energy sector.

Resource security (2)

FOOD SECURITY IN SOUTH ASIA, A NEXUS APPROACH

South Asia has just 3 per cent of the world's land but around 25 per cent of the world's population. Thus, water and food security are vital. South Asian countries are home to 40 per cent of the world's poor population, and over half the population is food-energy deficient. Moreover, about 20 per cent of the population lack access to safe drinking water. Thus there are major challenges in providing food and water security to the South Asian population.

Key indicators related to agriculture security in South Asia

Indicator	2007	2050 projection
Population (millions)	1520	2242
Undernourished population (%)	21.8	4.2
Arable land (million ha)	204	213
Irrigated area (million ha)	104	135
Cultivated land (ha per person)	0.12	0.08
Agricultural growth rate (%)	2.4	1.3

Key features of, and challenges for, food security in South Asia

Key characteristics	Adaptation challenges	Interface among food, water and energy resources and adaptation to climate change
Huge chronically undernourished population		
About half of the world's poor (46%) and 35% of the world's undernourished live in South Asia.	Provision of food, water and energy to a large malnourished population without degrading the natural resource base and environment.	To meet the nutritional needs of all, food production to double in the next 25 years.
Burgeoning human population		
About 25% of the world's population) lives in just 3% of the world's land area.	To feed the growing population, agricultural production will have to increase by 70%.	Increased pressure on land, water and energy to meet demand.
Declining cropland		
Per capita arable land continually declining due to population growth, urbanization, and increasing biofuel cultivation to meet energy demand.	Limited option for growing more food grain by expanding crop area.	Competing demand for land for food, bio-energy production, and ecosystem services.
Intensive food production		
Food production becoming increasingly water- and energy-intensive.	Adapting to the declining groundwater table.	Agricultural growth is constrained due to shortage of enegy and water.
Changing food preferences towards meat		
The meat production process requirs more energy and water.	About 7 kg of grain equivalent is required to produce 1 kg of meat.	Increased pressure on water to meet the food requirement.
Sensitivity to climate change		
Food production is highly vulnerable to climate change due to rising temperatures, accelerated glacial melting, increased evapotranspiration, and erratic rainfall.	Uncertainty in water availability due to rapid glacier melt and changes in monsoon pattern in the Himalayas.	Climate change is likely to be a critical factor in increasing water and energy demand for food production and land demand for biofuel production.

CHECK YOUR UNDERSTANDING
15. Explain why South Asia has a major problem with food production.
16. Outline the reasons why arable land/head is declining in South Asia.

Resource security (3)

ENERGY SECURITY IN SOUTH ASIA, A NEXUS APPROACH

Just as food and water are essential for human existence, energy is key to human development. The nexus approach stresses the need for cooperation among the water, food and energy sectors, despite the competition for scarce resources. The ecosystem services provided by the Hindu Kush Himalayan (HKH) region are vital for the security of all three sectors.

Key characteristics and challenges for energy security in South Asia

Key characteristics	Adaptation challenges	Interface among food, water and energy resources and adaptation to climate change
High energy poverty		
About 63% of the population without access to electricity; 65% use biomass for cooking.	Providing adequate and reliable energy to a large population without increasing pollution.	Growing demand for water and land for energy production.
Under-utilised potential for hydropower and clean energy		
Hydropower in the Himalayas is limited in places due to the risk of causing landslides.	Adaptation options are restricted.	Energy diversification to meet the growing demand for food, water and economic growth.

The challenges that face South Asia include population growth, rapid urbanization, industrialization as well as the uncertainties of climate change. These changes are leading to increased demand for, and pressure on, resources. Most ecosystem services are used and managed at a variety of scales and by a variety of stakeholders, for example, farmers, politicians, industrialists, water engineers and urban populations.

The HKH is the source of water for hydroelectric power. However, the region is experiencing deforestation, land degradation, soil erosion, overgrazing and declining productivity. Soil erosion has led to an increase in the frequency and severity of flooding. Water quality and quantity are adversely affected by land-use changes. Without proper ecosystem management in the HKH, water, food and energy security are all at risk.

THE IMPACT OF LANDSLIDES IN NEPAL

In 2014 heavy rain caused a massive landslide from the hillside in Jure, in Nepal's central region. It created a high artificial dam across the Saptakoshi River, one of the main tributaries of the Koshi River, blocking the flow of water completely. Around 5,000 families were displaced and dozens of houses destroyed. It was the deadliest landslide in Nepal in a decade. The landslide also covered the main highway to China in mud and debris 20 metres deep, and blocked the Sun Koshi River roughly 80 km east of Kathmandu. Water quickly pooled behind the rubble, forming a lake that submerged a small hydropower station three kilometres upstream. The lake posed a flood risk for at least 400,000 people in two countries. The landslide damaged a hydroelectric power station downstream and cut electrical transmission lines along the valley. In all, nearly a tenth of the nation's hydroelectric capacity, some 67 megawatts, was severed, leading to power cuts in the capital and elsewhere in the country.

Nepal's undeveloped hydropower potential is second highest in the world, behind Brazil. Nepal's deep narrow canyons would be an ideal site for HEP, if it were not for monsoonal rain and the risk of landslides. Nepal's energy ministry set a goal in 2010 of building 37,000 megawatts of new hydropower capacity within 20 years. Heavy rain from 14 to 16 August, caused massive floods and several landslides in 18 districts throughout the country. It destroyed crops, contaminated water and once again cast doubts on whether Nepal should try to develop its HEP project, despite abundant raw materials.

COMMON MISTAKE

✗ *HEP does not have any negative environmental impacts.*

✓ Although HEP is a clean form of energy, it has negative environmental impacts. For example, it may increase the risk of landslides and earthquakes, it disturbs natural habitats, and much CO_2 is released during the construction of the dam.

CHECK YOUR UNDERSTANDING

17. Suggest reasons why access to biomass may be limited to many (poor) people in future.
18. Explain why it may be difficult for Nepal to develop its HEP potential.

Climate change

INFLUENCE ON THE NEXUS

Climate change could influence the water–food–energy nexus in many ways. In some areas it may reduce agricultural productivity, whereas in other areas it may increase it. Water supplies will diminish in some areas and increase in others. Climate change is expected to increase the frequency of climate-related shocks, and these will have an impact on food, water and energy supplies. Moreover, due to their interdependence, an impact on one part will have an effect on the other two.

Climate change will influence food availability, crop yields, water availability, and the distribution of pests and diseases. Higher temperatures and evapotranspiration rates will reduce water supplies and increase the need for irrigation. Energy demand may rise due to the increased need for the pumping or treatment of water. There may be increased competition between the agricultural and energy sectors for scarce water resources.

Moreover, attempts to limit climate change may have an impact on the water–food–energy nexus. For example, the production of biofuels may create new demands for water resources. The use of drip irrigation and desalinization of seawater, are very energy intensive. Increased groundwater use would also require extra pumping and therefore more energy resources.

There are a number of benefits of adaptations to climate change that have knock-on effects for the nexus.

Synergies between the climate change adaptation and nexus approaches

Sector-specific adaptation measures	Positive implications for the sector	Potential for synergies across the nexus
Water		
Increasing water use efficiency	Reduced water per capita	Increased availability of water for energy and agriculture
Switching from use of freshwater to waste water	Reduces freshwater use per capita	Increased availability of freshwater for food, energy and other uses
Switching from wet to dry cooling at thermoelectric power plants	Reduces water use and associated thermal pollution	Increased availability of water for energy and agriculture
Desalinization	Increase in brackish and freshwater supplies	Increased availability of freshwater and overall water supply for energy and agriculture and other uses
New storage and conveyance of water to serve new demands	Increased water supplies to meet demand	Increased availability of freshwater and overall water supply for energy and agriculture and other uses
Watershed management	Increased water suplies to meet demand	Increased water supply for energy and other uses, improved water quality, reduction in flood potential
Land		
Switching to drought-tolerant crops	Increased/maintained crop yields in drought areas	Reduced water demand
Using waste or marginal lands for biofuels	Increase in renewable energy	Reduced pressure on non-renewable energy as some fossil fuels are replaced with biofuels
Energy		
Increasing transmission capacity	Reduced economic and social impacts	Potential for reduced emissions if new transmission and wind/solar power supplied to the grid
Increasing renewable energy, for example, solar, wind, biogas, bioenergy	Increased clean energy and reduced pressure on energy	Reduced GHG emissions, reduced water demand for cooling, thermal power

CHECK YOUR UNDERSTANDING

19. Outline one advantage and one disadvantage of desalinization as a potential source of freshwater.
20. Outline two benefits of increasing supplies of renewable energy.

The disposal and recycling of waste

SOLID DOMESTIC WASTE

Waste management options for solid domestic waste

Waste management options	How it works
Reduce the amount of waste	• Producers think more about the lifespan of goods and reduce packaging • Consumers consider packaging and lifespan when buying goods
Reuse goods to extend their lifespan	• Bring-back schemes where containers are refilled (e.g. milk bottles) • Refurbish/recondition goods to extend their useful life, e.g. use of old car tyres to stabilize slopes/reduce soil erosion • Used goods put to another use rather than thrown out (e.g. plastic bags used as bin liners; old clothes used as cleaning cloths) • Charity shops pass on goods to new owners
Recover value	• Recycle goods such as glass bottles and paper • Compost biodegradable waste for use as fertilizer • Incinerate (burn) waste – collect electricity and heat from it
Dispose of waste in landfill sites	• Put waste into a hole (natural or the result of quarrying) or use to make artificial hills

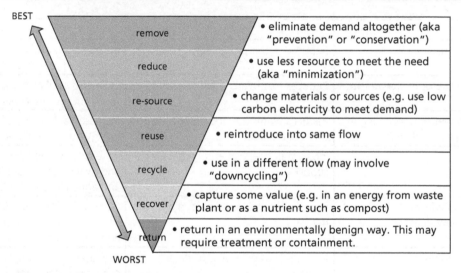

BEST

remove	• eliminate demand altogether (aka "prevention" or "conservation")
reduce	• use less resource to meet the need (aka "minimization")
re-source	• change materials or sources (e.g. use low carbon electricity to meet demand)
reuse	• reintroduce into same flow
recycle	• use in a different flow (may involve "downcycling")
recover	• capture some value (e.g. in an energy from waste plant or as a nutrient such as compost)
return	• return in an environmentally benign way. This may require treatment or containment.

WORST

Methods of waste disposal

INTERNATIONAL FLOWS OF WASTE

China imports more than 3 million tonnes of waste plastic and 15 million tonnes of paper and cardboard each year. Containers arrive in many countries, with goods exported from China, and load up with waste products for the journey back. A third of the UK's waste plastic and paper is exported to China each year. Low wages and a large workforce mean that this waste can be sorted much more cheaply in China, despite the distance it is transported.

CHECK YOUR UNDERSTANDING
21. Suggest why "remove" is considered to be the best form of waste disposal.
22. Outline two positive and two negative impacts of the disposal of e-waste in Guiyu, China.

E-WASTE

In 2012 China generated over 11 million tonnes of e-waste, followed by the US with around 10 million tonnes. However, per capita figures were reversed: on average, each American generated around 30 kg of e-waste, compared to less than 5 kg per person in China. The European Environment Agency estimates that between 250,000 tonnes and 1.3 million tonnes of used electrical products are shipped out of the EU every year, mostly to west Africa and Asia. In 2010, the US discarded around 260 million computers, monitors, TVs and mobile phones, of which only 66% were recycled. The life of a mobile phone is now less than two years. In 2011 in the US, although 120 million phones were bought, only 12 million mobile phones were collected for recycling.

Guiyu in China has been described as the e-waste capital of the world. The industry is worth $75 million to the town each year, but Guiyu's population has high rates of lead poisoning, cancer-causing dioxins, and miscarriages.

Contrasting views of population and resources (1)

NEO-MALTHUSIANS

In 1798 the Reverend Thomas Malthus produced *An Essay on the Principle of Population Growth*. He believed that population growth would potentially outstrip the growth of food production, leading to a decline in the standard of living and to "war, famine and disease".

THE LIMITS TO GROWTH MODEL

This study, also referred to as the Club of Rome model (1970), examined the five basic factors that determine and therefore ultimately limit growth on the planet:
- population
- agricultural production
- natural resources
- industrial production
- pollution.

The model suggests that food output and population grow exponentially until the rapidly diminishing resource base forces a slowdown in industrial growth. Because of delays in the system, both population and pollution continue to increase for some time after the peak of industrialization. Population growth is finally halted by a rise in the death rate due to decreased food production.
   The team concluded that:
- if present trends continued, the limits to growth would be reached in 2070, probably resulting in a sudden and uncontrollable decline in population and industrial production
- it is possible to alter these growth trends and establish a condition of environmental and economic stability that is sustainable.

However, there are a number of criticisms of the Limits to Growth model:
- It is a world model and does not distinguish between different parts of the world.
- It ignores the spatial distribution of population and resources, of agricultural and industrial activity, and pollution.

- The model emphasizes exponential growth and not the rate of discovery of new resources or technologies.

Original limits to growth

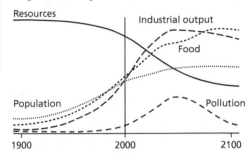

Sustainable limits to growth

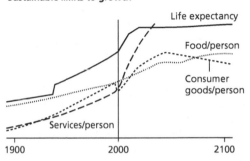

In a revised model, the authors introduced a number of new features:
- deliberate constraints on population growth
- trade-offs between population size and levels of material well-being
- technologies to conserve resources
- reductions in levels of pollution
- the use of renewable resources does not exceed their rate of regeneration
- the use of non-renewable resources does not exceed the rate at which sustainable renewable resources are developed
- the rate of pollution does not exceed the tolerance of the environment.

PAUL EHRLICH

Paul Ehrlich is an American biologist who wrote *The population bomb* (1995). The opening sentence of the first edition stated: "The battle to feed all of humanity is over. In the 1970s hundreds of millions of people will starve to death in spite of any crash programs embarked upon now. At this late date nothing can prevent a substantial increase in the world death rate."
   He is a neo-Malthusian who believes that population control, an increase in food supply, and the redistribution of wealth are needed in order to address the world's population problem.

He argued that population growth was outstripping the growth of food and resources. In 1980 he had a wager (bet) with the economist Julian Simon in which Ehrlich argued that the cost of raw materials would rise in future, whereas Simon argued that their price would come down. They wagered over five raw materials. The wager started in 1980 and lasted until 1990. Ehrlich lost the wager as all five commodities fell in price during the 1980s. However, had they chosen different commodities Ehrlich would have won, and had they chosen a longer period Ehrlich would have won. Thus, it is possible to reach very different conclusions based on different resources and timescales.

EXAM TIP
The Limits to Growth model and Paul Ehrlich are considered to be neo-Malthusians since they shared the same pessimistic view as Malthus, but they differed in terms of their proposed solutions.

CHECK YOUR UNDERSTANDING
23. Outline the differences between the Malthusian view and the neo-Malthusian view.
24. Identify the main limits to population growth.

Contrasting views of population and resources (2)

ESTHER BOSERUP'S THEORY OF POPULATION

Esther Boserup believed that people have the resources to increase food production, namely knowledge and technology. When a need arises, someone will find a solution.

Whereas Malthus thought that food supply limited population size, Boserup suggests that an increase in population stimulated a change in agricultural techniques so that more food could be produced. Population growth has thus enabled agricultural development to occur.

She examined different land-use systems, which varied in terms of intensity of production. At one extreme was the forest fallow association with shifting cultivation; at the other extreme was the multi-cropping system, which had more than one harvest per year. She believed that any increase in the intensity of productivity would be unlikely unless population increased. Thus, population growth will lead to agricultural development and the growth of the food supply.

Boserup's theory was based on the idea that people knew the techniques required by a more intensive system, and adopted them when the population grew. If that knowledge was not available, the agricultural system would limit the population size in a given area.

CARRYING CAPACITY

The concept of a population ceiling is of a saturation level, where population equals the carrying capacity of the local environment. There are four models of what might happen as a growing population approaches carrying capacity:

OPTIMUM, OVER- AND UNDERPOPULATION

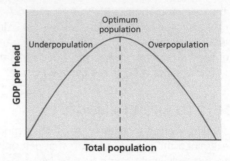

Optimum, over- and underpopulation

Optimum population is the number of people who, when working with all the available resources, will produce the highest per-capita economic return. It is the highest standard of living and quality of life. If the size of the population increases or decreases from the optimum, the standard of living will fall.

Overpopulation occurs when there are too many people, relative to the resources and technology locally available, to maintain an adequate standard of living, for example, South Sudan and Somalia.

Underpopulation occurs when there are far more resources in an area than can be used by the people living there. Countries such as Canada can export their surplus food, energy and mineral resources.

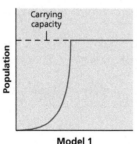

The rate of increase may be unchanged until the ceiling is reached, at which point the increase drops to zero. This highly unlikely situation is unsupported by evidence from either human or animal populations.

Model 1

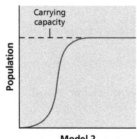

Here the population increase begins to taper off as the carrying capacity is approached, and then to level off when the ceiling is reached. It is claimed that populations that are large in size, with long lives and low fertility rates, conform to this S-curve pattern.

Model 2

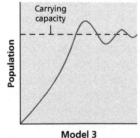

The rapid rise in population overshoots the carrying capacity, resulting in a sudden check, for example famine, birth control; after this, the population recovers and fluctuates, eventually settling down at the carrying capacity. This J-shaped curve appears more applicable to small populations with short lives and high fertility rates.

Model 3
Carrying capacity

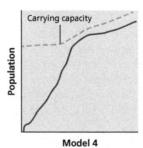

New innovations (irrigation, terracing, wetland drainage, high yielding varieties and chemical fertilizers etc.) allow the world to support a larger population, and the carrying capacity increases.

Model 4

CHECK YOUR UNDERSTANDING
25. Outline the main differences between model 4 and models 1–3 in terms of carrying capacity.
26. Outline ways in which carrying capacity can be increased.

Resource stewardship (1)

Resource stewardship suggests that humans can use resources in such a way that they will be available to future generations. It suggests that there will be not only environmental sustainability but also social equity over access to resources. The global commons refer to areas that lie outside the political reach of any nation state: the high seas, Antarctica, the atmosphere and outer space. The global commons require management and protection. Renewable resources such as fish need not be depleted, provided the rate of use does not exceed maximum sustainable yield. If resources become over-exploited, depletion and degradation will lead to scarcity.

THE TRAGEDY OF THE COMMONS

Garrett Hardin, an American ecologist, used the term "the tragedy of the commons" to explain the lack of control over the way common resources are used and how the selfish acts of a few countries/individuals can destroy the resource for others. In any given ocean, for example, many nations may be fishing. Apart from the seas close to land, where there is an exclusive economic zone (EEZ), no country owns the oceans or the resources they contain.

If one country takes more fish from the oceans, their profits increase. The "tragedy" is that other countries feel compelled to increase their catch in order to match the scale of the country that initiated the increase. Thus the rate of use may exceed maximum sustainable yield and the resources may become depleted. The tragedy of the commons explains the tendency to over-exploit shared resources and the need for agreements over common management.

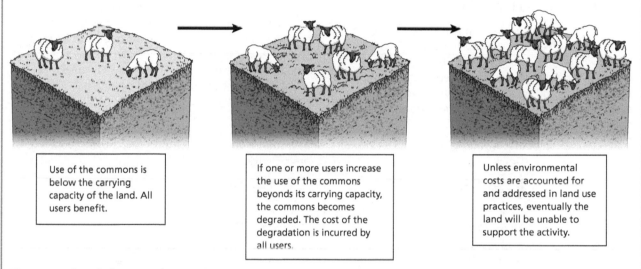

Use of the commons is below the carrying capacity of the land. All users benefit.	If one or more users increase the use of the commons beyonds its carrying capacity, the commons becomes degraded. The cost of the degradation is incurred by all users.	Unless environmental costs are accounted for and addressed in land use practices, eventually the land will be unable to support the activity.

The tragedy of the commons

The US Department of Agriculture has developed a resource stewardship evaluation programme to help farmers identify their conservation goals and improve their outcomes. It enables them to monitor progress in achieving their goals in relation to soil management, water quality, water quantity, air quality and wildlife habitat. An example of cropland stewardship achievement and goals is shown:

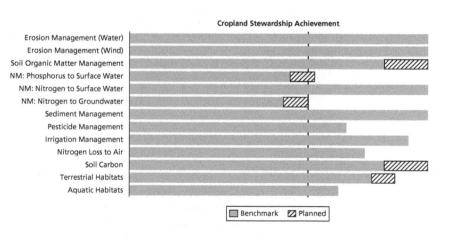

Cropland Stewardship Achievement

Erosion Management (Water)
Erosion Management (Wind)
Soil Organic Matter Management
NM: Phosphorus to Surface Water
NM: Nitrogen to Surface Water
NM: Nitrogen to Groundwater
Sediment Management
Pesticide Management
Irrigation Management
Nitrogen Loss to Air
Soil Carbon
Terrestrial Habitats
Aquatic Habitats

Benchmark   Planned

CHECK YOUR UNDERSTANDING
27. Define the term "resource stewardship".
28. Explain the meaning of the term "optimum population".

Resource stewardship (2)

THE CHARACTERISTICS OF A CIRCULAR ECONOMY

A **circular economy** is one that preserves natural capacity, optimizes resource use and reduces loss through managing finite stocks and renewable flows. It is an economy that restores and regenerates resources, and keeps products, materials and components at their highest utility and value always. A circular economy aims to rebuild capital, whether it is financial, manufactured, natural, social or human. In a completely circular economy, consumption only occurs in bio-cycles, in which resources can be recovered and restored.

There are three principles behind the circular economy:
- Preserve and enhance natural resources by controlling non-renewable resources and balancing renewable resources.
- Optimize resource yields by recycling and remanufacturing products, materials and components.
- Improve effectiveness by eliminating negative externalities such as pollution and climate change.

In the circular economy, waste is minimized by biological materials. Artificial materials are designed for repeated use. Systems are designed to run on renewable energy. For example, agriculture could be run on solar energy.

CASE STUDY

RESOURCE USE: MOBILE PHONES

In most HICs, many people change their phone every few years. In 2010 the volume of waste electrical and electronic equipment (WEEE) in the EU was 750,000 tonnes. Although mobile phones are light (less than 160 g), their materials and components have considerable economic value, since they contain resources such as gold, silver and rare earths. In the EU, some 160 million mobile phones are discarded annually, representing a loss of materials of $500 million a year. Only about 15% of phones are currently collected and recycled. Collecting reusable components and remanufacture could be made easier if the design of certain parts of a phone were standardized. The main parts that could be remanufactured include the charger, battery, camera and display. Recycling of phones would generally occur close to the market and result in reduced imports of phones, which are mainly produced outside the EU.

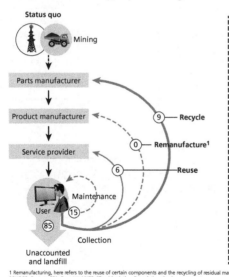

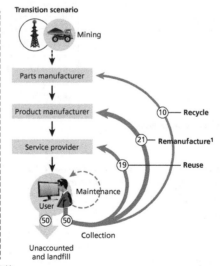

1 Remanufacturing, here refers to the reuse of certain components and the recycling of residual materials
SOURCE: Gartner; EPA; Eurostat; UNEP; Ellen MacArthur Foundation circular economy team

Resource use: Car manufacturing

The car industry accounts for 60% of the world's lead use, which is predicted to run out by 2030. In Cuba many old cars have been remanufactured using spare parts from other machines to keep them viable. At Choisy-le-Roi, in south-east Paris, the savings from the remanufacture of cars include 80%t less energy, 88% less water, 92% fewer chemical products, and produces 70% less waste. Around 43% of car bodies are reusable and 48% are recycled for new parts.

COMMON MISTAKE
✗ Some students believe that there is no waste disposal in the circular economy.
✓ In a completely circular economy this is the case, but in most cases, there is some waste, which is minimized, and therefore there is limited disposal.

CHECK YOUR UNDERSTANDING
29. Define the term "circular economy".
30. Briefly explain how the mobile phone industry could be made more "circular".

The Sustainable Development Goals (SDGs)

The 17 SDGs were introduced in 2015, and are set to exist until 2030. They replaced the Millennium Development Goals (MDGs) that existed between 2000 and 2015.

The 17 SDGs

Goal 1 calls for an end to extreme poverty by 2030.

In 2015, around 10% of the world's population lived on less than $1.90 a day.

Goal 2 aims to end hunger and malnutrition by 2030, and seeks to achieve sustainable food production by 2030. There are still more than 800 million people who lack access to sufficient food.

Goal 3 focuses on improving reproductive, maternal and child health, reducing epidemics of infectious diseases, lowering the incidence of degenerative diseases, and improving access to health care for all. Between 2000 and 2015, there were 200 million cases of malaria.

Goal 4 aims to improve education and training. In 2013, around 60 million primary school children did not attend school. Most of these come from the poorest households.

Goal 5 is to empower women and girls to achieve their full potential. In Sub-Saharan Africa and Western Asia, girls still face barriers to entering primary and secondary school.

Goal 6 aims to ensure the availability and sustainable management of water and sanitation. Water scarcity affects 40% of the world's population and is predicted to rise.

Goal 7 aims to promote access to affordable energy, and to increase the use of renewable energy sources. There were still more than 1 billion people without access to electricity in 2012.

Goal 8 is to promote sustained, inclusive and sustainable economic growth, full and productive employment, and decent work for all. It also is attempting to eradicate forced labour, human trafficking and child labour. More than 200 m people are unemployed – 470 m new jobs will be required by 2030.

Goal 9 plans to promote infrastructure development, sustainable industrialization and innovation. Basic infrastructure like roads, sanitation, electrical power and water remains scarce in many LICs.

Goal 10 aims to reduce inequalities within and between countries. This includes inequalities in wealth, gender, age, race, class, ethnicity, religion and opportunity. A significant majority of households in LICs are living in societies which are more unequal than they were in the 1990s.

Goal 11 aims to make settlements more sustainable and to promote community cohesion, personal security, innovation and employment. Nearly one billion people live in slums, and the number if rising.

Goal 12 focuses on the need for sustainable consumption and production patterns, that is, the "circular economy" in which today's waste becomes tomorrow's raw materials and recycled products. By 2050, the equivalent of three planet Earths will be needed to provide the natural resources to sustain current lifestyles.

Goal 13 seeks to take urgent action to combat climate change and its impacts. By 2100 global temperatures are predicted to be 1.5°C warmer compared with the 1850–1900 average.

Goal 14 aims to promote the conservation and sustainable use of the oceans, seas and marine resources. Up to 40% of the world's oceans are affected by human activities including pollution, and depleted fisheries.

Goal 15 aims to manage forests sustainably, restore degraded lands, combat desertification, limit the degradation of natural habitats and reduce biodiversity loss. Around 1.6 billion people depend on forests for their livelihood, including 70 million indigenous people.

Goal 16 aims to promote peaceful and inclusive societies for sustainable development, provide access to justice for all and build effective, accountable and inclusive institutions at all levels. The rate of children leaving primary school in conflict affected areas is around 50% (some 28.5 million children).

Goal 17 calls for partnerships to strengthen the means of implementation of the SDGs, that is, the finance and technology. More than 4 billion people do not use the internet, 90% of them from LICs.

CHECK YOUR UNDERSTANDING

31. Identify the region of the world where there are high rates of poverty and food insecurity and most new cases of malaria/HIV.
32. Outline the potential benefits of improved education and training.

Exam practice

The diagram shows the interrelationships of water, energy and food.

(a) (i) Describe one way in which the use of water impacts on energy production. (2 marks)

(ii) Describe one way in which the use of energy production influences food production. (2 marks)

(iii) Describe one way in which food production influences water availability. (2 marks)

(iv) Explain the value of the nexus approach to the study of water, energy and food. (4 marks)

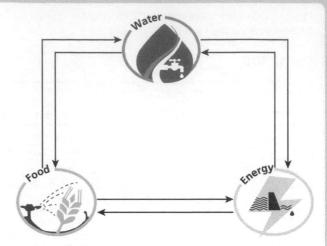

(b) The infographic shows a model of the circular economy.

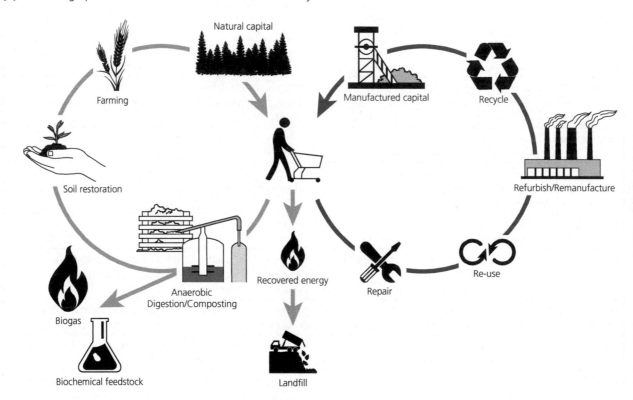

(i) Define the terms refurbish and recycle. (2 marks)

(ii) Suggest how natural capital can be re-used. (2 marks)

(iii) Suggest why repair, re-use and refurbish/re-manufacture are better options than recycling. (2 marks)

(iv) Outline the potential advantages of the circular economy. (4 marks)

(c) **Either**

Examine the effectiveness of ecological footprints as a measure of resource consumption. (10 marks)

Or

Discuss the value of the circular economy as a strategy for resource stewardship. (10 marks)

1 GLOBAL INTERACTIONS AND GLOBAL POWER

Globalization indices

DEFINITION

Globalization is the increasing interdependence of countries. This includes economic systems, physical systems (such as global warming), sociocultural systems (such as fashion, music and the film industry) and political systems. Globalization is not new – many countries had empires from which they sourced raw materials and labour – but the current form of globalization is more global, larger in scale and has developed at a much faster rate than in the past.

There are three main forms of globalization:
- economic: accelerated by the growth of transnational corporations (TNCs)
- social: the impact of western culture, art, media, sport and leisure activities around the world
- political: the growth of western democracies and their influence on poor countries, and the opening up of centralized economies.

THE KOF INDEX OF GLOBALIZATION

The **KOF Index of Globalization**, introduced in 2002, defines globalization as "the process of creating networks of connections among actors at multi-continental distances, mediated through a variety of flows including people, information and ideas, capital and goods. Globalization is conceptualized as a process that erodes national boundaries, integrates national economies, cultures, technologies and governance and produces complex relations of mutual interdependence".

The KOF Index covers the economic, social and political dimensions of globalization.
- The economic dimension includes long-distance flows of goods, capital and services, as well as foreign direct investment (36% of the Index).

- The social dimension includes the spread of ideas, information, images and people, international tourism, and cultural proximity (number of McDonalds and IKEA stores per capita) (38% of the Index).
- The political dimension includes the number of embassies in a country and membership of international organizations (26% of the Index).

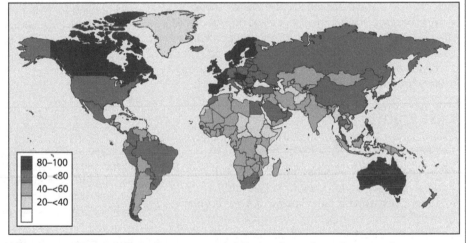

80–100
60 <80
40–<60
20–<40

The KOF Index of globalization, 2016 (based on data for 2013)

OTHER GLOBALIZATION INDICES

The EY Globalization Index measures the 60 largest countries/territories by GDP according to their level of globalization. The Index looks at openness to trade, capital flows, exchange of technology and ideas, labour movements and cultural integration.

The New Globalization Index is based on finance, trade and politics, and social factors. It differs slightly from the others in that it measures the distance of goods traded and counts the number of refugees in a country. Thus, for example, New Zealand and Argentina score higher on this measure than, for example, Belgium and the Netherlands. Lebanon and Turkey would appear to be more globalized due to the number of refugees they have taken in.

The New Globalization Index: dimensions and weights

Finance	Trade and politics	Social
37%	32%	31%

Source: Vujakovic, P. 2010. "How to measure globalisation? A New Globalisation Index." *Atlantic Economic Journal.*

CHECK YOUR UNDERSTANDING
1. Describe the main features of the 2016 KOF Index of globalization.
2. Identify two ways in which the New Globalization Index differs from the KOF Index.

Global superpowers

SUPERPOWERS AND SOFT POWER

Superpowers are countries that influence policy on an international scale, and often in different world regions at the same time. Superpowers have economic, cultural, military and geographical influence on a large scale, as seen, for example, in the USA and the former USSR and, more recently, in China and Russia.

"Soft power" refers to the ability to change individuals, communities and nations without using force or coercion.

Many countries achieve soft power through their culture, political values and foreign policies such as aid and investment. The USA is a favourite destination for overseas students, and the distribution of American films and TV programmes, such as *Friends*, has led to the Americanization of other cultures and languages – this is a form of soft power. Superpowers use soft power as well as hard power (force) to influence other countries.

RISING SUPERPOWERS AFTER 1991

After the collapse of communism in the USSR and its break-up, the USA was left as the world's only superpower. However, US military involvement in Iraq and Afghanistan following the 2001 terrorist attacks on the USA did not

achieve its desired ends, and some observers argue that the USA is losing its superpower status. Since the 2008 financial crisis, the USA has lost economic strength and other nations are gaining ascendancy. The Trump presidency introduced a more inward-looking government policy ("America first") and a potential decline in US overseas commitments.

THE USA: THE MILITARY-INDUSTRIAL COMPLEX

The USA has the world's largest and most technologically advanced fleet of warplanes, ships, tanks and artillery systems. These give it dominance over air, sea and land. Control of space and information are key aspects of military strategy. The USA's global military presence includes overseas bases, ships and aircraft that allow the USA to apply force to many parts of the world, and to supply weapons and military training to a wide range of countries.

The US defence industry employs over 2 million people. Approximately one in six households in the USA have someone employed in the military-industrial complex. Annual spending on defence exceeds $100 billion a year and funding for military research is $40 billion a year, twice what is spent on health, energy and the environment combined. The budget for environmental research was cut heavily during the Trump presidency, while that of defence was raised.

CASE STUDY

CHINA – A RISING SUPERPOWER

China's economic growth has also had a major impact on trade among the world's wealthiest countries.

China's demand for raw materials for this economic growth had a major impact on world markets in oil, iron ore, metals, petrochemicals and machinery.

Despite rapid change, the structure of the Chinese economy still has the character of an NIC with almost one-third of the workforce employed in agriculture, 30%

employed in industry which accounts for 40% of GDP. This reflects the fact that China has become the world's leading manufacturing centre, although much of the value-added output is from foreign-owned plants in China's special development zones. According to the IMF, in 2014 China produced 17% of the world's gross domestic product, surpassing the USA's 16%. However, China remains well below the USA in terms of GNI/head, with a GNI of $15,400 compared with the US's $57,300.

Superpowers

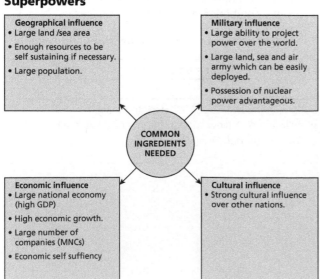

Geographical influence
- Large land /sea area
- Enough resources to be self sustaining if necessary.
- Large population.

Military influence
- Large ability to project power over the world.
- Large land, sea and air army which can be easily deployed.
- Possession of nuclear power advantageous.

COMMON INGREDIENTS NEEDED

Economic influence
- Large national economy (high GDP)
- High economic growth.
- Large number of companies (MNCs)
- Economic self suffiency

Cultural influence
- Strong cultural influence over other nations.

COMMON MISTAKE

✗ *The USA will always remain as a superpower.*

✓ It is easy to think of superpowers as static, remaining as superpowers. However, they grow and decline – just as empires did in ancient Greece and Rome, China in the Middle Ages, the UK in the 19th/20th centuries, and the decline of Russia in the late 20th century. The USA's position as a superpower is not guaranteed. The USA may well be replaced as the world's main superpower in the 21st century.

CHECK YOUR UNDERSTANDING

3. Explain, using an example, the meaning of the term "soft power".
4. Outline the evidence that the USA rather than China, is the world's leading superpower.

Global organizations and groups (1)

THE G7 AND THE G8

The G7 (Group of Seven) is a group of powerful HICs – the USA, France, Germany, Italy, the UK, Japan and Canada – that meets annually to discuss matters such as the global economy, global governance, energy policy and international security. They were joined by Russia in 1998 to form the G8 (Group of Eight). However, Russia was suspended following its activities in Ukraine.

The G7 is not a formal institution but acts in an advisory capacity. The small number of similar countries that make up the G7 should, in theory, enable decisions to be made, but critics argue that it does not speak for any emerging economy.

Without Russia, the G7 has a more similar outlook but represents a smaller proportion of the world's population, and without China its claim to be a global organization is limited.

THE G20

The G20 (G-20 or Group of Twenty) is an international grouping for the governments of 20 major economies. It was established in 1999 to discuss policy issues related to global financial stability.

The G20 countries account for about 85% of gross world product, 80% of world trade, and about 65% of the world's population. The G20's main interest is global economic governance, but it has also discussed issues such as the impacts of an ageing population, reform of the World Bank and the IMF, energy security and resource depletion. Although the G20 claims that its broad membership and high economic profile makes it legitimate, some critics argue that Africa is under-represented. Over 170 countries are not represented in the G20.

THE OECD

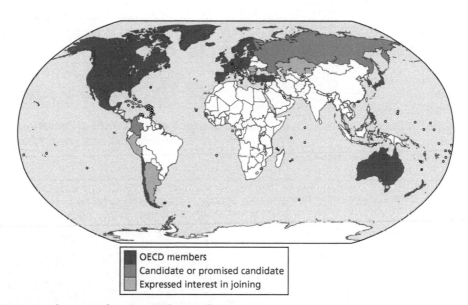

OECD members
Candidate or promised candidate
Expressed interest in joining

OECD members and potential members
Source: https://en.wikipedia.org/wiki/Organisation_for_Economic_Co-operation_and_Development

After the Second World War, the Organization for European Economic Cooperation (OEEC) was established to run the Marshall Plan for the reconstruction of Europe. It recognized the need for cooperation between countries. In 1960 Canada and the USA joined the OEEC to form the Organization for Economic Cooperation and Development (OECD). Other countries joined, and in 2017 the OECD had 35 members. The OECD aims to identify, analyse and discuss problems, and find policies to solve them. Its stated aim is to help countries around the world to:

- restore confidence in markets
- re-establish healthy public finances
- foster and support new sources of growth through innovation, environmentally friendly "green growth" strategies
- ensure that people of all ages can develop the skills to work productively.

?

CHECK YOUR UNDERSTANDING
5. Outline one advantage and one disadvantage of the G7.
6. Comment on the membership of the OECD.

Global organizations and groups (2)

OPEC – AN OIL CARTEL

The Organization of Petroleum Exporting Countries (OPEC) was established in 1960 to tackle oil price cuts by American and European oil companies. Founder nations included Iran, Iraq, Kuwait, Saudi Arabia and Venezuela. By 1979 the OPEC countries produced 65% of world petroleum, but only 36% by 2007. As early as 2003, concerns that OPEC members had little excess pumping capacity sparked speculation that their influence on crude oil prices would begin to slip. However, OPEC continues to have the major share of world crude oil reserves.

The development of OPEC and the control of oil has had important implications. As OPEC controlled the price of oil and much of the production in the 1970s and 1980s, Middle Eastern countries gained economic and political power. This provides an incentive for the old industrialized countries to increase energy conservation or develop alternative forms of energy.

The importance of oil means that countries need to maintain favourable relationships with OPEC countries and that the Middle East will be involved in economic cooperation and development with industrialized countries.

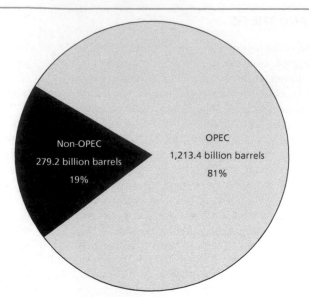

Non-OPEC	OPEC
279.2 billion barrels	1,213.4 billion barrels
19%	81%

OPEC proven crude oil reserves, at end 2015 *(billion barrels, OPEC share)*

Venezuela	300.88	24.8%	Nigeria	37.06	3.1%
Saudi Arabia	266.46	22.0%	Qatar	25.24	2.1%
Iran	158.40	13.1%	Algeria	12.20	1.0%
Iraq	142.50	11.7%	Angola	9.52	0.8%
Kuwait	101.50	8.4%	Ecuador	8.27	0.7%
United Arab Emirates	97.80	8.1%	Indonesia	3.23	0.3%
Libya	48.36	4.0%	Gabon	2.00	0.2%

OPEC share of world crude oil reserves, 2015

Source: http://www.opec.org/opec_web/en/data_graphs/330.htm

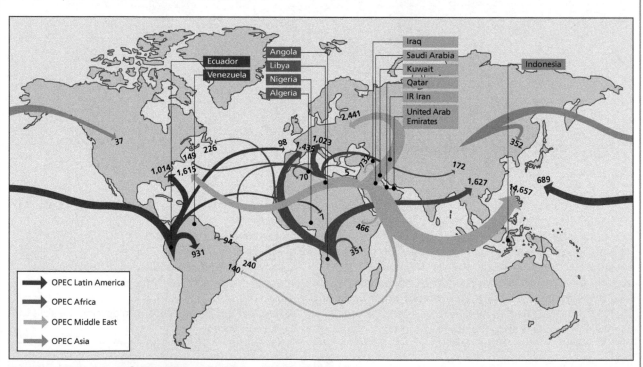

Source OPEC Annual Statistical Bulletin, 2016, p.68

http://www.opec.org/opec_web/static_files_project/media/downloads/publications/ASB2016.pdf

CHECK YOUR UNDERSTANDING

7. State the approximate amount of the world's crude oil reserves that OPEC accounts for.
8. Describe the distribution of OPEC members.

Global lending institutions

THE WORLD BANK

The World Bank, which was established in 1944, is a source of financial and technical assistance to LICs and NICs. Its mission is to fight poverty by providing resources, sharing knowledge and building capacity.

The World Bank's current focus is on the achievement of the Sustainable Development Goals (SDGs), lending mainly to middle-income countries (MICs). The Bank's mission is to aid these countries and their inhabitants to achieve development and reduce poverty. This includes achievement of the SDGs, and previously MDGs, by helping countries develop an environment for investment, jobs and sustainable growth.

Critics of the World Bank

Critics of the World Bank have claimed that its free-market reform policies are harmful to economic development. Another critism is that there is an assumption that LICs cannot modernize without money and advice from abroad. Moreover, although the World Bank represents 186 countries, it is run by a small number of rich countries.

THE INTERNATIONAL MONETARY FUND (IMF)

The International Monetary Fund (IMF) is the international organization that oversees the global financial system.

Member states with balance of payment problems may request loans to help fill gaps between what they earn and/or are able to borrow from other official lenders and what they must spend to operate. In return, countries must usually launch certain reforms such as structural adjustment programmes (SAPs).

Critics of the IMF

One of the main SAP conditions placed on borrowers is that the governments sell as much of their national assets as they can, normally to western corporations at heavily discounted prices. Moreover, the IMF sometimes advocates "austerity programmes" – increasing taxes/reducing social spending even when the economy is weak – to generate government revenue/reduce spending.

The IMF is for the most part controlled by the major western nations.

THE NEW DEVELOPMENT BANK

The New Development Bank (NDB) was established by Brazil, Russia, India, China and South Africa in 2014. Formerly the BRICS Development Bank, the NDB supports public or private projects through loans, guarantees, equity participation and other financial instruments.

Its main focus for lending will be infrastructure and sustainable development projects, such as clean energy.

Brazil, Russia, India, China and South Africa initially contributed $10 billion to the fund. The bank plans initially to fund one project from each member. The bank has its headquarters in Shanghai.

According to the NDB, although multilateral banks provide up to $100 billion/year in development loans, this is insufficient to meet the infrastructural development needs of emerging economies.

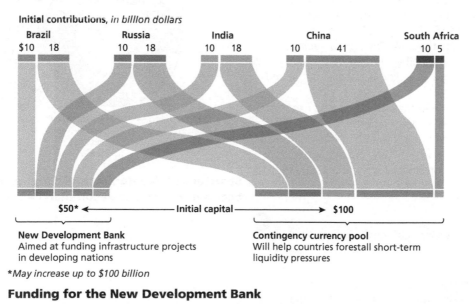

Initial contributions, *in billion dollars*

Brazil	Russia	India	China	South Africa
$10 18	10 18	10 18	10 41	10 5

$50* ←————— Initial capital —————→ $100

New Development Bank
Aimed at funding infrastructure projects in developing nations

Contingency currency pool
Will help countries forestall short-term liquidity pressures

*May increase up to $100 billion

Funding for the New Development Bank

EXAM TIP
Try to be balanced in your comments about global organizations – point out some of the positive aspects as well as some of the negative aspects.

CHECK YOUR UNDERSTANDING
9. Outline one advantage and one disadvantage of the IMF.
10. Outline one advantage and one disadvantage of the New Development Bank.

Global trade in materials, manufactured goods and services

GROWTH OF WORLD TRADE

The value of world trade in materials, manufactured goods and services roughly doubled between 2005 and 2015. However, growth reduced between 2012 and 2014. Asia, Europe and North America continue to account for the bulk of trade, although the share in merchandise exports from NICs increased from 33% in 2005 to over 40% in 2015.

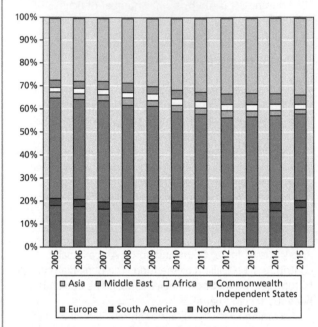

Merchandise trade by region, 2005–15 (% share)

Moreover, the merchandise trade between emerging economies increased from just over 40% to over 50% of their global trade between 2005 and 2015. Merchandise trade was worth over $16 trillion in 2015, and was dominated by China and the USA. The top 10 trading nations accounted for over half of the world's trade in merchandise in 2015, and emerging economies accounted for over 40% of it.

In 2015, China remained the world's leading exporter and the USA the world's leading importer. China's exports in 2015 were valued at over $2 trillion, followed by the USA at $1.5 trillion. The USA's imports were valued at over $2.3 trillion, followed by China's at around $1.7 trillion.

SERVICES

Developing countries' share in commercial services continued to rise, accounting for nearly one-third of global exports. The increase was mainly due to China, India, Hong Kong and South Korea. Travel and tourism account for the major share of commercial services in NICs and LICs.

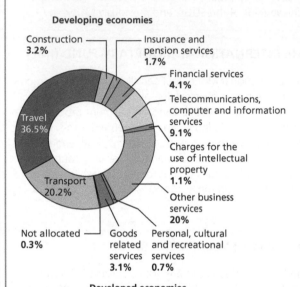

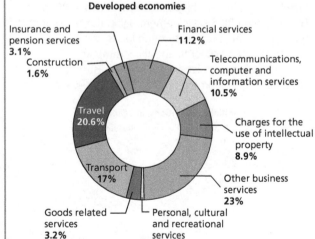

Structure of developing and developed economies' exports in commercial services, 2015

International aid, loans and debt relief (1)

There are many types of aid, including emergency/relief aid, development aid, and short- and long-term aid. Two of the most well-known are top-down and bottom-up development.

Top-down development	Bottom-up development
Usually large in scale	Small in scale
Carried out by governments, international organizations and "experts"	Labour intensive
Done by people from outside the area	Involves local communities and local areas
Imposed upon the area or people by outside organizations	Run by locals for locals
Often well funded and responsive to disasters	Limited funding available
Does not involve local people in the decision-making process	Involves local people in the decision-making process
Emergency relief can be considered top down	Common projects include building earthen dams, creating cottage industries

DEVELOPMENT AID

The main donors of development aid are HICs. In contrast, the main recipients are NICs and LICs. Highest levels of aid would appear to be go to much of sub-Saharan Africa, Eastern Europe and Russia and South East Asia. The largest donors are the USA and Japan, although as a percentage of their GNP each donates less than 0.25 per cent of GNI. France and the UK are the next largest donors, donating less than 0.5 per cent of their GNI. The largest donors in relation to GNI are the Scandinavian countries, Norway, Denmark and Sweden.

LOANS

A loan is a transfer of money or skills that require repayment over a set time. The main pattern of loans is a transfer from richer countries to poorer countries. In 1970 the OECD adopted the target for donors to spend 0.7 per cent of their GNI on overseas development assistance (ODA), but by 2012 only five donors had met this target.

DEBT RELIEF

Sub-Saharan Africa includes most of the countries classified as heavily indebted poor countries (HIPCs), and 25 of the 32 countries rated as severely indebted. In 1962 sub-Saharan Africa owed $3 billion. In 2017 its debt was about $230 billion. The most heavily indebted country is Nigeria (around $35 billion).

As a result of debt relief, Tanzania introduced free schooling, built more schools and employed more teachers. Critics of debt relief state that it does not help the poor, that it does not help countries that do not get into debt, and that it encourages countries to overspend.

STRUCTURAL ADJUSTMENT PROGRAMMES

Structural adjustment programmes (SAPs) are loans from the IMF requiring the borrowing country to cut its government expenditure, reduce state intervention in its economy, and promote liberalization and international trade. SAPs encourage international trade and long-term economic growth. They have four main requirements:
- greater use of a country's resources
- social and economic reforms to increase economic efficiency
- diversification of the economy and increased trade to earn foreign exchange
- a reduction in the active role of the state in the economy.

EXAM TIP
Be sure to specify whether you are writing about the absolute amount of aid that a country donates (usually in US$) or the relative amount (as a percentage of GNI). For example, the USA gives the largest amount in absolute terms, but not in relative terms.

CHECK YOUR UNDERSTANDING
13. Outline two advantages of top-down development and two advantages of bottom-up development.
14. Outline one advantage and one disadvantage of debt relief.

International aid, loans and debt relief (2)

REMITTANCES

Many of the world's migrants travel from poorer to richer countries, but people from rich countries also migrate, to oil-rich countries, for example. The largest regional migrations are from South East Asia to the Middle East, lured by employment in the oil economy, the boom in the construction industry and the demand for domestic servants. The largest flow between two single countries is from Mexico to the USA.

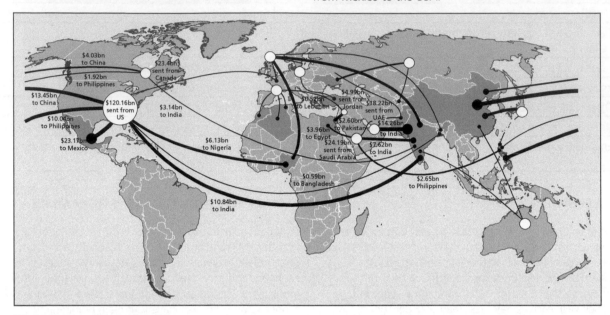

Main flows of remittances

The value of remittances ($ million)

1	India	70,389
2	China	62,332
3	Philippines	28,403
4	Mexico	24,968
5	France	24,460
6	Nigeria	20,921
7	Egypt	19,570
8	Germany	17,629
9	Pakistan	17,066
10	Bangladesh	14,969

Source: *The Economist*. 2017. Pocket World in Figures. London, UK. Profile Books

The region that receives the most in remittances is south Asia, in particular India, Pakistan and Bangladesh. In these countries the value of remittances is said to be greater than the amount of international aid that they receive. Countries in South East Asia, such as Indonesia, Malaysia and Vietnam, receive a considerable amount of money through remittances. In contrast, most of Africa and the Caribbean receive a relatively small amount of remittances. Sub-Saharan Africa appears to be worst off. The pattern is different from the usual rich–poor divide in a number of ways; for example, the low value of remittances received in eastern Europe and in an arc of countries through Turkey to Kazakhstan makes this pattern unusual.

Some countries are very dependent on remittances. Nearly one-quarter of Haiti's GDP comes from remittances and in Jordan it is over one-fifth. In the Philippines, not only do remittances bring in a huge amount of income, about $28 billion, they also account for around 13% of GDP. In India and China, the two largest recipients, remittances in 2014 accounted for around 3% and 1% of GDP respectively.

The great advantage of remittances over other forms of financial assistance is that they go directly to the migrants' family/household, and can be used for whatever purpose the household wishes. However, many money transfer companies charge a substantial amount to transfer money back to migrants' homes, and in some cases, migrants' wages have been withheld by the employer and/or transferred back to the government, for example, from the UAE to North Korea.

CHECK YOUR UNDERSTANDING

15. Describe the main pattern of global remittances.
16. Briefly explain the main advantage of remittances.

Illegal flows

There are many forms of illegal flows, such as trafficked people, counterfeit goods, fraudulent medicines, counterfeit food and drink and drugs.

TRAFFICKED PEOPLE

Trafficking of people is a crime on a global scale. The UN Office on Drugs and Crime (UNODC) 2016 *Report on Trafficking in Persons* stated that between 2012 and 2014 there were over 63,251 victims of trafficking, from data provided by 106 countries. Trafficking can be domestic or international. About 60% of victims are international, although fewer than 30% are inter-regional. Women accounted for 51% of those trafficked, men 21% and children 28% (girls made up 20% and boys 8%). Most trafficking occurs within the same geographical sub-region. Frequently, victims come from relatively poorer countries and are exploited in relatively richer countries. The Middle East has the highest share of inbound trafficked people from other regions.

COUNTERFEIT GOODS

Counterfeit goods are believed to generate more than $250 billion each year, and are responsible for labour exploitation, environmental damage and health implications for consumers. In addition, there are links between counterfeit goods, money laundering, illicit drugs and corruption.

Corruption and bribery are linked to the trade in counterfeit goods, especially when they are transported internationally.

The trade in counterfeit goods reduces tax revenues for governments. It raises extra costs with the need for more surveillance and policing. The size of the market for online goods is unknown, but likely to be substantial and increasing.

FRAUDULENT MEDICINES

According to the UNODC, the trade in fraudulent medicines from East Asia and the Pacific to South East Asia and Africa is worth about $5 billion a year. The World Health Organization claims that about 1% of medicines in HICs and up to 30% in LICs are fraudulent. According to *The Lancet*, one-third of malaria medicines in sub-Saharan Africa and East Asia were fraudulent.

COUNTERFEIT FOOD AND DRINK

According to the UK Food Standards Agency, up to 10% of the food bought in the UK is fraudulent. For example, in 2013 a "horse meat scandal" broke out, as meat passed off as beef was found to be horse meat.

FLOWS OF DRUGS

Drug trafficking is a major global trade involving the cultivation, manufacture, distribution and sale of substances that are prohibited by law. The global drug trade is worth more than $300 billion, or 1% of total global trade.

The map shows the main flows in cocaine around the world.

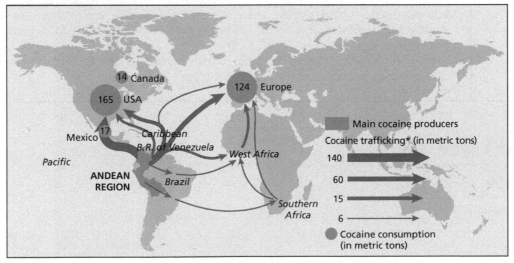

The flow of cocaine
Source: https://www.unodc.org/unodc/en/drug-trafficking

CHECK YOUR UNDERSTANDING
17. Outline why trade in counterfeit food and drink and fraudulent medicine has negative impacts.
18. Describe the main flows of cocaine around the world.

Foreign direct investment (FDI) (1)

FDI is the investment by a company into the structures, equipment or organizations of a foreign country. It does not include investment in shares of companies of other countries.

FDI fell after the financial crisis of 2008–09 but by 2015 had improved. Much of the growth was due to investment in HICs, especially in the USA and Europe. In addition, FDI in NICs reached a new high in 2015. Asia remained the main focus for FDI in NICs/LICs, whereas FDI into Africa, Latin America and the Caribbean slowed. One exception was Cuba, which re-established diplomatic ties with the USA, and FDI there is expected to increase.

Outward FDI flows from HICs increased by a third in 2015. Nevertheless, this was still 40% short of the peak in 2007. Europe became the world's largest investing region. Investment in agriculture declined while that in manufacturing increased. FDI to Africa fell by 7% in 2015, partly because of low commodity prices. In contrast, FDI to Asia increased but is predicted to slow down as many economies mature. FDI to transition economies such as the CIS remained low because of low commodity prices, declining domestic markets and the impact of restrictive measures (such as sanctions) and geopolitical tension. Flows to landlocked LICs and developing small-island states had major declines.

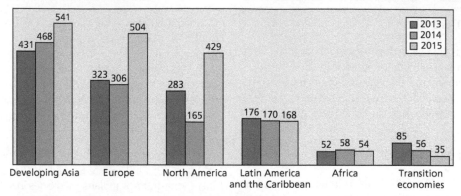

FDI inflows by region, 2013–15 ($ billion)

The positive and negative impacts of the global shift

	Positive	Negative
In HICs	• Cheaper imports. • Growth in LICs may lead to a demand for exports from HICs. • Greater worker mobility to areas with relative scarcities of labour. • Greater industrial efficiency.	• Rising unemployment, especially of unskilled workers. • Large gaps develop between skilled and unskilled workers. • Job losses are invariably concentrated in certain areas and certain industries. • Branch plants are particularly vulnerable.
In NICs and LICs	• Higher export-generated income. • The effects can spread to local areas with many new, highly paid jobs. • Negative trade balances can be reduced. • Employment growth in relatively labour-intensive manufacturing spreads wealth.	• Jobs tend to be concentrated in the core region of urban areas, leading to in-migration. • TNCs may be exploitative and establish sweatshops. • Overdependence on a narrow economic base can result. • Food supplies may be reduced as people give up agriculture.

CHECK YOUR UNDERSTANDING
19. Outline two potential advantages of foreign direct investment.
20. Outline two potential disadvantages of foreign direct investment.

Transnational companies

TRANSNATIONAL COMPANIES

A transnational company (TNC) is an organization that operates in a number of countries. Generally, TNC headquarters are in HIC cities, with research and development (R&D) and decision-making concentrated in growth areas of HICs (the core), and assembly and production located in LICs and NICs, and depressed parts of the HICs (the periphery).

CASE STUDY

THE TATA GROUP

The Tata Group is comprises over 100 companies, covering sectors such as cars and consulting, software and steel, tea and coffee, transport and power, chemicals and hotels.

Tata Steel is India's largest steelmaker and Tata Consultancy Services is Asia's largest software company. Tata Power is the country's largest private electricity company, while Tata Global Beverages is the world's second-largest maker of branded tea.

Tata was founded in 1868, creating India's first Indian-owned steel plant, and it is now one of the stars of India's globalization. Tata operates in over 80 countries and employs about 600,000 people. Over half of its revenue comes from outside India.

Tata created many of India's greatest institutions, such as the Tata Memorial Hospital. Tata prides itself above all on its culture, which is defined by three characteristics: loyalty, dignity and what is now called corporate social responsibility (CSR). As early as 1912, the company introduced an eight-hour working day, and it introduced paid leave in 1920. Tata trusts fund worthy causes, from clean-water projects and literacy programmes, all of which cost over $105 million annually.

Jamshedpur, the home of Tata Steel, is, perhaps, the world's most successful company town. Tata Steel runs almost all the city's institutions, including a 980-bed hospital, a giant sports stadium and the local utility company.

The Group is currently committed to "frugal innovation": new products that are affordable to poor people and the rising middle class. These include the Tata Nano (a $2,300 car), a cheap water filter using rice husks, and a prototype of a $500 house that can be bought in a shop.

APPLE INC

Apple Inc. is one of the richest corporations in the world, valued at around $250 billion in 2015. However, the Apple supply chain has received much criticism on account of human rights and ethical issues in China.

For the manufacture of its iPhone, Apple has some 785 suppliers – 349 of them in China. In its Supplier Code of Conduct, Apple states that "suppliers are required to provide safe working conditions, treat workers with dignity and respect, act fairly and ethically, and use environmentally responsible practices wherever they make products or perform services for Apple".

Foxconn, the world's largest electronic manufacturing services company, employs approximately 1.6 million people in China. It is Apple's principal supplier. Since 2006, there have been allegations of poor working conditions at Foxconn, where workers working 15 hours a day earn just $50/month. In 2010 demands for improved working conditions and higher wages culminated in 18 attempted suicides, 13 of which succeeded. After the suicides, Foxconn put up mesh netting around its buildings to prevent people jumping from them, provided counselling for its workers and increased wages.

Apple's social actions have been described as "reactive", while Chinese labour laws and lack of enforcement do little to protect workers.

CHECK YOUR UNDERSTANDING

21. Outline how the Tata Group has pursued its corporate social responsibility.
22. Suggest why Apple Inc. chose Foxconn as its principal supplier of electronic manufacturing services.

Multi-government organizations

DEFINITION

Multi-government organizations (MGOs) operate across a number of different states. Some are international, such as the World Bank, the IMF and the UN, whereas others are regional, such as the North American Free Trade Agreement or the European Union. Most MGOs focus on economic matters in an attempt to increase trade and interactions, although there is increasing nationalism and calls for protectionism in many countries.

TRADING BLOCS

A trading bloc is an arrangement between countries to allow free trade between member countries but to impose tariffs (charges) on external countries that wish to trade with them.

- **Free trade areas** are areas where members abolish tariffs and quotas on trade between member nations but restrict imports from non-member countries. NAFTA is a good example of a free trade area.
- **Customs unions** are a closer form of economic cooperation, such as Mercosur within South America. As well as having free trade between members, all members operate a common external tariff on imports from abroad.
- **Common markets** are customs unions which, in addition to the free trade in goods and services, allow free movement of people and capital.
- **Economic unions** are groups of nations that not only allow free trade and free movement of people and capital, but also require members to have common policies on such sectors as agriculture, industry and regional development, for example, the European Union.

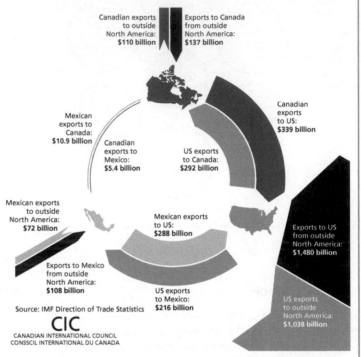

Canadian exports to outside North America: $110 billion

Exports to Canada from outside North America: $137 billion

Canadian exports to US: $339 billion

Mexican exports to Canada: $10.9 billion

Canadian exports to Mexico: $5.4 billion

US exports to Canada: $292 billion

Mexican exports to outside North America: $72 billion

Mexican exports to US: $288 billion

Exports to US from outside North America: $1,480 billion

Exports to Mexico from outside North America: $108 billion

US exports to Mexico: $216 billion

US exports to outside North America: $1,038 billion

Source: IMF Direction of Trade Statistics

CIC
CANADIAN INTERNATIONAL COUNCIL
CONSSCIL INTERNATIONAL DU CANADA

The state of North American trade, 2012 merchandise trade between North American countries and beyond

EXPORT PROCESSING ZONES AND FREE TRADE ZONES

Export processing zones (EPZs) are areas that offer incentives to foreign companies to develop export-orientated industries. Free trade zones (FTZs) are areas where goods can be stored, manufactured and re-exported without customs duties (taxes). Both are important parts of the so-called new international division of labour, and they are relatively easy paths to industrialization. By the end of the 20th century more than 90 countries had established EPZs and FTZs as part of their economic strategies.

Their popularity is due to three factors that link the economies of LICs and NICs with HICs. These are:
- problems of indebtedness and foreign exchange shortages in LICs since the 1980s
- economic incentives that encourage open economies, foreign investment and non-traditional exports
- the search by TNCs for cost-saving locations, particularly in terms of wage costs, to shift manufacturing, assembly and component production from locations in HICs to LICs and NICs (global shift).

CHECK YOUR UNDERSTANDING

23. Distinguish between a free trade area, a customs union and a common market.
24. Describe the main trading patterns among North American countries.

Economic migration

CONTROLS AND RULES

There are four main trends in international migration:
- Migration is becoming more global and the diversity of areas of origin is increasing.
- Migration is accelerating – the number of migrants is growing in all major regions.
- Migration is becoming more diversified – migrants include combinations of permanent settlers, refugees, skilled labour, economic migrants, trafficked people, forced migrants, retirees, arranged marriages, and so on.
- Migration is increasing among women, who are playing a much fuller part in their own right, notably as economic migrants.

There are new challenges for governments to provide for migrants and refugees, but there is also increased hostility in receiving countries. Increasing globalization and the growing diversity of migrants make it harder for governments to restrict migration, although many countries have attempted to restrict migrants.

Migration is important for the growth of an economy. In the USA, economic prosperity is associated with the country's ability to attract labour – both skilled and unskilled. Due to the ageing population, migrants are needed to increase the size of the working population. Nevertheless, many people still want to control migration.

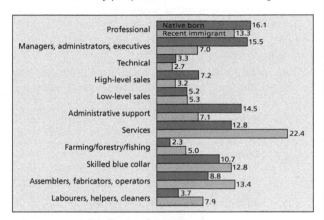

Migrants and jobs in the USA

CASE STUDY

Migration control in the USA

Illegal immigration to the USA refers to foreign nationals voluntarily residing in the country in violation of US immigration and nationality laws. Illegal immigration carries a civil penalty – punishment can include fines, imprisonment and/or deportation.

There are believed to be around 12 million illegal immigrants in the USA, entering the country in two ways:
- Most enter illegally by evading immigration inspectors and border patrol.
- Others enter legally but over-stay their visa.

Visa over-stayers tend to be somewhat more educated and better off financially than those who cross the border illegally. People have also used sham marriages as a way to enter the USA.

Each year, an estimated 200,000 to 400,000 illegal immigrants try to cross the Mexican border to reach cities in the USA. In 2017, US President Donald Trump called for a wall between Mexico and the USA to be completed – and to be paid for by Mexico. He claimed it would be an "aesthetically pleasing" wall!

As many as 50,000 people are illicitly trafficked into the United States annually. Trafficking in women plagues the USA as much as it does underdeveloped nations. Indian, Russian, Thai and Chinese women have reportedly been brought to the USA under false pretences to be then used as sex slaves.

European migration the UK

A study on the economic impact of European migrants to the UK concluded that they pay out far more in taxes than they receive in state benefits. The research economists at University College London, showed that Britain was uniquely successful, even more than Germany, in attracting the most highly skilled and highly educated migrants in Europe.

Over 60% of new migrants from western and southern Europe were university graduates, and 25% of eastern Europeans who come to Britain had a university degree, compared with 24% of the UK-born workforce. They calculated that European migrants made a net contribution of £20bn to UK public finances between 2000 and 2011. Those from the 15 countries that made up the EU before 2004, including Germany, France, Italy and Spain, contributed 64% – £15bn more in taxes than they received in welfare – while eastern European migrants contributed 12%, equivalent to £5bn more.

The educational qualifications of new migrants to Britain, especially from western and southern Europe, was now extraordinarily high and higher than any other EU country. The UK would have had to spend £6.8bn on education to build up the same level of "human capital". The study showed that not only are European migrants more highly educated than the UK-born workforce but they are less likely to be in receipt of state benefits – 43% less likely among migrants in the past decade – and more likely to be in employment. They were also less likely to live in social housing.

CHECK YOUR UNDERSTANDING
25. Compare the employment structure of migrants to the USA with that of native-born Americans.
26. Comment on the four main trends in migration. Suggest how political events from 2016 onwards may have affected some of these trends.

The "shrinking world": data flow patterns and trends

The world is more interconnected than ever before. While the physical flow of goods and services characterized the 20th century, the flows of data and information characterize the 21st century. Digitization has changed globalization in many ways. It is driving down the cost of cross-border communications and transactions, and allowing businesses to connect with other businesses and customers in most parts of the world. HICs are the most globally connected countries, although NICs and LICs are closing the gap slowly.

The digital age poses some challenges. Companies may be subject to pricing pressures, and security is a major issue. Social media may create new communities but it creates more social pressure on users, and is also a means for the spread of extremism.

When trade was dominated by physical goods, it was largely confined to TNCs in HICs. However, now global data flows are accelerating, more countries and smaller enterprises are increasingly able to compete on the global market. The volume of data being transmitted across borders has surged, connecting countries, companies and individuals. Although container ships are responsible for moving products to markets, goods are ordered, tracked and paid for online.

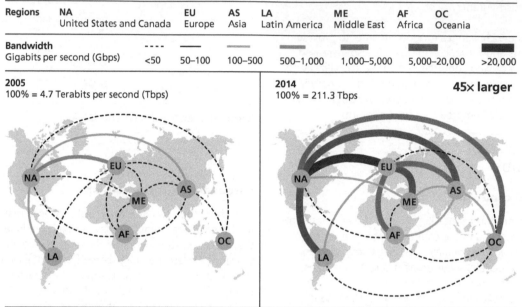

NOTE: Lines represent inter-regional bandwidth (for example, between Europe and North America) but exclude intraregional cross-border bandwidth (for example, connecting European nations with one another).

The surge in cross-border data flows connecting countries

About one-eighth of the global goods trade is carried out by e-commerce, and about half of the world's traded services are already digitized. Digitization allows the instantaneous exchange of virtual goods such as e-books, apps, online games, MP3 music files and streaming services, and software. Facebook suggests that there are around 50 million small and medium enterprises (SMEs) on its platform. Digital platforms allow companies to reach beyond the constraints of a small local market and reach a global audience.

Nevertheless, global flows are concentrated among a comparatively small number of users. The USA, Europe and Singapore are at the centre of the world's digital networks.

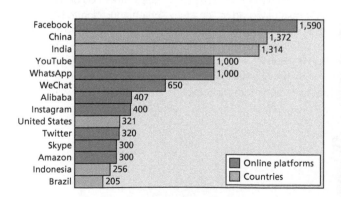

COMMON MISTAKE

✗ Everyone benefits from the "shrinking world".

✓ Not everyone benefits from the shrinking world – many elderly, infirm, immigrants and refugees, for example, do not have the same access and/or ability to use the internet for various reasons.

CHECK YOUR UNDERSTANDING

27. Explain the term "shrinking world".
28. Compare the volume of data flows in 2014 with that of 2005.

Transport developments over time

The **frictional effect of distance**, or **distance decay**, suggests that areas that are close together are usually more likely to interact with one another, whereas areas far apart are less likely to interact with one another. However, there has been a reduction in the frictional effect of distance, as improvements in transport have allowed people to travel greater distances in less time. In addition, improvements in ICT bring places in different parts of the world together almost instantaneously.

TIME–SPACE CONVERGENCE

For the development of transnational corporations (TNCs) and international trade, technologies are needed that overcome the frictional effect of distance and time. The most important of these are transport and communications. Neither of these technologies caused the development of TNCs or international trade, but they allowed such developments to occur.

Information communications technologies are the key technology in all economic activities. One of the most important technologies to enhance global communications is satellite technology. Satellite technology has made possible remarkable levels of global communication and the transmission of data. Another technology is optical fibre cables (OFCs) – these are now challenging satellite communications. These systems have a huge carrying capacity, and transmit information at very high speed and with a high signal strength.

However, only very large organizations – whether business or government – have the resources to utilize fully the new communications technologies. For example, Texas Instruments, the electronics TNC, has approximately 50 plants located in some 19 countries. It operates a satellite-based communications system to coordinate production planning, cost accounting, financial planning, marketing, customer services and personnel management.

CONTAINERS

Containers are the backbone of the modern global economy. About 90% of non-bulk cargo worldwide is transported in containers stacked on articulated lorries, trains and freight ships. Cargo shipped is now measured in TEUs (20 foot equivalent units) or FEUs (40 foot equivalent units). A TEU is the measure of a box 20 feet long and 8 feet wide, with a maximum gross mass of 24 tonnes.

However, geography and topography limit the ever-increasing size of ships. For example, their size is restricted by the depth of the Straits of Malacca, linking the Indian Ocean to the Pacific Ocean. This limits a ship to dimensions of 470 metres long and 60 metres wide. The Panama Canal is being expanded to allow ships up to 12,000 TEUs to pass. Many of the larger container ships can now carry 17,000 TEUs.

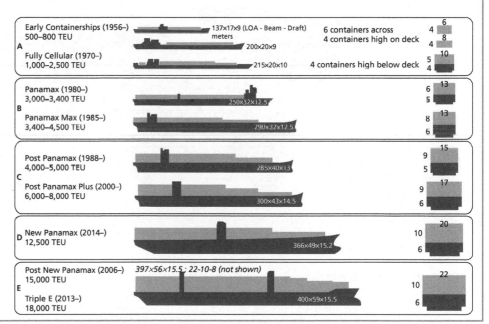

THE JET ENGINE

The jet engine is perhaps the most significant innovation in long-distance transport ever. The jet is safer, easier to maintain, better suited for longer distances, and more fuel efficient than the propeller. Jet aircraft have a much higher power-to-weight ratio, which enables longer range, faster travel and bigger payloads. For example, the Airbus A380, the world's largest passenger plane, can carry about 555 people.

While most global trade is by maritime shipping, air transport fills an important niche in just-in-time production systems. In Brazil, known for its primary goods exports, air cargo in 2000 accounted for 0.2% of total export volume by weight, but almost 19% by value.

Inexpensive and frequent air services have allowed countries like Chile, Colombia and Kenya to sell agricultural and horticultural products to markets in Europe, the Middle East and North America. A prime example is Kenya, which today has a third of the global market for cut flowers.

CHECK YOUR UNDERSTANDING
29. Explain the meaning of the "frictional effect of distance".

30. Explain the importance of containers.

Patterns and trends in communication infrastructure

There are major variations in international access to fixed telephone lines, mobile subscriptions, the internet and broadband.

CHANGING TRENDS

Every year, ICT technology brings faster processing speeds, greater storage capacity and more advanced software. Facebook had 50 million users in just one year, and Twitter took even less time to reach that number.

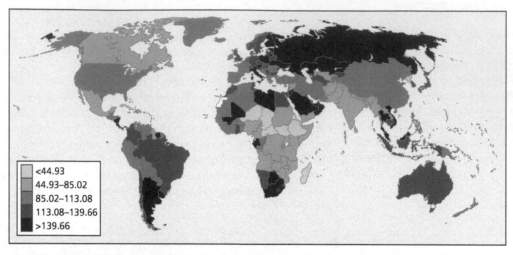

Mobile cellular subscriptions per 100 people

Legend:
- <44.93
- 44.93–85.02
- 85.02–113.08
- 113.08–139.66
- >139.66

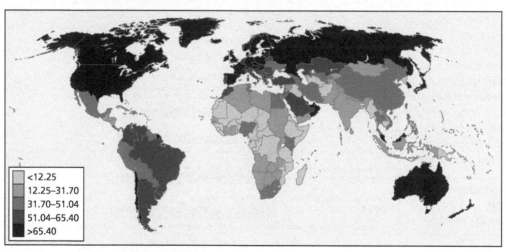

Internet users per 100 people

Legend:
- <12.25
- 12.25–31.70
- 31.70–51.04
- 51.04–65.40
- >65.40

By 2025, there will be 3 billion new users of the mobile internet and half of the growth is predicted to be in NICs and LICs. By 2015, some 1.8 billion people will move up to the global consumer class – those with enough money to purchase goods and services, having met their basic needs.

The biggest rise will be in smartphones. Since 2005, sales of mobile handsets have increased by 20% per year in Africa and by 15% per year in Asia and Latin America. China and India each have more than 1 billion mobile phones. Smartphones enable people to have access to internet functions, such as financial services. Safaricom's M-Pesa has 15 million accounts in Kenya and 9 million in Tanzania.

Mobile phones can improve access to health care. For example, in Bangladesh, where 90% of births occur outside hospitals and clinics, a scheme to notify midwives when labour starts has resulted in almost 90% of births taking place with a midwife present. Education courses can also be delivered over the internet. MOOCs (massive open online courses) are creating learning opportunities for people of all ages.

Nevertheless, despite all these advances, not everybody has access to the internet. This includes large numbers of poor people, the elderly, those with certain disabilities, and those living in areas with no service.

CHECK YOUR UNDERSTANDING
31. Describe the global variations in access to the internet.
32. Describe the global variations in mobile cellular subscriptions.

The influence of the physical environment on global interactions

NATURAL RESOURCE AVAILABILITY

The availability of natural resources is a significant factor in world trade. For example, the Middle East countries dominate the export of oil. Countries endowed with other raw materials, such as food products, timber, minerals and fish, also have the potential to trade.

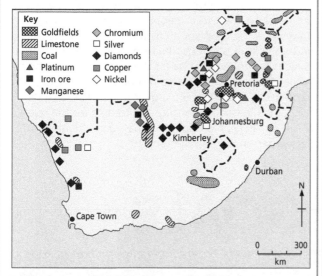

Key
- ▦ Goldfields
- ▨ Limestone
- ▥ Coal
- ▲ Platinum
- ■ Iron ore
- ◆ Manganese
- ◇ Chromium
- ☐ Silver
- ◆ Diamonds
- ▪ Copper
- ◇ Nickel

Resource availability in South Africa

South Africa, Australia and Canada are rich in resources and they continue to exploit them. Other countries, such as Germany and the UK, may have used the bulk of their resources but have since diversified. Many HICs have developed due to their export of raw materials. MICs and LICs rich in raw materials, such as Brazil and South Africa, have been trying to follow a similar path – using the wealth from exporting their raw materials to diversify and produce a more broad-based economy. Countries with a single resource product – for example, Ghana with cocoa and Zambia with copper – are more vulnerable to climate extremes, demand, competition, disease and currency fluctuations because they have fewer alternatives to earn foreign currency.

GEOGRAPHIC ISOLATION

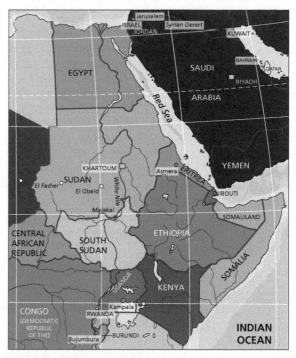

Location of South Sudan

Isolation from world markets can be a major hindrance to development. Increased transport costs and times may reduce access to markets and profitability. Countries that are landlocked may also have to pay substantial tariffs to export their goods. For example, South Sudan has to pay either Sudan or Kenya to get its oil to the coast. Landlocked countries may also have to pay for the use of another country's air space. Access to market is essential for economic development. Locations such as Singapore, on the tip of the Malay peninsula, at the entrance to the Straits of Malacca, has had a flourishing port as early as the 13th century. It continues to thrive, partly due its geographic location.

EXAM TIP

In an exam, make it clear that not everyone benefits from new technologies. For example, despite the innovations in ICT, there are some areas that are too remote, inaccessible and isolated due to physical geography, and technological developments have not been able to overcome these difficulties.

✓ CHECK YOUR UNDERSTANDING ❓

33. Outline the problems that landlocked countries face.
34. Briefly explain how resource endowment helps countries to develop.

Exam Practice

(a) Examine the value of globalization indices for measuring global interactions. (12 marks)

(b) Discuss the view that technological innovations are making the world smaller. (16 marks)

1 DEVELOPMENT OPPORTUNITIES

The UN Sustainable Development Goals

In 2015, once the Millennium Development Goals (MDGs) had finished, the 193 countries of the United Nations adopted the 2030 Agenda for Sustainable Development. This includes 17 Sustainable Development Goals (also called Global Goals), and they came into force on 1 January 2016. At the COP22 (Conference of the Parties) meeting in Marrakech in 2016 United Nations Secretary-General Ban Ki-moon said that the SDGs were the only way forward for the world as "we don't have plan B because there is no planet B".

The new goals continue – and extend – the efforts to end poverty, reduce inequalities and tackle environmental degradation, including climate change. The goals are wide ranging and include 169 targets. They are not legally binding, but governments are expected to develop systems in order to achieve the goals.

Some countries, including Japan and the UK, have reservations about the SDGs, and believe that there should be fewer of them. There are also doubts about how the funding will be made available within all countries to achieve them. "The new agenda is a promise by leaders to all people everywhere. It is a universal, integrated and transformative vision for a better world. It is an agenda for people, to end poverty in all its forms. An agenda for the planet, our common home. An agenda for shared prosperity, peace and partnership. It conveys the urgency of climate change. It is rooted in gender equality and respect for the rights of all. Above all, it pledges to leave no one behind." UN Secretary-General Ban Ki-moon.

UN Sustainable Development Goals (SDGs)
Source: Adapted from https://sustainabledevelopment.un.org/?menu=1300

CHECK YOUR UNDERSTANDING
1. Suggest why access to clean water and sanitation is an important goal.
2. Outline the evidence that some children do not have a high quality of life.

The Human Development Index and the Gender Inequality Index

THE HUMAN DEVELOPMENT INDEX (HDI)

The Human Development Index (HDI) is a composite measure of development. It includes three basic components of human development:
- life expectancy
- mean years of schooling, expected years of schooling
- income adjusted to local cost of living, that is, purchasing power.

The United Nations (UN) has encouraged the use of the HDI, as it is more reliable than single indicators such as gross national income (GNI) per head. It is a composite index so that the importance of any one factor is reduced.

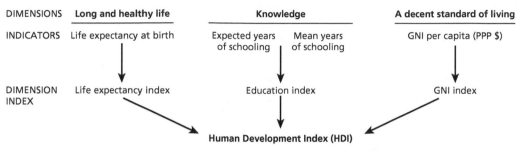

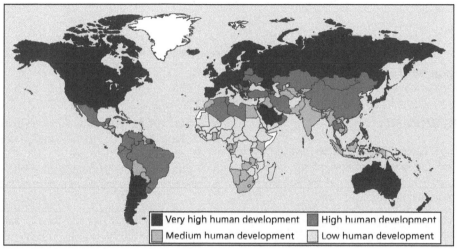

Global HDIs for 2015
Source: Based on data in *The Human Development Report 2016*. http://www.undp.org/content/undp/en/home/librarypage/hdr/2016-human-development-report.html

Countries with a very high human development index generally have a relatively high index for each component. A country with a high purchasing power, for example, but a low life expectancy and low education index will only have a medium HDI.

The countries at the top of the HDI are all HICs, and include Norway, Australia, Switzerland, Germany and Denmark. In contrast, the five countries at the bottom of the HDI are LICs – namely Central African Republic, Niger, Chad, Burundi and Burkina Faso. Countries at the top of the HDI have a similar level of development to one another whereas those at the bottom have more varied levels. Although sub-Saharan Africa has the most countries ranked low on the HDI, a number of countries there, such as Namibia and Zambia, have moved up to the medium level. In 2014, the countries with the greatest relative drop in HDI rank were Syria and Libya.

THE GENDER INEQUALITY INDEX (GII)

In 2010, the UNDP introduced the Gender Inequality Index. The countries with the lowest gender inequality (meaning women are better off) in 2015 were, in rank order, Switzerland, Denmark, the Netherlands, Sweden and Iceland. The USA, ranked 10 on the HDI, was ranked 43 on the GII, and Qatar, ranked 33 on the HDI, was ranked 127 on the GII. In contrast, at the bottom end of the scale (lowest first) were Niger, Chad, Mali, Cote d'Ivoire and Afghanistan. The latest statistics for gender inequality can be found at http://hdr.undp.org/en/composite/GII.

CHECK YOUR UNDERSTANDING
3. Describe the global variations in the HDI.
4. Describe the global variations in the GII.

Measuring gender inequality

The Gender Inequality Index measures gender inequalities in three aspects of human development:
- reproductive health – measured by maternal mortality ratio and adolescent birth rates
- gender empowerment – measured by the proportion of parliamentary seats held by women, and the proportion of adult females and males aged over 25 years with some experience of secondary school
- economic status – measured by labour force participation by males and females aged 15 and over.

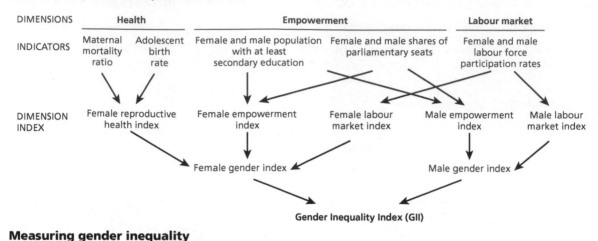

Measuring gender inequality
Source: UNDP Gender Inequality Index

EMPOWERING WOMEN

UN Women and UN Global Compact have developed the following Women's Empowerment Principles:
- Establish corporate leadership for gender equality.
- Treat all women and men fairly at work.
- Ensure the health, safety and well-being of all workers.
- Promote education, training and professional development for women.
- Promote equality through community initiatives.
- Measure and report on progress to achieve gender equality.

CASE STUDY

EMPOWERING WOMEN IN COLOMBIA

One of the longest armed conflicts in Latin America took place in Colombia, from 1964 onwards. During that time, women became mobilized. Colombia now has more women in decision-making positions than ever before – by 2011, 32% of cabinet members were women, up from just 12% in 1998. Girls' enrolment in secondary and tertiary education exceeds that of boys, and women's participation in the labour force increased from 30% in 1990 to over 40% in 2012.

Empowering indigenous and minority groups

There are some 370 million indigenous people around the world, accounting for a disproportionate amount of its poor. Indigenous people are generally marginalized and isolated, in worse health than the general population, and much less able to participate in economic and political processes. Most live in rural areas and depend on agriculture and related activities for their livelihoods.

The International Fund for Agricultural Development (IFAD) is a UN agency that aims to eliminate rural poverty by financing projects to aid agricultural development in LICs. In Panama, for example, IFAD established a sustainable rural development project for the Ngobe–Bugle Territory to restore land rights to indigenous communities. It provided financial and technical support to the communities, and improved local leaders' planning and administrative skills. The aim of the project is to support natural resource management based on traditional practices.

COMMON MISTAKE

✗ *The Gender Inequality Index always compares conditions for women and men.*

✓ Not all aspects of the Gender Inequality Index compare conditions for men and women. Data for health look at the maternal mortality ration and the adolescent birth rate to produce a female reproductive health index.

EXAM TIP

Some of the examples that you can use for gender inequality may also be useful as examples for empowering women, and vice versa.

CHECK YOUR UNDERSTANDING

5. Outline ways in which female empowerment in Colombia has progressed.
6. Describe an example of a project to empower indigenous/minority groups.

Social entrepreneurship and human development (1)

MICROFINANCE

Around 2.5 billion adults have no access to financial services, and 80% of people living on less than $2 a day have no bank account. Microfinance has grown to meet the needs of these people. Microfinance lending schemes aim to reduce poverty, address social issues including gender discrimination, and enable market access for the poor. Microfinance has grown dramatically since 1974, when they were first introduced. This led to the formation of the Grameen Bank and other organizations that provide small business loans to poor people.

Most people who use microfinance are people in rural areas, mainly farmers, who cannot access other forms of finance. They need to borrow money to buy seeds and fertilizers and improve their farms. In the past, their lack of access to banks may have forced them to use "loan sharks" (money lenders charging very high rates of interest).

Microfinance schemes often focus on women, who in some societies are unable to own land or borrow money. In most microfinance schemes, members are part of the

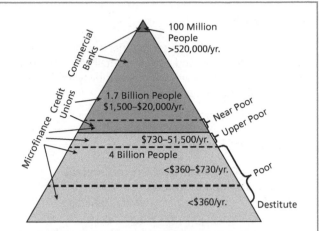

The hierarchy of banks and their customers
Source: VISA International, World Bank, C.K. Prahalad

community, and the community lends out money to its members. This makes repayment of the loan much more likely.

WMI: Women's microfinance initiative

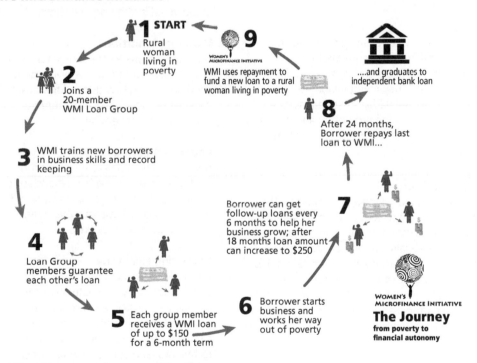

Microfinance schemes – the theory

Microfinance schemes have the potential to improve living standards for many poor people. However, the schemes have their critics:
- Their interest rates are higher than those of commercial banks.
- Some people will use the loans to pay for food or health care rather than for starting or improving their business.

- Not all poor people are entrepreneurs and so the loans may be wasted.
- Microfinance loans may be used to pay off other loans rather than for business purposes.

Some also claim that microfinance does not tackle the root causes of poverty but actually makes poverty worse.

CHECK YOUR UNDERSTANDING
7. Outline one advantage of microfinance schemes.
8. Outline one disadvantage of microfinance schemes.

Social entrepreneurship and human development (2)

FAIR TRADE	
Fair trade can be defined as trade that attempts to be socially, economically and environmentally responsible. It is trade in which companies take responsibility for the wider impact of their business.	Fair trade is an attempt to address the shortcomings of the global trading system and ensure that producers in poor countries get a fair deal, including a fair price for goods and services, decent working conditions and a commitment from buyers that there is reasonable security for the producers.

Principles of Fair Trade

Creating Opportunities for Economically Disadvantaged Producers

Fair Trade is a strategy for poverty alleviation and sustainable development. Its purpose is to create opportunities for producers who have been economically disadvantaged or marginalized by the conventional trading system.

Transparency and Accountability

Fair Trade involves transparent management and commercial relations to deal fairly and respectfully with trading partners.

Trade Relations

Fair Trade organisations trade with concern for the social, economic and environmental well-being of marginalized small producers and do not maximize profit at their expense. They maintain long-term relationships based on solidarity, trust and mutual respect that contribute to the promotion and growth of Fair Trade. An interest-free pre-payment of at least 50% is made if requested.

Payment of a Fair Price

A fair price in the regional or local context is one that has been agreed through dialogue and participation. It covers not only the costs of production but enables production which is socially just and environmentally sound. It provides fair pay to the producers, and takes into account the principle of equal pay for equal work by women and men. Fair Traders ensure prompt payment to their partners and, whenever possible, help producers with access to pre-harvest or pre-production financing.

Child Labour

Fair Trade organisations respect the UN Convention on the Rights of the Child, as well as local laws and social norms in order to ensure that the participation of children in production processes of fairly traded articles (if any) does not adversely affect their well-being, security, educational requirements and need for play.

Organisations working directly with informally organised producers disclose the involvement of children in production.

Non-Discrimination, Gender Equity and Freedom of Association

The organisation does not discriminate in hiring, remuneration, access to training, promotion, termination or retirement based on race, caste, national origin, religion, disability, gender, sexual orientation, union membership, political affiliation, HIV/Aids status or age. Fair Trade means that women's work is properly valued and rewarded. Women are always paid for their contribution to the production process and are empowered in their organisations.

Working Conditions

Fair Trade means a safe and healthy working environment for producers. Working hours and conditions comply with conditions established by national and local laws and International Labour Organization conventions.

Capacity Building

Fair Trade is a means to develop producers' independence. Fair Trade relationships provide continuity, during which producers and their marketing organisations can improve their management skills and their access to new markets.

Promoting Fair Trade

Fair Trade organisations raise awareness of Fair Trade and the possibility of greater justice in world trade. They provide their customers with information about the organisation, the products, and in what conditions they are made. They use honest advertising and marketing techniques, and aim for the highest standards in product quality and packing.

The Environment

Fair Trade actively encourages better environmental practices and the application of responsible methods of production.

Source: https://www.trade aid.org.nz/index.php/page/wfto

CHECK YOUR UNDERSTANDING

9. Define "fair trade".
10. Outline the ways in which Fair Trade impacts working conditions.

Social entrepreneurship and human development (3)

CORPORATE SOCIAL RESPONSIBILITY

Corporate social responsibility (CSR) refers to the practices of companies to manage the social, economic and environmental impacts of their activities, and take action to reduce these impacts if necessary. Companies vary widely in their responsibilities towards workers and the environment. In general, companies involved in fair trade take more care of their workers and the environment than many large TNCs. Most TNCs have a code of CSR, but whether they follow that code is another matter.

McDonald's Corp. CSR 2020

In 2014 McDonald's Corp. announced its 2020 Corporate Social Responsibility Framework. Its goals include:
- supporting sustainable beef production
- sourcing only coffee, palm oil and fish that are sustainably produced
- increasing in-restaurant recycling to 50% and minimizing waste in nine of its top markets
- increasing energy efficiency in company-owned restaurants by 20% in seven of its top markets.

Benefits of CSR

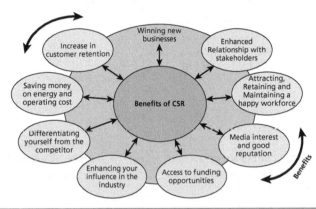

CASE STUDY

THE RANA PLAZA DISASTER, BANGLADESH

On 24 April 2013 an eight-storey garment factory in Rana Plaza on the outskirts of Dhaka, Bangladesh, collapsed, killing more than 1,100 people. Over half the victims were women and children. It was one of the worst industrial accidents in South Asia since the Bhopal disaster of 1984. Local police had warned that the building was unsafe but the owners allegedly threatened to fire those who did not carry on as usual (*The Economist*, 4 May 2013). Planning permission had been given for a five-storey building, not an eight-storey building.

As a result of the Rana Plaza disaster, many clothing companies have made efforts to improve their CSR.

According to pressure groups, CSR audits had looked at working hours and child labour, but not the structural soundness of buildings or fire exits.

CASE STUDY

UNILEVER

Unilever is a TNC in the food and beverage sector, with a comprehensive CSR strategy. It has been ranked "Food Industry leader" in the Dow Jones Sustainability World Indexes for 11 consecutive years and ranked 7th in the "Global 100 Most Sustainable Corporations in the World".

Unilever has a "sustainable tea" programme. It sources all of its Lipton and PG Tips tea bags from Rainforest Alliance Certified™ farms. The Rainforest Alliance Certification offers farms a way to differentiate their products as being socially, economically and environmentally sustainable.

Unilever plans that by 2020, 100% of their tea, including loose tea, will be sustainably sourced.

Unilever's 10-year Sustainable Living Plan shows how businesses can reduce environmental impact, while also maintaining sales growth. The Plan, which launched in 2010, has three main goals: (1) improving the health and well-being of the company's customers and consumers; (2) reducing the company's overall environmental impact; (3) enhancing the livelihoods of millions of people around the globe. In addition to these goals, the company hopes to double the size of its business.

COMMON MISTAKE

✗ *Workers in LICs are exploited by TNCs, and derive no benefits.*

✓ Although it is very easy to criticize large TNCs for the exploitation of workers and resources in LICs and NICs, many of the workers want to work for TNCs, as it will provide an income, however small.

CHECK YOUR UNDERSTANDING

11. Explain the meaning of the term "corporate social responsibility".
12. Outline the conditions that may have led to the Rana Plaza disaster.

What is "culture"?

Culture is the way of life of a particular society or group of people. Among other factors, it includes beliefs, behaviours, customs, traditions, rituals, dress, language, art, music, sport and literature. Culture is a complicated concept with a range of meanings, and it is important to all human populations. Culture varies from region to region, with some areas being relatively similar and others offering greater diversity.

AN EMERGING GLOBAL CULTURE

The world is changing fast, and the rate of this change is probably greater than ever before. New technologies such as the internet and satellite communications, are making the world more global and more interconnected. The increased speed of transport and communications, the increasing interactions between economies and cultures, and the growth of international migration, are among the factors that have changed everyday lives in recent decades.

Proponents of an emerging global culture suggest that different places and cultural practices around the world are converging and becoming more similar. A global culture might be the product of two very different processes:

- The export of supposedly "superior" cultural traits and products from advanced countries, and their worldwide adoption ("modernization", "westernization", "Americanization").
- The mixing, or hybridization, of cultures through greater interconnections leading to a new universal cultural practice.

A McDonald's restaurant in Ulsan, South Korea

LANGUAGE

A number of languages have more than 100 million native speakers. These include English, Mandarin, Hindi, Arabic and Bengali. English has become one of the dominant world business languages, and the main language of the internet, but it has major variations in vocabulary and accent from country to country and from region to region.

RELIGION

There are five major global religions: Christianity, Islam, Hinduism, Chinese folk religion and Buddhism. While Christianity and Islam can claim to be truly global, the remaining three are more regional in their distribution.

MUSIC

Transnational corporations control the global music industry, with the USA and the UK dominating domestically generated popular music.

TELEVISION

Until recently, television programmes tended to be produced primarily for domestic audiences within national boundaries, and could be subjected to rigorous governmental control. However, with the advent of cable, satellite and digital technologies, several television channels are now globally disseminated, and to some extent circumvent national restrictions. The USA, France, Germany and the UK are major exporters of television programmes. Companies such as Al Jazeera, BBC, CNN and ABC have an international reach.

SPORT

Sports are forms of cultural expression that are becoming increasingly globalized, as well as increasingly commodified. Football/soccer is the most obvious example, but similar trends can be observed in US major league baseball.

TOURISM

Tourism is a form of international cultural exchange that allows large numbers of people to experience other cultures and places. It also locks specific destinations into wider international cultural patterns.

CHECK YOUR UNDERSTANDING
13. Define the term "culture".
14. Suggest why cities are often culturally diverse.

Cultural traits

CULTURAL DIFFUSION

Cultural diffusion, the spread of cultural traits, occurs when two cultures intermingle. This occurred historically when members of different cultures interacted with one another through trade, intermarriage or warfare, and it happens today when, for example, different countries share an interest in a particular sport. Cultural diffusion may also be forced, as when one culture defeats another and forces its beliefs and customs on to the conquered people. This is known as cultural imperialism. It reflects an assumption that the culture being imposed is somehow superior to the one being replaced. Cultural diffusion today is taking hold around the world as cultural ideas spread through information communications technology and the mass media. Globalization is regarded as a key factor in driving culture towards a global model. Media TNCs and the movement of workers and tourists aid this process.

CULTURAL IMPERIALISM

Global cultural imperialism today has resulted from economic forces, as when the dominant culture (usually the USA) captures markets for its commodities and thereby gains influence and control over the popular culture of other countries. The export of entertainment is one of the most important sources of cultural diffusion. In the political sphere, cultural imperialism plays a major role in dissociating people from their cultural roots, replacing them with media-created desires. The political effect is to alienate people from traditional community bonds and from one another. As countries are attracted by and brought under the influence of the dominant world system, they are pressured into shaping their culture to follow the values and structures of the dominant system. Some of the means by which this happens occur through language, tourism, global brands, the media and democracy.

HOW HAS CULTURAL IMPERIALISM CHANGED?

Scholars of cultural imperialism believe that it started during the industrial colonialism phase. Colonialism reached its peak just before the First World War – the British Empire reached its maximum territorial extent. It is usually the case that the cultural imperialist is a large, economically or militarily powerful nation and the victim country is a smaller or less affluent one. The end of formal colonialism in the second half of the 20th century did not spell the end of cultural imperialism. However, the nature of cultural imperialism changed. The world is becoming more uniform and standardized, through a technological, commercial and cultural process originating from the West. Cultural imperialism has become a social and economic process as well as a political one. It is facilitated by TNCs that represent the interests of the rich and powerful, especially those of the West.

CASE STUDY

CULTURAL DIFFUSION IN SINGAPORE

Port cities are characterized by a constant mixing of people, ideas and goods. Port cities in Asia, such as Singapore, have many multiracial communities, and are characterized by global trade links, and the import and export of primary products and manufactured goods. Socially, it has led to intermarriages and the creation of altered cultural forms. Mixed-race communities are perhaps the most visible aspect of cultural integration in port cities. For centuries Asian people have been mixing with other Asian nationalities. For example, the Peranakam people in Singapore are a mixed-race people based on the intermarriage of Malay women and Indians creating Chitty Melaka Peranakam and Jawi Peranakam, and intermarriage of with Chinese leading to Straits Chinese Peranakam.

The convergence of people and goods created cities characterized by diverse architecture and a demand for imported goods. For those who became wealthy as a result of trading, much of their income was spent on luxury goods and fashion goods.

Ex-pats in Singapore

There are around 2.5 million non-permanent residents (ex-pats) living in Singapore, who account for about 40% of the island's population. This is up from 25% in 2000. One in four skilled workers in Singapore are from overseas. There are over 40,000 British nationals and other large populations of overseas communities include Australians, Americans French and Japanese. There are many popular ex-pat neighbourhoods including Orchard, Holland Village, Bukit Timah and the east coast. To cater for the ex-pat communities, there are many sporting and social clubs, international schools, expat websites and magazines. Singapore continues to be a place of great cultural mixing.

CHECK YOUR UNDERSTANDING
15. Explain the term "cultural diffusion".
16. Explain the changing nature of cultural imperialism.

Glocalization of branded commodities and cultural hybridity

GLOBAL BRANDS

The rise of global consumer culture has seen many brands become truly global. Coca-Cola is sold in nearly every country, yet it is very much linked with US culture. Pringles are found worldwide – although some of the flavours vary from country to country, such as in jalapeno flavour in Cuba and seaweed flavour in Singapore. McDonald's, for example, operates over 35,000 outlets in more than 100 countries.

COCA-COLA

Coca-Cola is the biggest-selling soft drink in history and one of world's best-known products. It was first offered as a soda fountain beverage in Atlanta. The Coca-Cola trademark is now one of the most recognized in the world and the word "Coca-Cola" itself is thought to be the second most widely understood word in the world after "OK".

The Coca-Cola Company is still based in Atlanta but it employs 49,000 people worldwide, operating in over 200 countries. The production and distribution of Coca-Cola uses a franchising model, so that local people with local resources produce the drinks. Coca-Cola produces a syrup concentrate, which it sells around the world.

There are a number of variations on the original Coca-Cola formula including:

- Coca-Cola with lime – available only in the USA, Canada, Singapore, the UK, Belgium and the Netherlands
- Coca-Cola Blak (2006–8) – available only in the USA, Canada, France, the Czech Republic, Bosnia and Herzegovina, Bulgaria and Lithuania
- Coca-Cola Citra (2005–present) – available only in Bosnia and Herzegovina, New Zealand and Japan
- Coca-Cola clear (2016) – available only in Canada, France and Australia.

Since 2012, Coca-Cola has been officially available in every country of the world except Cuba and North Korea. In 2012 it was widely rumoured that Bolivia was about to ban Coca Cola, however, the government called on its population to drink less cola and have more healthy drinks instead.

ZAMZAM

Zamzam (previously Zamzam cola) is a type of soft drink produced by the Iranian company Zamzam Group. It is popular in parts of the Middle East, where it is seen as an alternative to Coca-Cola and Pepsi. It was created in 1954 as a subsidiary of Pepsi, but became independent in 1979 following the Islamic Revolution. The product's name is derived from the Well of Zamzam in Mecca. The Zamzam group produces other products in Iran and the Middle East.

MCDONALD'S RESTAURANTS

McDonald's has been considered an emblem of globalization. On an average day, more than 70 million customers are served at one of 35,000 McDonald's restaurants. Over 1.9 million people work in McDonald's and its franchises. The first McDonald's restaurants were located in the USA and Canada and more opened in Europe, Australia and Japan during the early 1970s. By the end of the 1970s McDonald's was consolidating its position in Europe and New Zealand and had opened restaurants in South America, namely in Brazil. The 1980s saw further expansion and consolidation in South America, Mexico, parts of Europe and South East Asia. China, Russia and parts of the Arab world were reached in the 1990s. Over half these restaurants are in the United States, but the UK has over 600 outlets, Brazil over 250, China nearly 200 and Thailand nearly 50.

McDonald's is famed for its corporate uniformity; it has largely the same decor and the same service style around the world. Nevertheless, McDonald's has been glocalized into some traditional cultural forms and practices.

McDonald's menus vary around the world. The basic menu is similar but there are national specialities. Japanese McDonald's offers green tea-flavoured milkshakes, as well the gratin korokke burger – a sandwich of mashed potato, filled with shrimp, macaroni and shredded cabbage.

Some other variations you may not have heard of include:

- **McLobster:** Lobster roll (New England and Canada)
- **Croque McDo:** Melted Emmental cheese and ham on hamburger buns (Europe)
- **McRice:** Asia's steamed rice buns with beef or chicken patty (Asia)
- **McAloo Tikki Burger:** Vegetable burger made out of potatoes, peas and spices (India)
- **McTurco:** Fried pita with two burger patties, cayenne pepper sauce and vegetables (Turkey)
- **Ebi Filet-O:** Shrimp burger (Japan)
- **McSpaghetti:** Spaghetti dinner meal (Indonesia and the Philippines)
- **Kiwiburger:** Beef patty with cooked beetroot, egg, tomato, lettuce, cheese, onion and mustard

EXAM TIP

For answers about glocalization, make sure that you have examples. The examples do not have to include McDonald's!

CHECK YOUR UNDERSTANDING

17. Suggest why branded commodities may become "glocalized".
18. Outline the reasons for the growth of hybridized commodities.

The changing urban cultural landscape

EVOLUTION

The evolution of uniform urban landscapes is the result of a variety of factors:
- improvements in communications technology, so that people in cities around the world are aware of opportunities and trends in other cities
- increased international migration and the spread of ideas and cultures
- the desire of global brands such as Costa and Starbucks to reach new markets
- improvements in standards of living
- aspirations to be part of a global network of urban centres
- globalization of economic activity and culture
- attempts to create smart cities.

Aspects of cultural landscape changes in the built environment

Many urban landscapes in different countries look very similar. Tall towers are a feature of many cities. Industrial estates and science parks are a familiar sight in many countries, as TNCs outsource their activities to access cheap labour, vital raw materials and potential markets. Many cities have similar retailing characteristics – pedestrianized shopping centres and out-of-town supermarkets.

CASE STUDY

SEOUL, SOUTH KOREA

Seoul is a good example of the debate on the homogenization of urban landscapes. On the one hand, it fits the theory of a homogenized landscape – global firms (such as McDonald's) are located in Seoul, just as Korean firms like Hyundai and Samsung are located in other countries. The CBD is characterized by skyscrapers and international firms such as Barclays and Tesco. There are high-rise apartments and edge-of-town developments.

On the other hand, a massive urban redevelopment project has centred on the restoration of the Cheong Gye Cheon River in central Seoul. This restoration has not been just of a river but also has historical, cultural and environmental value. Murals along the side of the river recount some of the most important events in Seoul over the last 600 years, and the river has become an important focus for Seoul residents and visitors – rather like Times Square in New York – partly because it is stressing the individuality and uniqueness of Seoul, and of Korea.

What type of urban area is unique?

All urban areas have something in common. All have something unique. Urban areas that are less westernized or less globalized might be expected to be more different from one another than those that are more globalized. Some Arab cities are different in structure and environment from western cities. Many large western cities have mosques. Western and Arabic cultures may be merging in some urban areas, or co-existing peacefully. For example, the urban landscape of Bandar Seri Begawan (Brunei) is dominated by a mosque. Equally unusual is that a high proportion of people live in Bandar's water village, yet there are elements of a western culture.

CHECK YOUR UNDERSTANDING
19. Suggest ways in which the built environment is becoming more similar around the world.
20. Suggest why the built environment is becoming more similar around the world.

Diasporas and cultural diversity

WHAT IS A DIASPORA?

The scattering of a population, or diaspora, originally referred to the dispersal of the Jewish population from Palestine in 70 AD. It is now used to refer to any dispersal of a population formerly concentrated in one place. Examples include:

- the forced resettlement of Africans during the slave trade
- professional and business diasporas, such as the movement of Indians and Mexicans overseas
- cultural diasporas, such as the movement of migrants of African descent from the Caribbean.

Diasporas may bring their culture with them – some may marry people from other population groups, and so cultures may be hybridized.

THE CHINESE DIASPORA

Approximately 40 million people of Chinese origin live in sizeable numbers in at least 20 countries. Large concentrations are in Singapore (2.6 million), Indonesia (7.6 million), Malaysia (6.2 million), Thailand (7 million) and the USA (3.4 million). With globalization, Chinese migration for professional and business reasons has increased.

The Chinese diaspora has significant economic power, through remittances, as well as financial power invested in the host countries. Chinatowns are an important symbol of Chinese culture and identity.

THE SYRIAN DIASPORA

There are between 8 million and 15 million Syrians living outside Syria. Increasingly, more of them are refugees whereas earlier they were economic migrants. There have been three main waves of Syrian migration. In the late 19th century, many were attracted by jobs in South America. Many Syrians went to Brazil and Argentina, and both countries have a significant number of Syrian residents. During the late 19th and early 20th centuries, many Syrians went to the USA, in particular to New York, Boston and Detroit. Most were Christians fleeing persecution and in search of a better life. Many in New York lived in Manhattan in what was to become known as Little Syria. However, as they prospered they moved into more affluent neighbourhoods. Since the 1970s there has been an increase in the Syrian diaspora in the Middle East, initially due to the oil economies of the region, and later due to refugees fleeing Syria into neighbouring states.

In 2002 the Syrian government established the Ministry of Expatriate Affairs to encourage the diaspora to return (at least for a visit) and to invest in the country. Tourism and remittances are the main source of foreign exchange for Syria, worth $1–$2 billion annually. Civil war and conflict in Syria has had a negative impact on the number of diaspora returning for a holiday.

THE IRISH DIASPORA

The Irish diaspora consists of Irish migrants and their descendants in countries such as the USA, the UK, Australia, New Zealand, Canada and continental Europe. The Irish diaspora contains more than 80 million people, more than 14 times the population of Ireland.

Various aspects of Irish culture, such as sport (hurling and Gaelic football), traditional Irish music and dance (popularized by *Riverdance*), food and drink, are commonplace throughout many of the areas where the Irish have settled. In Australia, Gaelic football has been hybridized into Australian-rules football.

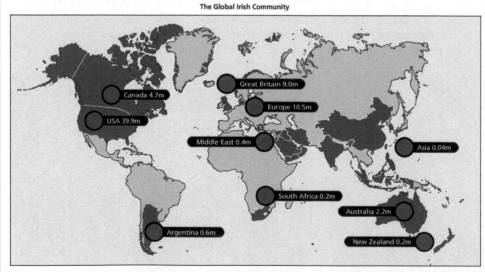

The Global Irish Community

Canada 4.7m
USA 39.9m
Great Britain 9.0m
Europe 10.5m
Middle East 0.4m
Asia 0.04m
South Africa 0.2m
Australia 2.2m
Argentina 0.6m
New Zealand 0.2m

COMMON MISTAKE

✗ *Migrants keep their home culture in a way that does not change.*

✓ Very often it is suggested that diaspora groups keep the culture of the country that they left. In many cases, however, migrants intermarry, and so cultures become hybridized or new cultures emerge.

CHECK YOUR UNDERSTANDING

21. Define the term "diaspora".
22. Suggest two ways in which migrant groups may increase cultural diversity.

The rejection of globalized production

ADVANTAGES AND DISADVANTAGES OF GLOBALIZED PRODUCTION

Globalized production has many advantages and disadvantages. Views vary widely between stakeholders.

The benefits and costs of globalized and localized food production

Local commercial production	Globalized production
Benefits	**Benefits**
Producer • Increased market access and sales • Possibly more farm-gate sales	*Producer* • Ability to produce foods cheaply and to a uniform standard
Consumer • Fresh food • Local products "in season" • Reuced air miles • Smaller carbon footprint	*Consumer* • Cheap food available year round • All types of products available year round • Competition between producers keeps main costs down
Local economy • Improved local farming economy • Multiplier effects, e.g. demand for fertilizers, vets, farm equipment	*Local economy* • May be able to provide large amounts of a single product to a major TNC • Specialization allows intensification and increased production
Costs	**Costs**
Producer • Increasing cost of oil makes cost of inputs higher • Greater emphasis on quality may make production less profitable	*Producer* • Increased air miles • Higher costs of inputs, especially fertilizers and oil • Profit margins increasing squeezed
Consumer • Higher cost of local farm products • Less choice "out of season"	*Consumer* • Increased costs are likely to be passed on to the consumer • Indirect costs such as pollution control, eutrophication of streams, soil erosion, declining water quality
Local economy • Cost of subsidies to maintain farming, e.g. payments to encourage farming in environmentally friendly ways	*Local economy* • Undercuts local farmers who may quit farming • Producers are vulnerable to changes in demand and are at the mercy of TNCs

ALTERNATIVES TO GLOBALIZATION

Global civil society is extremely wide ranging. Groups that comprise it can be liberal, democratic and peaceful, or the complete opposite. Some civil societies are very large organizations, such as Oxfam, whereas others are relatively small, such as Operation Hunger in South Africa.

THE RISE OF NGOs

The perception that some global institutions, such as the World Bank and the IMF, are undemocratic and do not help all people equally, has led to global civil society movements that are attempting to regulate the global system from below.

Several alliances have emerged within NGOs, such as the global environmental movement, the anti-globalization movement and the global women's movement, which are attempting to reform the global system. Well-known individual NGOs include Greenpeace, The Fair Trade Network, Stop The War Coalition, Globalize Resistance, CAFOD and Amnesty International. Each of these has different aims and methods, but all agree that major globalizing bodies such as the World Bank, the IMF and the G7/G8 countries are pushing an agenda that favours rich western countries more than others.

Important areas global civil societies have focused on include:
- creating a more level playing field for the global south
- supporting free access to information
- making global civil society more accountable and transparent
- developing a new relationship with global institutions.

EXAM TIP
Try to be balanced in your responses. For example, consider an advantage and a disadvantage of locally produced goods. Locally produced goods may have less food miles, but they may be more expensive than globally produced goods – their sales may fall during times of economic slowdown or a recession.

CHECK YOUR UNDERSTANDING
23. Outline two advantages of locally produced goods and two advantages of globally produced goods.
24. Suggest why campaigns against TNCs develop.

The rise of anti-immigration movements

THE RISE OF ANTI-IMMIGRATION GROUPS

There are many reasons for the rise of anti-immigration groups including the perceived threats over competition for jobs, and the cost of housing, education and health care. In some cases, notably in LICs and NICs, environmental issues may also be a concern, as a result of rapid population growth. Some argue that certain immigrant groups isolate themselves from society and refuse to integrate into mainstream society.

On the other hand, there is a lot of evidence to suggest that migration benefits the host country. A 2014 study by the University College London suggested that the UK benefited from EU migrants to the amount of £20 billion per year. However, the euro crisis and the increase in the number of migrants from Syria, the Middle East and North Africa have made many Europeans anxious about migration. There has been an increase in right-wing political parties opposed to immigration in many European nations. Anti-migration issues were considered to be a very important reason for the UK voting to leave the EU ("Brexit"). Following the Brexit referendum there was a reported rise in race crime in the UK.

In 2017, US President Trump prevented migrants from seven, mainly Muslim, countries from entering the USA, although his decision was overturned by a number of US courts. He reiterated his intention of having a wall built between the USA and Mexico to reduce migration.

In 2017 Buddhist nationalist groups protested at Yangon, Myanmar, against the arrival of a Malaysian ship carrying aid for thousands of Rohingyas. The government initially refused the ship permission to enter its waters, and also demanded that the aid be distributed to both Rohingya and Buddhist ethnic Rakhine people. Buddhist nationalists claim that Rohingyas are illegal migrants from Bangladesh.

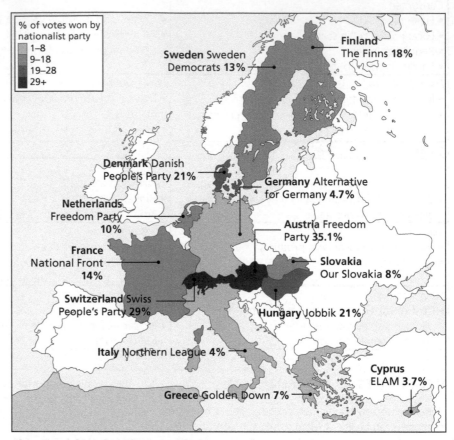

% of votes won by nationalist party
1–8
9–18
19–28
29+

Finland The Finns **18%**

Sweden Sweden Democrats **13%**

Denmark Danish People's Party **21%**

Germany Alternative for Germany **4.7%**

Netherlands Freedom Party **10%**

Austria Freedom Party **35.1%**

France National Front **14%**

Slovakia Our Slovakia **8%**

Switzerland Swiss People's Party **29%**

Hungary Jobbik **21%**

Italy Northern League **4%**

Cyprus ELAM **3.7%**

Greece Golden Down **7%**

The rise of nationalism in Europe

CASE STUDY

IMMIGRATION LAWS IN DENMARK

Denmark has some of the tightest immigration laws in Europe. Denmark introduced a points system designed to make it more difficult for "family reunions" that allowed non-Danes into the country for marriage, imposing a minimum age of 24 years for both the Danish spouse and the immigrant, proof of financial independence, and evidence of an active commitment to Danish society.

In 2016 the Danish government voted in favour of seizing asylum seekers' assets in order to help pay for their stay while their application for asylum is considered.

However, Denmark is not alone in claiming assets. Switzerland and some German states also take funds from asylum seekers.

EXAM TIP
When referring to the rise of anti-immigration movements, try to refer to recent elections and their results.

CHECK YOUR UNDERSTANDING
25. Identify the European country with the highest votes won by a nationalist party.
26. Outline the measures taken by the Danish government to control migration.

Geopolitical constraints on global interactions (1)

GOVERNMENT AND MILITIA CONTROL

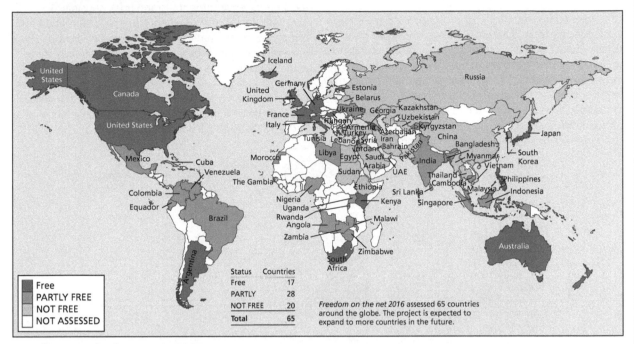

Status	Countries
Free	17
PARTLY	28
NOT FREE	20
Total	**65**

Freedom on the net 2016 assessed 65 countries around the globe. The project is expected to expand to more countries in the future.

Legend:
- Free
- PARTLY FREE
- NOT FREE
- NOT ASSESSED

Freedom on the internet, 2016

Internet access is highly controlled in communist countries such as China, North Korea, Cuba and Vietnam.

China has a "Great Firewall" which blocks critical websites and social media. In 2014 a government paper, Document 9, was leaked. It stated that the role of the media was to support government rule.

Eritrea, along with North Korea, is one of the most censored countries in the world. It has banned independent journalists. A UN inquiry into human rights in Eritrea claimed that there were systemic human rights violations, widespread detention and indefinite military service. Most Eritreans in national service earn $1–$2 per day.

Eritrea also has one of the lowest figures worldwide for cellphone users, at 5.6%. Some Eritrean exiles try to provide independent online websites and broadcasts, although these are blocked by the state-run telecommunications company The internet is available only through slow dial-up connections, and less than 1% of the population go online.

Up to 320,000 Eritreans have fled the country, many making the dangerous journey to Europe to start a new life away from the "pervasive state control". In 2015 they were the second largest nationality, after the Syrians, to arrive on Italian shores.

North Korea largely prevents its population from travelling around the country or abroad, and only the political elite own vehicles. Emigration and immigration are strictly controlled. Only political supporters and the healthiest citizens are allowed to live in Pyongyang. Less than 10% of the population have cellphones, although some phones are believed to be smuggled in from China. Some schools and other institutions have access to a highly-controlled intranet. According to Reporters Without Borders, radio and TV sets bought in North Korea are only able to receive government frequencies.

In 2017 Kim Jong-un, North Korea's dictator allegedly used assassins to kill his half-brother Kim Jong-nam at Kuala Lumpur airport, Malaysia, with VX nerve gas. VX is a banned, chemical-based weapon of mass destruction and one of the deadliest nerve gases ever created. Exposure to the liquid may lead to convulsions, loss of consciousness, paralysis and fatal respiratory failure. The USA, China, South Korea and Japan are believed to have tried to contain or ignore North Korea's illegal nuclear weapons and missile-building activities, but this incident was highly publicised. The dictator has had over 340 people killed (including an uncle and his half-brother) in just five years.

CHECK YOUR UNDERSTANDING
27. Suggest reasons for the large number of migrants leaving Eritrea.
28. Comment on North Korea's control on personal freedom to participate in global interactions.

Geopolitical constraints on global interactions (2)

NATIONAL TRADE RESTRICTIONS

Trade restrictions are a form of protectionism: most trade restrictions place an additional charge on traded goods to make home goods more competitive. Most economists would argue that trade restrictions increase inefficiency and lead to less choice for consumers, although they may help a country to industrialize. Trade barriers have been criticized as well, as they often affect LICs. Protectionism reduces trade between countries. This may be achieved through taxes on imports (tariffs), limits on the volume of imports (quotas), administrative barriers (for example, food safety, environmental standards), subsidies to home producers as well as anti-dumping legislation and campaigns to buy nationally produced goods. Most HICs have tried to embrace free trade, or at least free trade within a trading bloc. However, since 2008 70% of the 20 OECD nations have imposed restrictive trade policies in response to the global economic slowdown. For example, in 2015 the USA imposed a 256% tariff on Chinese steel and a 522% tariff on cold-rolled steel.

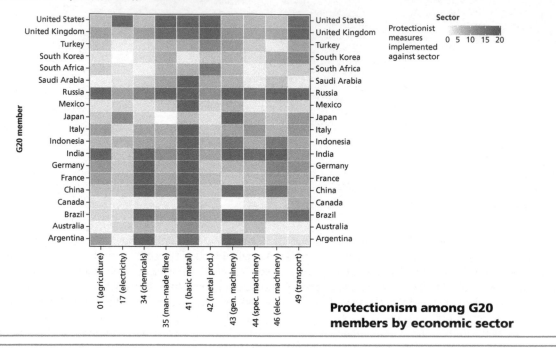

Protectionism among G20 members by economic sector

RESOURCE NATIONALISM

Resource nationalism occurs when a country decides to take all, or a part of, its natural resources under state ownership. Across much of Africa, governments have attempted to gain a larger share of the profits from mining. In South Africa, mineral wealth is estimated at $2.5 trillion, and the government is considering a 50% tax on mining profits and a 50% capital gains tax on the sale of prospecting rights.

In 2012 Ghana announced a review and possible renegotiation of all mining contracts, to ensure that mining profits are maximized for the benefit of the country. It intends to raise taxes on mining from 25 to 35%. In Zambia, the government doubled its royalties on copper to 6%. In Guinea, which contains some of the world's largest bauxite reserves, the government takes 15% of all profits from mining, and in Namibia, all new mining has been transferred to a state-owned company. In South Africa, following the end of the apartheid era, black-economic empowerment laws required mining firms to sell stakes of at least 26 per cent to black shareholders by 2014.

Some global mining companies are worried about resource nationalism. Mining is a capital-intensive industry and much of the equipment is extremely expensive. Companies worry that investments may cost billions of dollars, and take up to a decade to get a return.

Although resource nationalism holds many benefits for countries, it is important for the countries to give companies enough return on their investments that they will continue to invest in the future.

COMMON MISTAKE

✗ *Countries that pursue resource nationalism have complete control over the mining and production of raw materials.*

✓ Resource nationalism does not mean that a country has complete control of its resources – it has some control but may rely on TNCs for the equipment to develop the resources.

CHECK YOUR UNDERSTANDING

29. Identify the three countries with the most protectionist policies implemented.
30. Outline one advantage and one disadvantage of resource nationalism.

The role of civil society (1)

DEFINITION OF CIVIL SOCIETY

Civil society is composed of all the civic and social organizations or movements that form the basis of a functioning society, and work in the area between the household, the private sector and the state to negotiate matters of public concern. Civic societies include non-governmental organizations (NGOs), community groups, trade unions, academic institutions and faith-based organizations.

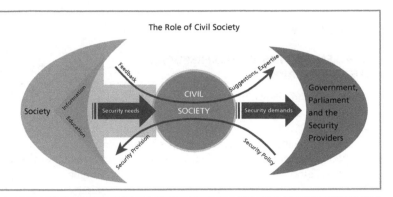

The Role of Civil Society

PEOPLE'S GLOBAL ACTION

People's Global Action (PGA) is a network for spreading information and coordinating actions between grassroots movements around the world. These diverse groups share an opposition to capitalism and a commitment to direct action and civil disobedience as the most effective form of struggle.

The PGA is an organized network for spreading information. It is an informal support group and it helps with fundraising; there is a website and numerous email lists.

CIVIL SOCIETY, SOCIAL MEDIA AND INCREASED USE OF THE INTERNET

The Global Internet Freedom Program is a civil society that aims to strengthen policy for internet freedom in the global south, as well as in internet forums. For example, the majority of countries in West Africa do not yet have laws in place that govern or regulate online activities. It also aims to increase the capacity of wider civil society in the global south to understand and influence relevant policies in support of internet freedom at the national, regional and global levels.

In 2005, the World Summit on the Information Society (WSIS) affirmed the commitment of all countries present to "develop a people-centred, inclusive and development-orientated information society". However, many Arab countries have limited civil society engagement in internet policy, preferring government approaches to setting internet policy. The challenges that civil societies face include the monitoring of violation of internet rights, pushing for the protection of activists, and developing fairer laws for use of the internet. For example, 70% of "prisoners of opinion" in the Arab world are imprisoned on charges of unlawful internet use, and 70% of those are prosecuted because of social media content.

The OECD Development Centre's Wikigender platform has been has been engaging with a cross-section of gender equality actors, from civil society to governments, as a means of promoting women's voices in policy-making. Social media has proven potential for mobilizing attention and accountability to women's rights, and challenging discrimination and stereotypes. Strategies to enhance social media's potential for women's empowerment include facilitating their access to technology; increasing women's representation in public life and media; and working with a cross-section of actors. The explosion of social media and unprecedented use by women of new technologies represents important opportunities to bring gender equality and women's rights issues to the forefront of both policy making and media attention.

Three areas where social media have enabled women's political activism include:
- hashtag activism bringing women's issues to the forefront of political agendas
- tackling violence against women through social media tools
- public accountability towards gender equality.

GLOBAL INTERNET POPULATION BY 2016 FOTN (Freedom of the Net) STATUS
FOTN assesses 88 percent of the world's internet user population.

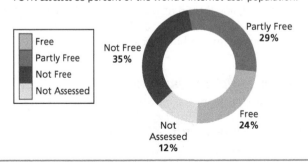

Free
Partly Free
Not Free
Not Assessed

Partly Free 29%
Not Free 35%
Free 24%
Not Assessed 12%

❓

CHECK YOUR UNDERSTANDING
31. Define the term "civil society".
32. Briefly explain how the use of social media can help reduce gender inequalities.

The role of civil society (2)

CHALLENGING RESTRICTED FREEDOM: THE ARAB SPRING

The Arab Spring refers to the demonstrations, protests, riots and civil wars that spread through the Middle East and North Africa after 2010. Most of the revolutions and protests were over by 2012, although the conflict in Syria is an important exception. The only country involved in the Arab Spring to become a democracy was Tunisia. Many factors lay behind the Arab Spring: dissatisfaction with governments, dictatorships, corruption, economic decline, unemployment, inequalities in wealth, food shortages and escalating food prices. Widespread access to social media networks made the Arab Spring possible in countries such as Tunisia and Egypt, whereas in other countries (for example, Yemen and Libya) people communicated through the traditional forms of media.

Following the protest and changes of the Arab Spring came the so-called Arab Winter, a wave of violence, instability and economic decline. The Arab Spring has thus had mixed success. For some, there has been greater freedom compared with the restrictions before, as in Tunisia. For others, the Arab Spring has led to a collapse of law and social order, as in Syria.

A number of reasons have been put forward to explain what has determined success in some areas but not others. They include:

- strong civil societies – countries such as Tunisia with strong civil societies were more successful than those without because they were able to transform the country after political change
- the degree of state censorship – in countries such as Egypt where Al Jazeera and the BBC provided widespread coverage, mass violence by the government and the military was suppressed, in contrast to countries such as Syria, where there was less television reporting
- social media – countries with greater access to social media were able to mobilize support for the protests
- support of the national military – in Egypt and Tunisia, the military supported the protesters in removing the government, whereas in Syria the military have contributed to civil war
- the mobilization of the middle class – countries with a strong, vocal middle class were more likely to see political change than countries with a weak or limited middle class.

POLITICAL CHANGE IN MYANMAR/BURMA

There has been considerable political change in Myanmar, from the decades of authoritarian military rule to the democratic election of Aung San Suu Kyi of the National League for Democracy party in November 2015, and her swearing in as de facto "president" in March 2016. Nevertheless, there are still many important challenges ahead, notably for Myanmar's ethnic communities and for minority religious groups, such as the Rohingyas. The military still have a disproportionate share of parliamentary seats. Many important, difficult decisions have been left for the new government, and there is no guarantee that they will succeed.

In 2017 the leader Aung San Suu Kyi urged all the country's armed ethnic groups to sign a ceasefire agreement. Ms Suu Kyi has come under criticism from human rights groups for ignoring the plight of the Rohingya minority, amid allegations of human rights abuses against the Rohingyas living in Rakhine state. Myanmar has suffered more than 50 years of armed conflict involving ethnic rebels since its independence.

In 1947 General Aung San (Ms Suu Kyi's father) reached an agreement with ethnic leaders to grant a level of autonomy to major ethnic groups if they cooperated with the federal government to separate from Britain. Five months after the deal was signed, General Aung San was assassinated, and ethnic groups have claimed that, since then, successive governments have failed to adhere to the agreement.

The Rohingyas became stateless in 1982 under the military government's Citizenship Act.

CHECK YOUR UNDERSTANDING

33. Suggest ways in which the Rohingyas have restricted freedom.
34. Suggest why the Arab Spring was successful in some countries but not in others.

Exam practice

(a) Analyse the validity and reliability of development indicators and indices. (12 marks)

(b) Examine the reasons for the growth of anti-immigration groups. (16 marks)

1 GEOPOLITICAL AND ECONOMIC RISKS

Threats to individuals and businesses (1)

HACKING

Hacking is a major threat to all internet users, from individuals, TNCs, national governments and even internet providers. For example, in 2016 there was a major attack on companies such as Twitter and Paypal, and many websites were affected including the *New York Times*, CNN and the *Wall Street Journal*. Hackers used thousands of interconnected devices that had been infected with a malicious code, known as a "botnet", to disable the internet. It was one of the largest internet attacks ever. The complexity of the attack made it difficult to deal with. Such attacks are becoming more common although not all are on such a large scale.

On a smaller scale, mobile devices and PCs are also vulnerable to attack. Often the source of the attack is a PDF or JPEG graphic that appears harmless until it is opened. It is likely that such attacks will increase in complexity in future, as more people own computers and mobile devices.

IMPLICATIONS OF SURVEILLANCE FOR PERSONAL FREEDOM

In 2013 the United Nations published a report, "The link between state surveillance and freedom of expression", concerning the threat of state surveillance to the right to privacy. According to the report: "State surveillance of communications is ubiquitous, and such surveillance severely undermines citizens' ability to enjoy a private life, freely express themselves and enjoy their other fundamental human rights." (https://www.privacyinternational.org/node/392)

It was the first time the UN has emphasized the centrality of "the right to privacy to democratic principles and the free flow of speech and ideas".

Modern surveillance technologies enable states to intrude into individuals' private lives. For example, by placing taps on the fibre optic cables, states can achieve almost complete control of telecommunications and online communications. Such systems were reportedly adopted, for example, by the Egyptian and Libyan governments in the lead-up to the Arab Spring.

In May 2017, there was a global cyber-attack, with more than 200,000 victims in 150 countries. The culprit was ransomware, a malicious program that locks a computer's files until a ransom is paid. In the UK, the National Health Service (NHS) was hit hard, although by the next morning the majority of the 48 affected health trusts in England had their machines back in operation. The WannaCry virus infected only machines running Windows operating systems. Unlike many other malicious programs, ransomware has the ability to move around a network by itself. Most others rely on humans to spread by tricking them into clicking on an attachment harbouring the attack code.

IDENTITY THEFT

Identity theft is one of the fastest-growing white-collar crimes. Cybercrime differs from traditional crimes in that the offender and the victim can be geographically separate.

The incidence of identity theft is expected to rise in the future because of the increase in the number of computers, many of which have insufficient protection. In addition, the very low prosecution rates and lenient sentencing make identity theft appealing to some criminals.

In a study of identity theft in the USA, regional differences and hotspots were identified. Certain populations are at increased risk of identity theft. These include people in close groups such as university students and military personnel, medical patients and even the deceased. In a study of urban areas in Florida, whites were most likely to be the victims of identity theft (72%) and Hispanics the least (1%).

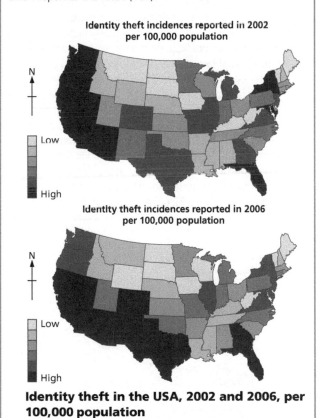

Identity theft incidences reported in 2002 per 100,000 population

Identity theft incidences reported in 2006 per 100,000 population

Identity theft in the USA, 2002 and 2006, per 100,000 population

CHECK YOUR UNDERSTANDING
1. Suggest why hacking is becoming more frequent.
2. Outline one advantage and one disadvantage of increased surveillance by governments.

Threats to individuals and businesses (2)

RISKS TO THE GLOBAL SUPPLY CHAIN

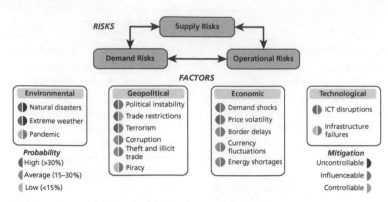

Risks to global supply-chain flows

There are many risks to global supply-chain flows including risks on the supply and demand sides and operational risks.

Supply risks refer to the ability to meet the demand for goods whereas demand risks refer to the changes in demand, possibly due to currency fluctuations or political unrest. Operational risks refer to the transport infrastructure required for delivering the goods.

Supply-chain risks are affected by many factors, including environmental, geopolitical, economic, or technological:

- Environmental factors include natural disasters, extreme weather events and epidemics. Floods in Bangladesh regularly disrupt the supply chain there, while the outbreak of Ebola in West Africa limited the movement of people and goods to and from the region. The 2011 tsunami that affected Japan had a major impact on the motor industry: up to 150,000 fewer cars were produced in the USA as a result of disruption of the supply chain for parts.
- Political factors include protectionism, trade restrictions, and conflict, all of which have a major impact on the supply chain. Supplies of goods from the Middle East and North Africa were severely disrupted at times following the Arab Spring protests. Governments with poor political or economic stability may take decisions (such as resource nationalism) that cause much uncertainty among investors. The UK's decision to leave the EU may lead to major changes in the supply chain.
- Economic factors include currency fluctuations and trade restrictions imposed by governments. Debt crises in Europe among the PIIGS countries (Portugal, Ireland, Italy, Greece and Spain) provided much economic uncertainty to risk managers.
- Technological factors include disruption to transport networks and ICT networks. Transport infrastructure failures are unusual, but ICT reliability is an issue.

Managing supply chains is difficult. A survey in 2013 found that more than three-quarters of companies had experienced supply-chain disruption within the previous two years.

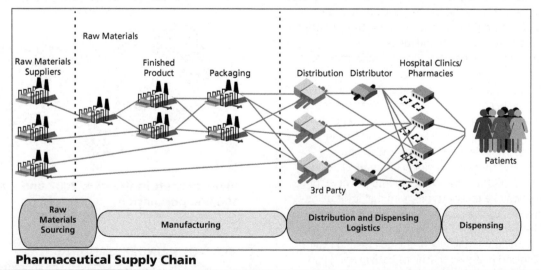

Pharmaceutical Supply Chain

EXAM TIP
Responses to questions on risks to supply chains benefit from named examples connecting the risk/causes of risk with the disruption, for example, a named event/product chain/supply chain being affected.

CHECK YOUR UNDERSTANDING
3. Briefly explain how physical factors may disrupt global supply chains.
4. Briefly explain how political factors may disrupt global supply chains.

New and emerging threats to the sovereignty of states (1)

PROFIT REPATRIATION AND THE AVOIDANCE OF TAX

Repatriation of profits is the movement of profits made in a business or investment in a foreign country back to the country of origin. Profits are normally repatriated to protect against expropriation or to take advantage of currency fluctuation. Profit repatriation is an important factor in determining whether foreign direct investment (FDI) in another country is actually profitable for the parent firm.

Profit repatriation laws vary from country to country. For example, when the Volkswagen Group earns profits anywhere in the world, it takes a share back home to Germany. China has slowly and successfully liberalized its economy and laws to suit FDI needs. It is possible legally to repatriate up to 90% of annual profits from China.

Some companies choose to open subsidiaries in other countries. They do this because the corporation tax in their home country is extremely high, and significant savings can be made by being based elsewhere.

Tax avoidance is not confined to HICs. Oxfam estimates that LICs could be losing up to $124 billion each year as a result of tax avoidance by TNCs and wealthy individuals. Action Aid reported on tax avoidance by Associated British Foods (ABF) in Zambia. ABF has operations in 46 countries and an annual turnover of around £11 billion.

Between 2007 and 2012, Zambia Sugar, a subsidiary of ABF, made profits of $123 million but paid less than 0.5% tax. It also repatriated over one-third of its profits to countries such as Ireland, Mauritius and the Netherlands, which offer very low tax rates. The Zambian government estimates that it loses over $2 billion per year from tax avoidance by TNCs and wealthy people.

TAX AVOIDANCE BY THE WEALTHY

There are a number of ways in which wealthy people avoid paying tax. One way is to live in countries with lower rates of tax. For example, the former Formula 1 world champion Lewis Hamilton lives in Switzerland where there is a much-reduced tax rate.

The Panama Papers is the name given to 11.5 million documents that were leaked, in 2015, from the database of the world's fourth largest offshore law firm, Mossack Fonseca. The records show the complex ways in which wealthy individuals exploit secretive offshore tax regimes. Twelve national leaders were among 143 politicians, their families and close associates known to have been using offshore tax havens.

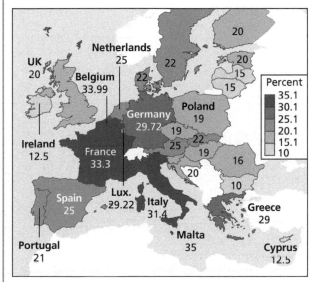

Corporation tax in the European Union

CASE STUDY

TAX AVOIDANCE: APPLE

In 2016 Apple Inc. was ordered by the European Union to pay €13 billion worth of back taxes to Ireland. Apple claimed that such a move could lead to less investment in the EU by TNCs. The European Commission ruled that the tax deal between Apple and the Irish government was illegal, since the same deal was not open to other companies. The rate of corporation tax in Ireland is one of the most competitive in Europe at 12.5%, but Apple had an arrangement whereby the maximum tax rate it paid in Ireland was just 1%; in 2014 it paid just 0.005%.

Apple has avoided paying tax on most of its profits from sales of its iPhone and other goods. It transferred the profits to Ireland rather than keep them in the country in which the phones were sold. It does not repatriate the money to the USA, as their corporation tax is around 35%.

Tax avoidance and US TNCs

For the USA, the share of overseas profits remaining overseas and being taxed at lower rates than in the USA is increasing, and this is happening in many areas of the world.

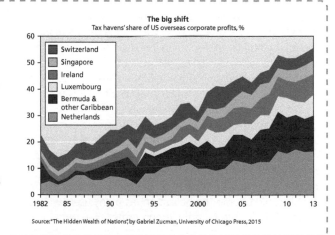

The big shift
Tax havens' share of US overseas corporate profits, %

Source: "The Hidden Wealth of Nations", by Gabriel Zucman, University of Chicago Press, 2015

CHECK YOUR UNDERSTANDING

5. Identify the European countries with the (i) highest and (ii) lowest corporation taxes.
6. Briefly explain why the European Commission was against Apple's tax deal with Ireland.

New and emerging threats to the sovereignty of states (2)

DISRUPTIVE TECHNOLOGICAL INNOVATIONS

Drones

Drones are unmanned aerial vehicles (UAVs) operated by remote control, either by an operator or by an on-board computer. Drones have a number of advantages: they can be used for surveillance in natural and man-made disasters to survey damage, locate victims, help the police search for lost children and monitor large crowds.

Growth of the drone industry between 2015 and 2018 is expected to generate over $13 billion, and by 2025 over $80 billion. Moreover, it is expected that the industry will create some 30,000 manufacturing jobs and more than 70,000 technical jobs.

However, concerns exist about the use of drones in war and for surveillance. Drones used in war are operated far from the conflict zone and may thus desensitize military personnel to war and killing. In addition, drones have caused many civilian fatalities.

Drones have also been linked with an invasion of personal privacy. Their powerful cameras and remote sensing imagery can be used to "spy" on people.

Other problems include cost, especially with drones used for military operations, breakdown or malfunction of computer software, and human error in their operation.

DRONE DEATHS

In a US survey, 56% of respondents indicated that they supported the use of drones and UAVs in war. Between 2006 and 2009 about 746 people were killed in attacks using drones. At least 147 of the victims were civilians, and 94 were children. Pakistan claimed that at least 400 civilians had been killed in US strikes in the country since 2006.

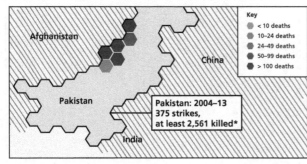

Key
- < 10 deaths
- 10–24 deaths
- 24–49 deaths
- 50–99 deaths
- > 100 deaths

Afghanistan

China

Pakistan

India

Pakistan: 2004–13
375 strikes,
at least 2,561 killed*

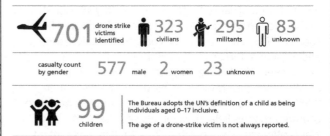

701 drone strike victims identified   323 civilians   295 militants   83 unknown

casualty count by gender   577 male   2 women   23 unknown

99 children

The Bureau adopts the UN's definition of a child as being individuals aged 0–17 inclusive.

The age of a drone-strike victim is not always reported.

Drone casualties in Yemen, Somalia and Pakistan
Source: https://storiesbywilliams.com/2013/08/07/drone-wars-new-leaks-reveal-costs-of-drone-use/

3D PRINTING TECHNOLOGY

3D printing technology, or "additive manufacturing", allows for the creation of physical objects from a digital model, by building them up in a sequence of layers. The technology is said to have great potential in engineering, medicine, the military, construction, architecture, education and the computing industries. In 3D printing it is possible to use a variety of different materials.

3D printing allows manufacturers to create complex 3D objects in a short time. There is little waste involved, as products are built up rather than reduced in size. 3D printing has been used to create human body parts, organs and tissues. Those in favour of 3D printing argue that it could counter globalization, as many users will do much of their own manufacturing rather than outsource or engage in trade. 3D printing technology could lead to the development of localized, customized production that responds to demand. Some believe that 3D printing could lead to a resurgence of manufacturing in HICs as highly skilled companies will be needed to develop the technology and access large amounts of capital to develop the technology needed for the industry.

However, 3D printing technology poses risks. The lack of legislation and regulations concerning the technology means that it can be used to create weapons and counterfeit goods. Critics argue that 3D printing will replace labour in increasingly complex tasks, as technology already has with ATMs and online banking, for example. There are also concerns that 3D printing can enable gangs to steal money from ATM machines.

CHECK YOUR UNDERSTANDING
7. Outline one advantage and one disadvantage of drones.
8. Outline one advantage and one disadvantage of 3D printing.

Increased globalization and renewed nationalism

According to Kearney, there are four possibilities in the evolution of globalisation.

The changing global economic order

1989–2000	2001–2008	2009–?
Globalization 1.0	Globalization 2.0	Global Hiatus
• Post-Soviet bloc countries and China begin to liberalize	• The BRICS and other emerging markets power the global economy	• Developed and emerging markets experience economic volatility

Four possible futures
- Globalization 3.0
- Polarization
- Islandization
- Commonization

Globalization 3.0, which corrects for deficiencies in current globalisation; "**Polarization**", in which rising geopolitical tensions and economic rivalries divide the global economy into competing blocs of countries; "**Islandization**", in which nationalism gains ground in key economies around the world, leading to protectionist measures and reduced global economic flows; "**Commonization**" – the rise of a new global commons through manufacturing and the sharing economy and a fall of the consumer capitalism.

By 2014–15 it was clear that the increase in globalization had been matched by an increase in nationalism. The increase in nationalism was due, in part, to poor economic growth following the financial crash of 2008, rising inequality, and, in Europe, rising immigration. The changes brought about by globalization led many people to look for stability in national or local features, such as a shared culture, history or language. This new nationalism takes many forms: protectionist policies such as trade barriers, policies favouring domestic workers, anti-immigration measures and resource nationalism. In the most extreme cases, economic failure and a lack of opportunities for young people are fuelling resentment of globalization and the West, and, in extreme cases, terrorism. The election of President Trump, the UK's vote for Brexit, and the growth of right-wing political parties in the EU are examples of renewed nationalism.

One country that did not share in the growth of nationalism was Singapore. In its 2017 strategy for the future, the Committee on the Future Economy (CFE) stated that Singapore must "remain plugged into global trade", and "must resist the threat of rising protectionism amid anti-globalization sentiment".

RENEWED NATIONALISM IN EUROPE

Nationalism in Europe is making a reappearance. In general, Europe is peaceful, prosperous and there is close cooperation among national governments. It gives substantial power to institutions such as the European Commission, the European Parliament and the European Court of Justice. However, there is growing nationalism with a stronger determination on the part of many governments to defend their national self-interest as well as the rise of right-wing politics. Europe has been affected by problems with the Euro (its currency), and the large number of refugees and migrants entering. The growth of right-wing politics has led to a greater focus on anti-immigration and the rise of Islamophobia.

However, the rise of nationalism in Europe does not necessarily provide many solutions to the problems that nationalists criticize. According to the *Financial Times*, renewed nationalism in Europe shows "no economic policies beyond an iconoclastic rage at the Euro, free trade and foreigners alleged to be parasites of the welfare state." Nationalism is not only in Europe. President Trump of the USA initiated a policy of "America first", and Narendra Modi, a Hindu nationalist, swept to power in India in 2014.

Nevertheless, *The Economist* claims that the biggest threat to Europe is Russian nationalism. They claim that it is the humiliation about the collapse of the former Soviet Union, and the nostalgia for its great-power status that creates a threat to Europe. The general rise in nationalism will increase international political tensions and make multilateral cooperation on climate, trade, taxation and development more difficult.

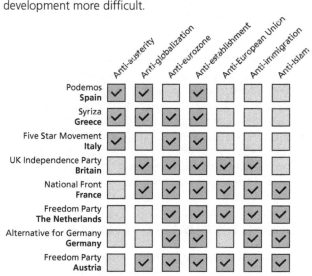

European populism

Source: http://timesofindia.indiatimes.com/photo/55925871.cms

CHECK YOUR UNDERSTANDING

9. Identify the impacts of renewed nationalism in countries.
10. Identify the two European countries that have the greatest number of populist policies.

Transboundary pollution (TBP)

DEFINITION

Transboundary pollution (TBP) is pollution affecting a large area or more than one country.

ACID RAIN

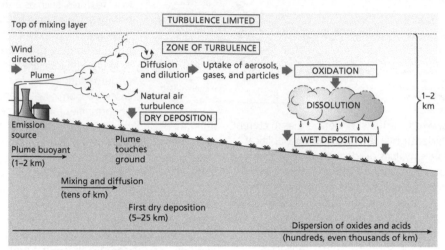

Dry and wet deposition

Acidic deposition can be wet or dry. **Dry deposition** of pollutants occurs typically close to the source of pollutants. **Wet deposition** occurs when the pollutants are dissolved in precipitation, and may fall at great distances from the sources. Wet deposition crosses international boundaries with disregard, and so it is a form of transboundary pollution.

Acid rain is rainfall that is more acidic than normal due to human activity. Acid rain has a pH of less than 5.5. Rainfall is naturally acidic because it absorbs carbon dioxide in the atmosphere and becomes a weak carbonic acid, with a pH between 5 and 6.

The major causes of acid rain are the sulphur dioxide and nitrogen oxides produced by burning fossil fuels such as coal, oil, and gas. When sulphur dioxide and nitrogen oxides are released into the atmosphere, they can be absorbed by the moisture and become weak sulphuric and nitric acids.

Coal-fired power stations are the major producers of sulphur dioxide, and vehicles, especially cars, are responsible for most of the nitrogen oxides in the atmosphere.

As a general rule, sulphur oxides have the greatest effect and are responsible for about two-thirds of the problem. However, worldwide emissions of SO_2 are declining while those of NOx are increasing, partly due to increased car ownership.

Acidification has a number of effects:
- weathering of buildings
- mobilization of metals, especially iron and aluminium, by acidic water which is carried into rivers and lakes
- aluminium damages fish gills
- tree growth is severely reduced
- soil acidity increases
- lakes become acidified and aquatic life suffers
- there are possible links (as yet unproven) to the increase in cases of senile dementia.

The effects of acid deposition are greatest in areas with high levels of precipitation (causing more acidity to be deposited on the ground) and those with acidic rocks that cannot neutralize the deposited acidity.

REDUCING THE IMPACTS OF ACID DEPOSITION

Various methods are used to try to reduce the effects of acid deposition, such as adding powdered limestone to lakes to increase their pH. However, the only really effective and practical long-term treatment is to curb the emissions of the pollutants. This can be achieved by:

- reducing the amount of fossil fuel used
- using less sulphur-rich fossil fuel
- using alternative energy sources that do not produce nitrate or sulphate gases
- removing the pollutants before they reach the atmosphere.

EXAM TIP

If you decide to write about acid deposition as a transboundary pollution, make sure that you distinguish between wet deposition (acid rain and snow) and dry deposition (particulate matter). Dry deposition (particulate matter) does not usually cause transboundary pollution whereas wet deposition (acid rain and acid snow) does.

CHECK YOUR UNDERSTANDING

11. Identify the two chemicals that cause acid rain.
12. Outline the main impacts of acid rain.

Environmental impacts of global flows (1)

SHIPPING

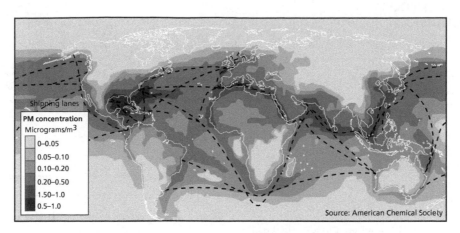

Shipping pollution and the world's main shipping routes

Shipping is one of the most important factors in the growth of globalization. However, there are high rates of pollution along certain shipping routes. There is an NO_2 track in the Indian Ocean between Singapore and Sri Lanka, and others in the Red Sea, the Gulf of Aden, the Mediterranean Sea and along the route from Singapore to China. Such tracks are less evident in the Pacific and Atlantic Oceans, where ships are not as concentrated in narrow zones.

Shipping causes a great deal of pollution and environmental damage, including:

- oil and chemicals released in deliberate discharges and accidental spills
- waste dumping, including sewage and garbage
- air pollution through the release of greenhouse gases
- physical damage through the use of anchors
- noise pollution, which disturbs large mammals such as whales.

Pollution is highly concentrated along the world's main shipping routes. Approximately 50,000 merchant ships are sailing these routes, carrying about 90% of the world's trade between countries. The International Maritime Organization and the UN Convention on the Law of the Sea set regulations to protect the oceans, but they have been criticized for being too slow and reactive rather than pushing for tighter environmental controls.

Rising levels of CO_2 are leading to increased acidification of the oceans. In addition, ships release sulphur and nitrogen oxides, which also leads to acidification.

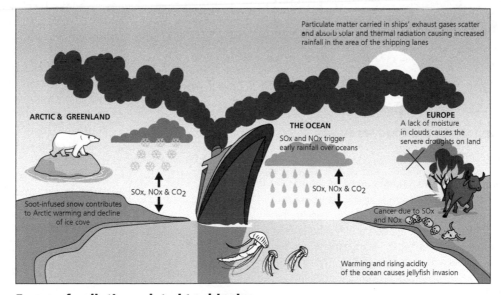

Forms of pollution related to shipping

❓

CHECK YOUR UNDERSTANDING
13. Outline the impacts of shipping on the natural environment.
14. Identify the areas where pollution by shipping is most intense.

Environmental impacts of global flows (2)

CARBON FOOTPRINTS FOR GLOBAL FLOWS OF FOOD AND OTHER GOODS

Food

The carbon footprint associated with food production is complex. The diagram below shows a model for food production, distribution and retailing. The flow of foodstuffs links the main producers and consumers. The carbon footprint depends on such factors as volume of produce, means of transport, refrigeration, packaging, storage, distribution and sales. However, the trade in food has other environmental impacts, such as the use of machinery, chemicals and pesticides.

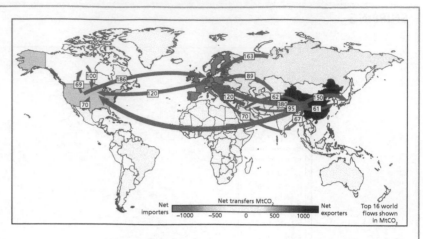

Flows of CO_2 associated with the export of goods

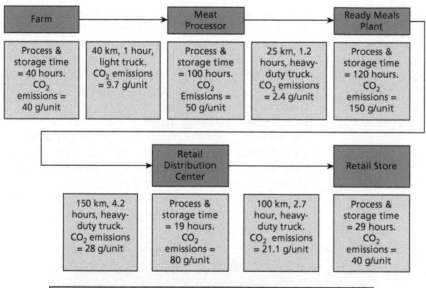

Farm	40 km, 1 hour, light truck. CO_2 emissions = 9.7 g/unit	Meat Processor	25 km, 1.2 hours, heavy-duty truck. CO_2 emissions = 2.4 g/unit	Ready Meals Plant
Process & storage time = 40 hours. CO_2 emissions = 40 g/unit		Process & storage time = 100 hours. CO_2 Emissions = 50 g/unit		Process & storage time = 120 hours. CO_2 emissions = 150 g/unit

	Retail Distribution Center		Retail Store
150 km, 4.2 hours, heavy-duty truck. CO_2 emissions = 28 g/unit	Process & storage time = 19 hours. CO_2 emissions = 80 g/unit	100 km, 2.7 hour, heavy-duty truck. CO_2 emissions = 21.1 g/unit	Process & storage time = 29 hours. CO_2 emissions = 40 g/unit

Emission calculations based on "current state" value stream

Transportation and storage CO_2 emissions per unit product = 421.2 g

(Production emissions = 142 g per unit product)

Carbon emissions in the food trade

Other goods

The flow of goods has a very different carbon footprint from the flow of food. By convention, these emissions are considered to belong to the producing nation, despite the consuming nations generating the demand for the goods. The diagram above shows only the 16 largest flows and identifies net exporters and importers of CO_2. The importance of China as an exporter of goods is clear, as is the importance of the USA and the EU as importers of goods.

Population flows

Human migration is at an all-time high. The migration of people from LICs to HICs will normally lead to an increase in global emissions. Migration from "low carbon" to "high carbon" countries will lead to an increase in greenhouse gas emissions, especially if the migrants' standard of living increases following the migration.

Immigrants to the USA

Research into the environmental impact of immigrants in the USA has shown that:
- CO_2 emissions of the average immigrant (legal or illegal) in the USA are 18% lower than those of the native-born American.
- Immigrants in the USA produce four times more CO_2 in the USA than they would have in their country.
- Legal immigrants have a much larger impact than illegal immigrants because they have higher incomes and higher resulting emissions.

CHECK YOUR UNDERSTANDING
15. Describe the main flows of carbon associated with the world's food trade.
16. Explain why legal migrants have higher carbon footprints than illegal migrants.

Environmental issues linked with the global shift of industry (1)

POLLUTING MANUFACTURING INDUSTRIES

HICs have long been siting their polluting industries in LICs, often with disastrous consequences. In 1984 the American-owned Union Carbide company released toxic gas from its pesticide plant in Bhopal, India, killing thousands. The shift of manufacturing industries from HICs to NICs has resulted in widespread pollution of air, water and soil, and impacts on the health of residents. HICs have stricter environmental laws, greater social supervision and more effective governments; pollution emissions are higher in NICs, where environmental regulations and their enforcement are weaker. These less-regulated environments give HICs a chance to export their waste and pollution. The environmental vulnerability of LICs and NICs to pollution is a result of their underdeveloped systems as well as their need for the economic benefits of the polluting industries.

MAQUILADORA DEVELOPMENT IN MEXICO

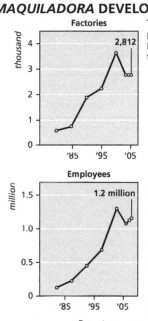

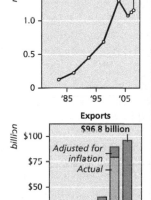

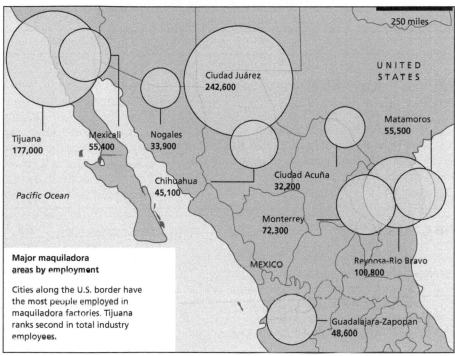

The maquiladora industry in Mexico

More than 2,800 maquiladora plants employing more than a million people were operating in 2005, with combined exports of nearly $100 billion. These plants are foreign-owned factories in Mexico where workers assemble imported parts into products for export.

Major maquiladora areas by employment

Cities along the U.S. border have the most people employed in maquiladora factories. Tijuana ranks second in total industry employees.

Relocations from the USA to Mexico

Mexico has attracted many US-owned companies to build low-cost assembly plants in places such as Ciudad Juarez, Nuevo Laredo and Tijuana. These factories, called *maquiladora* operations, are foreign-owned but employ local labour. Since 1989, more than 2,000 US firms have set up in Mexico's border cities. The main attractions are low labour costs, relaxed environmental legislation and good access to US markets.

Although Mexican law requires firms to transport hazardous substances back to the USA, illegal dumping in Mexico is common. Despite the environmental problems, many Mexicans are in favour of the *maquiladoras* because they bring investment, money and jobs to northern Mexico. Over 500,000 people are employed in these factories.

IS THE RELOCATION OF POLLUTING INDUSTRIES THE MAIN CAUSE OF POLLUTION IN MEXICO?

A study into *maquiladoras*, air pollution and human health in Paso del Norte found that particulate emissions from *maquiladoras* have impacts on human health, in particular on respiratory diseases. However, it found that *maquiladoras* are not the region's leading sources of particulates. Unpaved roads, vehicles and brick kilns were the main sources of particulate emissions. Given that vehicles and brick kilns emit far more combustion-related fine particulates than *maquiladoras*, they inflict more health damage.

CHECK YOUR UNDERSTANDING

17. Describe the main changes in maquiladora industries, as shown in the diagram.

18. Outline the main reasons for the relocation of US firms.

Environmental issues linked with the global shift of industry (2)

FOOD PRODUCTION SYSTEMS FOR GLOBAL AGRIBUSINESSES

The food market is truly global. Farming has become increasingly intensive, large in scale and globalized in the drive for cheaper food. Increasingly, modern farming methods are having a negative impact on the environment. The term "agro-industrialization" refers to the large-scale, intensive, high-input, high-output, commercial nature of much modern farming.

Modern farming has intensified, increased efficiency and adopted labour-saving technologies such as agro-chemicals and machinery.

Agro-industrialization has increased food production but is a major consumer of energy and a contributor to greenhouse gas emissions, air pollution, water pollution, land erosion and loss of biodiversity.

Gallons of water needed to produce 500 grams of various foods and selected items

Meat	Processed food	Fruit	Vegetables	Common goods
Beef 1857	Sausages 1382	Figs 379	Avocados 154	Pair of jeans 2900
Pork 756	Processed cheese 589	Plums 193	Corn 109	Hamburger 766
Chicken 469	Eggs 400	Bananas 103	Beans 43	Glass of milk 53
	Fresh cheese 371	Apples 84	Potatoes 31	Cup of coffee 37

IMPROVED YIELDS AND ENVIRONMENTAL IMPACTS

Food processors usually want large quantities of uniform-quality produce or animals at specific times. This is ideally suited to intensive farming methods. In HICs, since the 1960s, with increased inputs of fertilizers, insecticides and pesticides, for example, wheat yields have increased from 2.6 to 8 tonnes per hectare.

Unfortunately, intensive farming requires the heavy use of chemicals and methods that lead to land degradation and animal welfare problems. For example, apples receive an average of 16 pesticide sprays while lettuces can be sprayed 12 times. Air pollution and greenhouse gas emissions from farming cost the UK more than £1.1 billion annually.

LAND GRABS

Land grabs result in there being less land available for indigenous/national populations and give access to a foreign country/multinational. The new wave of land grabs is very different from the colonial land grabs of the 19th and 20th centuries. A large land deal used to be around 100,000 hectares (240,000 acres). Now the largest ones are many times that.

In Sudan alone, South Korea has signed deals for 690,000 hectares, the United Arab Emirates (UAE) for 400,000 hectares and Egypt has secured a similar deal to grow wheat.

It is not just Gulf states that are buying up farms. China secured the right to grow palm oil for biofuel on 2.8 million hectares of Congo, which would be the world's largest palm-oil plantation. In total, between 15 million and 20 million hectares of farmland in LICs have been subject to transactions or talks involving foreigners since 2006. The deals are worth $20–$30 billion.

When developed, the land will yield roughly two tonnes of grain per hectare and it will produce 30–40 million tonnes of cereals a year. That is a significant share of the world's cereals trade of roughly 220 million tonnes a year.

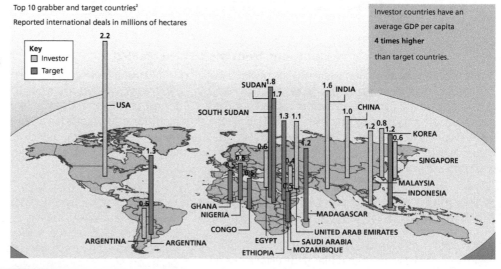

Top 10 grabber and target countries[2]
Reported international deals in millions of hectares

Investor countries have an average GDP per capita 4 times higher than target countries.

CHECK YOUR UNDERSTANDING
19. Define the term agro-industrialisation.
20. Outline two ways in which agriculture may contribute to global climate change.

Environmental issues linked with the global shift of industry (3)

The role of livestock in GHG emissions

Gas	Contribution to climate change (%)	Livestock emissions (billion tonnes carbon dioxide equivalent)	Livestock emissions as % of total anthropogenic
Carbon dioxide	70	2.70	9
Methane	18	2.17	37
Nitrous oxide	9	2.19	64

Cleaning up the chemical pollution and repairing the damage to habitats caused by industrial farming costs over £2 billion a year in the UK alone. It now costs water companies up to £200 million to remove pesticides and nitrates from drinking water. In the USA it is estimated that the costs of agriculture (pesticides, nutrient run-off, soil loss and so on) could be as high as $16 billion ($96 per hectare) for arable farming and $714 million for livestock.

The global food industry has a major impact on transport. Food distribution now accounts for 33–40% of all UK road freight. The food system has become almost completely dependent on oil, which means that food supplies are vulnerable, inefficient and unsustainable. A kilogram of blueberries imported by plane from New Zealand produces the same emissions as boiling a kettle 268 times.

Food miles and a Christmas dinner

The concept of food miles describes how far food has travelled before it appears on a plate. However, it is also important to consider how the food has been transported and even packaged; frozen food, for example, has higher energy costs.

The ingredients of a traditional Christmas meal bought from a supermarket in the UK may have cumulatively travelled more than 38,000 kilometres. Poultry imported from Thailand travels nearly 17,000 kilometres, runner beans could have come from Zambia (nearly 8,000 kilometres), carrots from Spain (1,600 kilometres), mangetout from Zimbabwe (over 8,000 kilometres), potatoes from Italy (2,400 kilometres), and sprouts from Britain (200 kilometres). By the time transport to and from warehouses to stores is added, the total distance the food had travelled was more than 38,000 kilometres. Transporting ingredients such great distances makes food supplies vulnerable.

CASE STUDY

WATER PROBLEMS AND GLOBAL FARMING IN KENYA

The shores of Lake Naivasha in the Happy Valley area of Kenya have been blighted. Environmentalists blame the water problems on pollution from pesticides, excessive use of water on the farms, and deforestation caused by migrant workers in the growing shanty towns foraging for fuel.

British and European-owned flower companies grow vast quantities of flowers and vegetables for export, but the official Kenyan water authority, regional bodies, human rights and development groups as well as small-scale farmers have accused flower companies near Mount Kenya of "stealing" water which would normally fill the river. Kenya's second largest river, the Ngiro, is a life-sustaining resource for nomadic farmers, but it also sustains big business for flower farms supplying UK supermarkets. The 12 largest flower firms may be taking as much as 25% of water normally available to more than 100,000 small farmers.

The flower companies are thereby exporting Kenyan water – this is known as "virtual water". A flower is 90% water. Kenya is one of the driest countries in the world and is exporting water to some of the wettest. The flower companies, which employ 55,000 labourers, are in direct competition with the peasant farmers for water, and the largest companies pay the same as peasant farmers for water.

In 2010, due to the eruption of the Eyjafjallsjokull volcano in Iceland, flights from Kenya to Europe were shut down. With most European airports closed, an estimated $8 million-worth of flowers had to be destroyed. The Kenyan horticultural industry lost around $3 million a day.

COMMON MISTAKE

✗ It is always best for the environment to source foods locally.

✓ It is usually thought that it is best to source food locally, because the amount of greenhouse gas (GHG) used to transport it is reduced. Critics also argue that transport cost is only part of the environmental impact of food production – there could be other costs such as reduction in biodiversity, eutrophication, decreased water quality and increased risk of flooding.

CHECK YOUR UNDERSTANDING

21. Evaluate the concept of food miles.
22. Suggest why there is conflict over water resources in Kenya.

International civil society organizations and risks related to global interactions (1)

AN ENVIRONMENTAL CIVIL SOCIETY: THE WORLD WIDE FUND FOR NATURE

The WWF logo

The World Wide Fund for Nature (WWF) is an international environmental non-governmental organization (NGO), or civil society. It was founded in 1961 with the aim of preserving wilderness areas and the species that inhabited them, and reducing the human impact on the environment.

The WWF is the world's largest conservation civil society, working in over 1,300 projects in more than 100 countries, and with over 5 million supporters worldwide. WWF's mission statement is "to stop the degradation of the planet's natural environment and to build a future in which humans live in harmony with nature".

The WWF's first manifesto stated that it needed "money, to carry out missions and to meet conservation emergencies by buying land where wildlife treasures are threatened, money … to pay guardians of wildlife refuges … [and] for education".

Its initial focus on protecting endangered species broadened into other areas, including preserving biodiversity, sustainable use of resources, reducing pollution, and climate change.

In the 1990s the WWF changed its mission statement to "stop the degradation of the planet's natural environment and to build a future in which humans live in harmony with nature" by:
* conserving the world's biological diversity
* ensuring that the use of renewable natural resources is sustainable
* promoting the reduction of pollution and wasteful consumption.

The WWF's current strategy is to:
* restore populations of keystone species, species that are important for their ecosystem or people, including elephants and whales
* reduce ecological footprints.

The WWF gets involved in many projects but also voices opinions on developments. For example, it has been a critic of the Canadian tar sands programme. From 2008–10 the WWF published reports that concluded that the tar sands programme could contribute significantly to climate change, that carbon capture and storage (CCS) technology was not as effective for tar sands as for other forms of oil, and that the development of the tar sands represented a major risk to caribou herds in Alberta.

CRITICISMS OF WWF

Critics argue that the WWF is too close to some large companies, such as Coca-Cola and IKEA. Wilfried Huismann's documentary *Silence of the Pandas* (2011) criticized WWF's involvement with TNCs responsible for destruction of the natural environment. In 2016 Survival International complained that the WWF was using eco-guards who had abused the rights of indigenous people in the Cameroon rainforest. WWF denied the allegations.

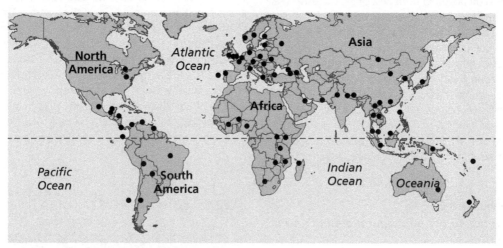

WWF offices around the world

CHECK YOUR UNDERSTANDING
23. Outline the mission statement of the WWF.
24. Comment on the criticisms about WWF.

International civil society organizations and risks related to global interactions (2)

A SOCIAL CIVIL SOCIETY: OXFAM

Oxfam is an international charity focused on the alleviation of global poverty. It was originally founded in Oxford, UK, in 1942, as the Oxford Committee for Famine Relief, with the aim of getting food to people in Greece during the Second World War.

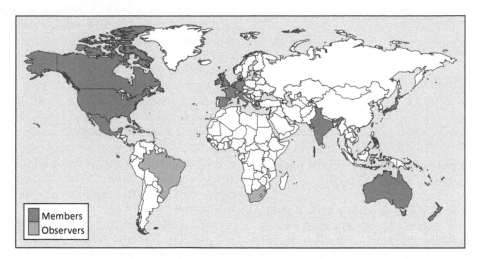

Countries in which Oxfam is present

Oxfam's aims have changed to address the causes of poverty and injustice, having human rights to the fore of its mission. The Oxfam International strategic plan states that everyone has the right to:
- a sustainable livelihood
- basic social services
- life and security
- be heard
- an identity.

Oxfam believes that poverty and powerlessness can be eliminated if there is political will and human action. Nevertheless, Oxfam continues to work in three main areas: development support, which aims to lift people out of poverty with sustainable projects, humanitarian work following natural disasters and conflict, and lobbying and campaigning.

Current focus

In its 2015–16 Annual Report, Oxfam claimed that the combined wealth of the world's richest 1% would overtake that of everyone else by 2016, given the trend of rising inequality. In its activities, Oxfam focuses on economic justice, essential services, rights in crisis and gender justice.
- Economic justice focuses on improving farming for farmers and labourers, fairer trade, and reducing shocks from energy changes and climate change.
- Essential services refers to the provision of health education, water and sanitation.
- Rights in crisis refers to assistance given during conflicts and after disasters.
- Gender justice seeks to support women's leadership and increase the number of women receiving an education.

Oxfam opened its first shop in 1948 and it now has about 1,200 shops worldwide, selling books, CDs, crafts, clothing and ethnic products. In 2013 Oxfam established a "Behind the brands" project, in which they provided information on the policies of the biggest food brands against the following criteria:
- transparency at a corporate level
- farm worker, including women, and small-scale producers in the supply chain
- small-scale farmers growing the commodities
- land rights and sustainable use of land
- water rights and the sustainable use of water
- methods of reducing climate change and adapting to climate change.

Oxfam's Make Trade Fair aims to eliminate a number of practices, including the dumping in LICs of highly subsidized foods produced in HICs; high import tariffs; unequal labour laws, in which women earn less than their male counterparts; and patent issues of seeds, medicines and software.

Criticisms of Oxfam

There have been some criticisms of Oxfam. These include allegations that Oxfam is politically motivated; that some of its trustees were tax avoiders, and that its stores have forced the closure of small specialist stores and other charity shops.

CHECK YOUR UNDERSTANDING
25. Outline Oxfam's main goals.
26. Outline the criticisms made about Oxfam.

Strategies to build resilience (1)

RESHORING OF ECONOMIC ACTIVITY BY TNCs

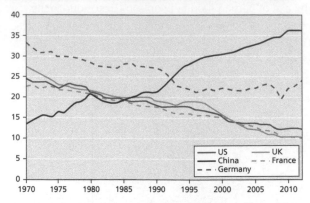

Changes in the manufacturing share of national output (% of national value added)

During the 1970s and 1980s, there was a major decline in manufacturing employment in HICs as a result of offshoring to LICs, which offered low-cost labour. However, that advantage is being eroded as the cost of labour in LICs rises, and concerns now exist in HICs over the source of goods, working conditions in LICs, delivery times and rising unemployment.

K'Nex Industries is a US company which makes Tinkertoys, Lincoln Logs and building sets. In the late 1990s it offshored, and by the early 2000s it had outsourced most of its toymaking to China. However, the long time required to ship toys to the United States placed a strain on the business. In 2017, 90% of K'Nex toys were manufactured in the United States.

Bringing manufacturing back to the United States is complex. Considerations include wages, energy costs and trade agreements. Between 2013 and 2014 the US moved from losing about 140,000 manufacturing jobs per year to gaining 10,000 or more annually. Increased automation reduces the wage advantage for countries like Mexico and China, even if those countries also automate. Moreover, the jobs that are automated are usually the lowest level.

The "Made in the USA" label has value for a toy company such as K'Nex – they can charge up to 15% more for toys because of the USA stamp.

Reasons for reshoring

Push factors from overseas	Pull factors to return home
• Rising global oil prices and increasing transport costs • A relative lack of skilled labour in LICs compared with HICs • Rising labour costs in LICs and NICs • Greater risks in the supply chain	• Increased demand for customization of products and smaller runs • A tradition of manufacturing, and public demand to maintain employment in HICs • Consumers increasingly demanding quick delivery times • Higher levels of R&D

Examples of reshoring include www.bathrooms.com, which is investing £2.5 million in the Midlands, UK, where it will manufacture 25% of its products. This will cut delivery times for customers.

Capital-intensive sectors with complex supply chains and rapidly changing markets are most likely to reshore. These include petroleum, chemicals and aerospace; pharmaceuticals and chemicals (R&D); textiles and leather goods.

Reshoring has several impacts:
• direct impacts on output, GDP and employment
• indirect impacts in the supply chain
• multiplier effects, when the employees of the reshored industries spend their earnings in the economy and increase demand for services.

The benefits of reshoring are geographically diverse. In the UK, the south-east region and the north-west benefit most, partly because of the concentration of the defence, electronics and aerospace industries there.

EXAM TIP
Try to be synoptic in your answers. For example, **political changes** due to the rise of nationalism and anti-globalization, have led to reshoring of **economic activity**, and this has a major impact on the **physical environment**.

CHECK YOUR UNDERSTANDING
27. Describe the trends in the share of manufacturing to national value added, as shown on the graph.
28. Identify two push factors and two pull factors that help explain the process of reshoring.

Strategies to build resilience (2)

CROWDSOURCING

Crowdsourcing is the process of sourcing ideas, services, finances and information from the public via the internet due to the benefit from the collective abilities of a large group of people. It has developed because top-down approaches, that is, government solutions to problems, have proved insufficient. Crowdsourcing empowers people.

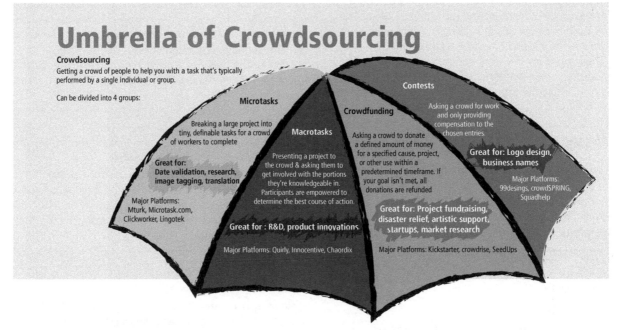

A model of crowdsourcing

Crowdsourcing is a way of using a community's best assets, namely its population, to find solutions, discuss ideas, fund projects etc. Crowdsourcing has been used in times of natural disasters. Following the earthquake in Nepal, crowdfunding raised over $20 million in 60 days. It also provided images of buildings, schools and hospitals daily, to help search, rescue and relief workers.

Crowdsourcing allows organizations to tap into the creativity of large numbers of people. However, plans and expectations must be clearly stated because there is a danger that different people may interpret plans differently.

The concept of resilience refers to the ability of individuals, communities or environments to respond to shocks and changes while continuing to operate under new circumstances. Crowdsourcing allows for rapid transmission of ideas and opportunities which would have been more difficult in the era before the internet. For example, Wikipedia is an example of a crowd-sourced effort that enables people in most places to obtain information on most subjects very quickly, and at very limited extra cost.

Crowdsourcing can be used to find new ways of providing services. For example, in 2011 the Bill and Melinda Gates Foundation introduced the Reinventing the Toilet Challenge, a US$100,000 challenge to design a toilet that could dispose of human waste without a septic tank or an outside water system. The winning design was a solar powered system that could break down waste and water to fertilizer and hydrogen.

Crowdsourcing offers many opportunities for people to interact with each other and find solutions to old and new problems. The increased interdependence and complexity of global interactions necessitates the involvement of many people from different stakeholder positions, to find solutions for problems as they arise. Crowdsourcing is a high-technology bottom-up approach of empowering communities around the world.

!

COMMON MISTAKE

✗ *Crowdsourcing focuses on fundraising.*

✓ Crowdsourcing is not just about raising money – that is crowdfunding. Crowdsourcing includes problem-solving, marketing, sharing knowledge, raising money, promoting open government, greater involvement in research and development, raising awareness, and disaster relief.

?

CHECK YOUR UNDERSTANDING
29. Outline the advantages of crowdsourcing.
30. Outline the disadvantages of crowdsourcing.

New technologies for the management of global flows of data and people

CYBERSECURITY

Cybersecurity, or computer security, is the protection of information systems, hardware and software from theft or damage, as well as the protection of information on computers and related technology. The need for cybersecurity is increasing as more people and organizations rely on computers and the internet.

There are many threats to computer security. One of the most common is "phishing" – the attempt to obtain personal or sensitive information such as user names, passwords, bank account details and credit card details. Most computers have some level of protection, but threats are becoming more sophisticated and protection systems need to keep ahead of the threats. Common targets are large organizations, government departments, military computer systems and airline carriers. The most common prevention systems are firewalls, which stop access to internal network systems and filter out different kinds of attack.

One of the main issues regarding cybersecurity is that there are no international regulations or common rules to abide by.

E-PASSPORTS

An e-passport has a computer chip in it which contains data about the owner. E-passports allow faster checking in at airports and border clearance, and they may also help in crime detection, as some contain biometrics such as fingerprints. E-passports are difficult to reproduce or forge, so security is improved. They also make it more difficult for one person to have several passports.

However, there are disadvantages. If the passport is stolen, the data could be used illegally. It would be possible for someone to hack into the system and change the data. Moreover, the person who owns the passport does not have access to the data.

CASE STUDY

CYBERSECURITY IN ISRAEL AND SINGAPORE

Israel, with a population of just 8.6 million people, is the world's second largest exporter of cybersecurity products and services – second only to the United States. Israel's cybersecurity exports are worth around $6 billion a year. Military experience is the key to Israel's experience in cybersecurity.

In February 2017, Singapore's Committee on the Future Economy (CFE) suggested using national service to train army personnel in cybersecurity skills to plug talent gaps for economic gains and defend the country against cyber attacks. For a country such as Singapore, that has aspirations to be a "smart nation", it is imperative that it has its own cyber-defence force.

Israel's progress in cyber-defence dates back to 1993 when it established Check Point Software Technologies. The company's three co-founders were in the army's elite intelligence unit before starting the company. Similarly, the two co-founders of another Israeli cybersecurity firm, CyberArk, also had military intelligence experience before founding the company in 1999.

CHECK YOUR UNDERSTANDING

31. Briefly explain why cybersecurity is an increasing problem.
32. Outline two advantages of e-passports.

Exam practice

(a) Examine the relationship between increased globalisation and the rise of nationalism. (12 marks)

(b) Discuss the view that global flows and global shifts inevitably lead to environmental risks. (16 marks)